ISBN 978-1-330-29062-0
PIBN 10015380

English
Français
Deutsche
Italiano
Español
Português

www.forgottenbooks.com

Mythology Photography **Fiction**
Fishing Christianity **Art** Cooking
Essays Buddhism Freemasonry
Medicine **Biology** Music **Ancient
Egypt** Evolution Carpentry Physics
Dance Geology **Mathematics** Fitness
Shakespeare **Folklore** Yoga Marketing
Confidence Immortality Biographies
Poetry **Psychology** Witchcraft
Electronics Chemistry History **Law**
Accounting **Philosophy** Anthropology
Alchemy Drama Quantum Mechanics
Atheism Sexual Health **Ancient History**
Entrepreneurship Languages Sport
Paleontology Needlework Islam
Metaphysics Investment Archaeology
Parenting Statistics Criminology
Motivational

The NATIONAL
GEOGRAPHIC
MAGAZINE

INDEX

—

January to June, 1921

VOLUME XXXIX

PUBLISHED BY THE
NATIONAL GEOGRAPHIC SOCIETY

CONTENTS

WASHINGTON, D. C.
PRESS OF JUDD & DETWEILER, Inc.
1921

INDEX FOR VOL. XXXIX (January-June), 1921 *v*

AN ALPHABETICALLY ARRANGED INDEX.   ENTRIES IN CAPITALS REFER TO ARTICLES AND INSERTS

"A"

VOL. XXXIX, No. 1   WASHINGTON   JANUARY, 1921

THE DREAM SHIP

The Story of a Voyage of Adventure More Than Half Around the World in a 47-foot Lifeboat

By Ralph Stock

WE ALL have our dreams. Without them we should be clods. It is in our dreams that we accomplish the impossible: the rich man dumps his load of responsibility and lives in a log shack on a mountain top, the poor man becomes rich, the stay-at-home travels, the wanderer finds an abiding place.

The difficulty is to turn one's dreams into realities, and as I happen to be enjoying that rare privilege at the present moment, it is too good to keep.

For more years than I like to recall, my dream has been to cruise through the South Sea Islands in my own ship. I can't help it; that has been my dream; and if you had ever been to the South Sea Islands, and love the sea, it would be yours also. They are the sole remaining spots on this earth that are not infested with big-game-shooting expeditions, globe-trotters, or profiteers; where the inhabitants know how to live, and where the unfortunate from distant and turbulent lands can still find interest, enjoyment, and peace.

A DREAM WITH A COMIC-OPERA PLOT

My dream was as impracticable as most. There was a war to be attended to and lived through, if Providence so willed. There was a ship to be bought, fitted out, and provisioned on a bank balance that would fill the modern cat's-meat man with contempt. There were the little matters of cramming into a chronically unmathematical head sufficient knowledge of navigation to steer such a ship across the world when she was bought, and of finding a crew that would work her without hope of monetary reward.

The thing looked and sounded sufficiently like comic opera to deter me from mentioning it to any but a select few, and they laughed. Yet such is the driving power of a dream, if its fulfillment is really desired, that I write this on the deck of my dream ship, anchored off the Isthmus of Panama, five thousand miles on the way to my goal.

HOW THE DREAM MATERIALIZED

Exactly how it all came about I find it difficult to recall. I have vague recollections of crouching in dug-outs in France, and while others had recourse during their leisure to letter-cases replete with photographs of fluffy girls, I pored with equal interest over sketches and plans of my dream ship.

In hospital it was the same, and when a medical board politely ushered me into the street, a free man, it took me rather less than four hours to reach the nearest seaport and commence a search that covered the next six months.

It is no easy matter to find the counterpart of a dream ship, but in the end I

FITTING OUT THE "DREAM SHIP" FOR ITS TRIP HALF AROUND THE WORLD

This staunch little 23-ton vessel, 47 feet in length and with a beam of 15 feet, was a Norwegian-built craft, designed as a lifeboat for service with the North Sea fishing fleet. It was equipped with an auxiliary engine, with which the skipper had trials which only he can describe (see pages 14 and 15).

found her—a Norwegian-built auxiliary cutter of twenty-three tons register, designed as a lifeboat for the North Sea fishing fleet, forty-seven feet over all, fifteen feet beam, eight feet draft.

Such was my dream ship in cold print. In reality, and seen through her owner's eyes, she was, naturally, the most wonderful thing that ever happened. A mother on the subject of her child is almost derogatory compared with an owner concerning his ship, so the reader shall be spared further details.

Having found her, there was the little matter of paying for her. I had no money. I have never had any money; but that is a detail that should never be allowed to stand in the way of a really desirable dream. It was necessary to m a k e some. How? B y conducting a stubborn offensive on the Army authorities for my war gratuity; by sitting up to all hours in a moth-eaten dressing-gown and a microscopic flat writing short stories; by assiduously cultivating maiden aunts; by coercion; by—but I refuse to say more.

A SISTER OF THE RIGHT VARIETY

The *Dream Ship* became mine. But what of a crew? I have a sister, and a sister is an uncommonly handy thing to have, provided she is of the right variety. Mine happens to be, for she agreed to forego all the delicacies of the season and float with me on a piece of wood to the South Sea Islands. So also did a recently demobilized officer, who, on hearing that these same islands were not less than three thousand miles from the nearest early-morning parade, offered his services with almost unbecoming alacrity.

Behold, then, the crew of the *Dream Ship*—Peter, Steve, and myself—and try not to laugh when I tell you that we learnt what we could of navigation inside of three weeks. On the first of July, 1919, we sailed from Devonshire, England, with a combined capital of £100 and a "clearance" for Brisbane, Australia.

We sailed, and have been sailing ever since, first across the dreaded Bay of

THE "DREAM SHIP," WITH THE DEVONSHIRE HILLS OF ENGLAND IN THE
BACKGROUND, READY TO START ON HER LONG VOYAGE TO THE SOUTH SEAS

Biscay, which treated us with the utmost kindness until off Finisterre, and then drove us before half a gale into Vigo, Spain.

Here we duly admired the soft-eyed senora and her charming children at play on the palm-bordered *alameda*, commiserated with the unfortunate Spanish mule, laid in ten gallons of *vino tinto*, and shaped a course for Las Palmas, Canary Islands.

THE ORDEAL OF COOKING IN A GIRATING FO'CASTLE

Four hours on and eight off was how we apportioned our watches, and, thanks to fair winds and the easy handling of the *Dream Ship*, it was never necessary for more than one of us to be on deck at a time. In fact, there were hours on end when the helmsman could lash the tiller and take a constitutional.

Cooking, a dreaded ordeal, we took week and week about. It is one thing to concoct food in a porcelain-fitted kitchen on *terra firma* and quite another to do it over a Primus stove in a leaping, gyrating fo'castle nine by five. Porridge

has been found adhering to the ceiling after "Steve's week." But hush! perhaps he may have something to say on the subject of Peter and myself.

There is always plenty to say about the other fellow, but in nine cases out of ten it is best left unsaid. Forbearance is as much the keynote of good-fellowship on a dream ship as elsewhere—perhaps more—and we are rather proud of the fact that we have covered half the world without battle, murder, or sudden death.

AMATEUR NAVIGATION CALCULATIONS PROVE CORRECT

At the end of ten days' pleasant routine and fair winds, we experienced the acute joy of finding land precisely where our frenzied calculations had placed it.

As the island of Grand Canary loomed ahead, Steve was seen to pace the deck with a quiet but new-born dignity—until hailed below to help wash dishes.

At Las Palmas we suffered a siege of bumboat-men, lost a good deal more than we could afford at roulette, laid in a fresh

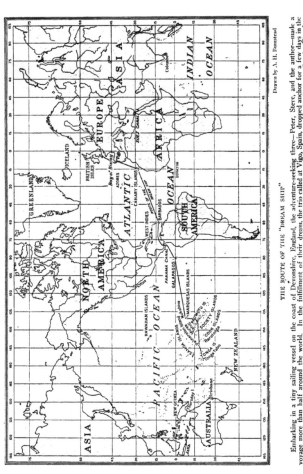

THE ROUTE OF THE "DREAM SHIP"

Drawn by A. H. Bumstead

Embarking in a tiny sailing vessel on the coast of Devonshire, England, the adventure-seeking three—Peter, Steve, and the author—made a voyage more than half around the world. In the fulfillment of their dream, the trio called at Vigo, Spain, dropped anchor for a few days in the Canaries, sailed smoothly across the Atlantic to Barbados, then experienced a series of thrills in passing through the Panama Canal. The first objective after entering the Pacific was San Cristobal Island, in the Galapagos Archipelago, off the coast of Ecuador. The Marquesas Group, the Society Islands, Rarotonga, Palmerston, and Tonga Tabu were visited and enjoyed in turn. After purchasing the *Dream Ship*, the combined capital of the voyageurs was only 100 pounds—less than $400 at the current rate of exchange.

supply of *vino tinto*, and set sail for the West Indies.

THE GREAT ADVENTURE REALLY BEGINS

The great adventure had now begun in earnest.

Three thousand miles of Atlantic Ocean lay ahead of us—a waste of waters holding we knew not what of new experience. For the first time since setting sail, our undertaking imbued us with a certain amount of awe.

At night, alone at the tiller, one began to think: Would the drinking water hold out? What if the chronometer broke down? Supposing— It was necessary to think of other things, but what?

Staring into the lighted binnacle, with its swaying compass card, or down at the phosphorescent water, swirling and hissing past the ship's stern, the helmsman became as one hypnotized. It seemed that he was not of this world, but an atom hurtling through space. The temptation was to surrender himself to the sensuous joy of it—a temptation only resisted by an almost painful effort and a knowledge that the lives of all aboard depend on his keeping his leaden eyelids from closing down.

A four hours' watch is too long. They do not allow it in the mercantile marine. But what were we to do? We kept a marlinspike handy, and when oblivion threatened we used it; that was all.

It will be seen that a dream ship is not all dream. If it were, such is the perversity of human nature, the dreamer would probably be tired of it inside of a month.

"I can promise you the northeast 'trades' the whole way across," said the skipper of a fine, six-masted schooner at Las Palmas, turning the pages of his log, and that may account for the fact that not for one day of the Atlantic passage did we encounter a northeast wind. We could have crossed in an open boat, for all of the weather, and three becalmed days in mid-ocean we occupied in swimming around the ship or diving down to scrape the barnacles off her copper.

Yet we made Barbados, West Indies, in thirty days, and gladly surrendered ourselves to the tender mercies of the most charming, hospitable people one could wish to meet.

My recollections of our two weeks' sojourn are a trifle vague, owing to the rapidity with which one pleasure succeeded another. I remember lying at anchor, with awnings up, in the most beautiful bay it is possible to conceive, and sleeping twenty-four hours on end. From then onward life consisted in "swizzles," car rides over a fairy island, and more "swizzles," pony races to the accompaniment of "swizzles," surf bathing followed by "swizzles," and evenings at the Savannah Club, where conversation was punctuated and sometimes drowned by the concoction of yet more "swizzles" by a hard-worked army of colored folk behind a gleaming mahogany bar.

There is no escaping the "swizzle" in Barbados, even if one wished to, which personally I did not. They are a delightful, healthful drink composed of the very best rum, Angostura bitters, syrup, fresh lime, nutmeg, and ice, the whole swizzled to the creamy consistency of— But I forget that I am addressing a country in the throes of total abstinence, and, whatever my faults, I have never been accused of making a man's mouth water without supplying the deficiency.

Hot-foot from a ball at one of the hotels, we literally fled aboard ship and sailed by stealth, otherwise I am convinced that we s ou be at Barbados still, imbibing "swh—.'ld

UNDAUNTED BY WEATHER PROPHETS

"Look out for the Caribbean Sea toward December," was another axiom of our six-masted-schooner friend at Las Palmas; but he proved no less fallible over the passage from Barbados to Colon than he had concerning the Atlantic. In fact, I am thinking of, in the future, asking advice of weather prophets and expecting the reverse.

A spanking, following wind, with mainsail and squaresail set, brought us within sight of land in seven days, a distance of twelve hundred miles. But what land? For a time we were at a loss. Comparing it with the chart and descriptions in "sailing directions" revealed nothing.

It was a low-lying, mist-enshrouded, sinister-looking land, and we sailed along its coast for a day and a night before we could tell whether we had passed Colon or hit the coast to the eastward.

STEVE WRESTLES WITH NAVIGATION : THEORY

Photographs by Ralph Stock

STEVE WRESTLES WITH NAVIGATION : PRACTICE

All the knowledge of the intricate problems of navigation possessed by the master and crew of the *Dream Ship* was acquired in three weeks of intensive study before weighing anchor off the Devonshire coast for the 12,000-mile voyage over trackless seas.

WHEN THE "DREAM SHIP" DROPPED THE PALMS OF THE CANARIES BELOW THE
HORIZON, THE VOYAGE TO THE SOUTH SEAS HAD BEGUN IN EARNEST

Ultimately, a light-house gave us the clue, and we found that, owing to a current that has the unpleasant knack of running at anything from a half to three knots an hour, we were still fifty miles from our objective; so we headed for sea and hove-to until daylight.

TERRIFYING REFLECTIONS OFF THE PANAMA CANAL

All night, as we lay rolling in a heavy swell, steamers passed us by, floating palaces of light, and with the dawn we joined the procession of giants making for the Panama Canal.

"We wished to go through the canal? Very well; a measurer would be sent off to determine our tonnage, and we must be ready to take the pilot aboard at five o'clock the next morning."

That, in effect, is what the Canal authorities said, and I answered it with a smile that I trust was sufficiently engaging to hide the fact that I was not at all sure we had enough money among us to pay the tolls.

It must be an expensive business, this passing from Atlantic to Pacific; I had never thought of that. There was quite a lot I had not thought about. What if

DRY GORGE, LAS PALMAS, CANARY ISLANDS

The Canary Archipelago was the first scheduled stop of the *Dream Ship* on its trip to the South Seas. The amateur navigators had been driven into Vigo, Spain, however, by heavy weather in the Bay of Biscay. In the Canaries the voyageurs replenished their stock of *vino tinto* and other ship supplies.

THE FAMOUS PELDE ROAD, LAS PALMAS, CANARY ISLANDS

Photograph by J. Percettello

Although the home of the canary, these islands derive their name from their large dogs rather than their birds. Writing of the expedition of a king of Mauretania before the Christian era, the elder Pliny records that the monarch called the islands "Canaria, from the multitude of dogs (canis) of great size" which roamed here.

Photograph by Ralph Stock

PAJAMAS CONSTITUTED THE SAILOR'S UNIFORM ON THE "DREAM SHIP"

Peter at the lookout. Leaving the Canaries, 3,000 miles of ocean stretch before her in the direction of Barbados, British West Indies.

the charges were altogether beyond us? It would mean Cape Horn! Cape Horn or the abandonment of the dream!

Which was worse for one who, after sixty below zero on the Canadian prairie, four below zero in France and Belgium, and sometimes far worse in coal-less London, has taken a solemn oath never again to leave the forties of latitude?

These terrifying reflections were cut short by a voice.

"I can't make it more than twelve tons."

"Twelve tons?"

The Canal official deigned to exhibit surprise by a slight elevation of the eyebrows, then smiled.

"The measurer has been aboard," he told me, "and you are twelve tons net. The tolls will be fifteen dollars. Will you pay now or at the other end?"

Such was my relief that I paid on the spot, reducing our united capital to £20, or, at the prevailing rate of exchange, $78.00.

This brief interview with officialdom is typical of Panama Canal methods.

Speed, silence, efficiency; nothing else "goes" in "the zone." Things are done in a few seconds and utter silence here that would take hours and pandemonium elsewhere. The entire miracle of passing a ten-thousand-ton liner from Atlantic to Pacific through six locks and forty miles of tortuous, ever-threatening channels has been performed in six and a half hours, and with a lack of fuss that is almost uncanny.

But the *Dream Ship* was twelve tons, and not ten thousand, and for that reason it is probable that she gave more trouble than any craft since the canal was opened. Yet on every hand we received the utmost courtesy and kindness.

Such treatment made us feel like a pestiferous mosquito being politely conducted to the door instead of squashed flat on the spot, as we deserved to be. But you shall see.

STARTING THROUGH THE CANAL

Punctually at 5 o'clock A. M. the pilot came aboard in his immaculate white drill uniform and, without a smile at his sur-

THE "DREAM SHIP" AT ANCHOR IN A WEST INDIES HARBOR

roundings, including ourselves in varie-gated costume, took up his position in the bows. I went below, and after a ten minutes' wrestle with the auxiliary en-gine, contrived to make three cut of the four cylinders "go" sufficiently to propel us at a dignified speed of three knots in the direction of the canal.

"Is that the best she can do?" inquired the pilot.

I lifted an apologetic, perspiring, and begrimed face to his and admitted that it was; moreover, that we were very lucky to be doing that.

"Ah, well, the day is young." he com-mented cheerfully. "What about an awn-ing? We shall be baked alive before we've done."

Did I tell him that the reason we had not rigged an awning was that I was more than half expecting the engine to break down, and that we should have to hoist sail? I did not. Whoever heard of sailing through the Panama Canal? An awning was rigged and we entered Gatun Lock in style, followed by two mere liners.

The giant gates closed. There was an eruption of water seemingly under our stern that caused the tiller to fly over

ON THE WATERFRONT, BRIDGETOWN, BARBADOS

It was a voyage of 30 days for the *Dream Ship* from the Canaries to this, the easternmost of the West Indies.

Photograph by Harold Stieg

SHARKS OF THE CARIBBEAN STRANDED ON THE BEACH OF ONE OF THE ANTILLES : WEST INDIES

Photograph by Harold Stieg

DIVING FOR COINS IN THE CLEAR TROPICAL WATERS AT BARBADOS

At a later period in their wanderings, the voyageurs on the *Dream Ship* were to have an opportunity of comparing this child's play with the activities of the pearl divers of the South Pacific.

and extract a groan of agony from Steve as it crushed him against the cock-pit wall; the aft warp snapped, and the *Dream Ship* commenced to rise, tearing her covering board to ribbons against the lock wall in the process.

There was nothing to be done. Our ascent was as inevitable as the sun's. We rose, and continued to rise, more like an elevator than a ship in a lock, until the blank, greasy wall ended, and above it appeared a row of grinning faces.

"That's that," said the pilot; and it was.

By some miracle the engine carried us to the next lock, where the same performance was gone through, with such slight variations as the loss of a hat, three fenders, and the remainder of the port covering board.

We passed out into Gatun Lake, a fairy place of verdure-clad islets and mist-enshrouded reaches, where cranes flew low over the water, and strange, wild cries came out of the bush.

It was also the place where our engine refused its office peremptorily, irrevocably.

THE DREAM SHIP'S ENGINEER SPEAKS FEELINGLY

I am engineer of the *Dream Ship*, probably the worst on earth, but still the engineer, and for an agonized hour I wrestled with lifeless scrap-iron. How the profession of marine motor engineering ever attracts adherents, it is beyond me to imagine. I know one man it has sent to the asylum, and many others who bear the marks of having trifled with it— finger nails that nothing short of cutting to the quick and gouging with a shovel will render clean; hands, clothes, and. for some unknown reason, face ingrained with ineradicable grime, a permanently furrowed brow, and a wistful expression that goes to the heart of the onlooker.

In order to avoid such a fate, I have made it a practice to try hard for one solid hour and, failing to gain a response from the atrocity, leave the matter in other, and perhaps more capable, hands.

I communicated this information to the pilot, and then and there the man's more human side came to the surface. It was raining as it knows how to rain on the Isthmus, and he was soaked to the hide; his natty uniform resembled nothing more closely than a dish rag; yet he smiled and proceeded to remove his jacket.

"Guess we'd better sail," he said.

Behold once more the *Dream Ship* sailing through the Panama Canal—alternately scudding before rain squalls, lying becalmed, and making tacks of fifty yards and less—a passage surely unique in the annals of "the zone."

The pilot said he enjoyed it, and by the way he swigged on halyards and gave us an old-time chanty to work by, I am inclined to believe him. We were lucky in our pilot.

Toward evening, and during a stark calm, Steve dived overboard and made us fast to a light-buoy, his jaw dropping perhaps half an inch and a thoughtful expression coming into his eyes when a little later a log on the muddy shore was suddenly imbued with life and slipped into the water with a whisk of a horny tail.

So it was that we had afternoon tea in comfort, some alleged music on piano and clarinet, and a pleasant chat with the pilot concerning the older and better days of the wind-jammer, while an ungainly pelican swooped and dived and somewhere ten-thousand-ton steamers were being hustled through the Panama Canal.

TAKEN IN TOW

We had no wireless; that was why it was impossible to summon a tug to take us on our way. Finally a monster steamer passed so close that it was possible to hail her, and a few hours later we were taken in tow by an apparition of noiseless engines, shining varnish, and gleaming brass.

It would cost us $6 an hour, the pilot told us, and I sat back to figure out just how long $78 would last under such an onslaught.

The result was alarming. We held a committee meeting about it in the bows and decided that there was nothing for it but to go on, and keep going on until we stopped. We had hoped to reach lands where money was of secondary importance, but we were not there yet; that was evident.

So we continued to race through the canal at the rate of $6 an hour until we reached the approaches to Pedro Miguel lock, where the apparition tied us up and steamed off, still at $6 an hour.

Something happened to us that night at Pedro Miguel. Looking back on it all, I can hardly persuade myself that it is not a dream. We met some canal officials—tall, sun-burned youths, with the mark of efficiency upon them, yet with a merry twinkle in the eye. We asked them aboard, and they came and marveled at what they saw. Their verdict was, as far as I can remember: "Some novelty!"

ENTERTAINMENT ON THE ISTHMUS

Then they asked us ashore, and it was our turn to marvel. One of our hosts was the chief operator of a lock, and we saw the miracle of the Isthmus of Panama from behind.

Futility overwhelms me at thought of trying to describe what we saw that night, over the lock, under the lock, at the sides of the lock; besides, you will find it all reduced to cold figures in technical journals if you are that way inclined.

It was the spirit of the thing that took hold of me—a pigmy man sitting at a lever! What was not possible after this?

We returned to the ship almost stupefied. One feels much the same if he attempts to think in Westminster Abbey. We were in the process of turning in when a cheerful head appeared through the skylight.

"We await your pleasure," quoth a voice.

I explained that the owner of the head was no doubt unconsciously violating, but still violating, the sanctity of my sister's bedroom. It made no difference.

I protested that at that moment my sister's costume consisted of a pair of ill-fitting pajamas and a kimona, and that Steve and I had nothing to our backs but what we had worn all day—an undershirt and a pair of football shorts; that we were all tired to death and literally ached for our pillows; that his kindness

Photograph by Ralph Stock

UNDER THE AWNING AT LAS PALMAS, CANARY ISLANDS

Buttons come off, rents occur in clothing, and holes in socks, even aboard a *Dream Ship.* Peter proves as handy with needle and thread as with frying-pan and tiller-rope.

been aptly termed (see leaflets). Fear not, lady" (this is an aside to Peter), "the man at the wheel values his life as much as yours, p e r h a p s more.

"*And now* we approach the historic city of Panama, passing cn our left the Union Club, otherwise known as t h e O n i o n Club, frequented solely by the nobility and gentry of the neighborhood.

"*A n d   o n   t h e right—*"

On the right was the blazing portico of a cabaret, and the car had come to a jarring full stop.

In vain we pleaded our costume, the hour of night, the utter degradation of exposing ourselves to the public gaze in such a condition. We literally *found* ourselves at a table drinking imitation lager beer and grape juice and listening to raucous-voiced, imported ladies rendering washy ballads to the accompaniment of tinkling ice and tobacco smoke.

It all sounds sordid enough, but it was vastly amusing to sea-weary wanderers, and will remain with us a memory of kindness and good-fellowship.

was overwhelming, but that— Nothing made any difference.

Somehow we found ourselves in a car, the chief operator's first car, that he had learnt to drive during the dinner hour the previous day.

Out into the moonlight we sped, or rather zigzagged at the rate of forty miles an hour, while between Peter and myself a youth named Bill—I shall never forget Bill—kept up a running flow of informative rhetoric.

"*On* the left we have the famous Isthmus of Panama, intersected by the still more famous Panama Canal—a miracle of modern engineering, as it has

Today we lie at anchor off Balboa, in the Pacific Ocean. We have come far and hope to go a great deal farther. To do so we have come to the conclusion it will be necessary to make some money.

How? Well, we have a ship; a group of pearling islands lies thirty miles to the westward, and—but of this anon.

A strange life, my masters, but one that I would not exchange with any man.

Chapter II

"Tomorrow," said Steve, mate of the *Dream Ship*, "we ought to raise Tower Island."

"Good," said I, with an indifference born of confidence in our navigating officer.

"Splendid!" s a i d Peter, who, owing to a professed and preferred ignorance of navigation, had not ceased to look upon the determination of a ship's position at sea as a species of conjuring trick.

After a seven-thousand-mile sail, we were approaching the ash-heap of the world.

At the time we had no notion that it was an ash-heap, but you shall judge.

Throughout t h a t night we took our appointed four-hour, single-handed watch; slept our four hours, as we had come mechanically so to do during the past four months, and went on deck at dawn to see Tower Island.

It was not there.

Photograph by Ralph Stock

THE COOK'S JOB ON THE "DREAM SHIP"

There was no system of castes on the *Dream Ship*. Master, mate, and "crew" took their turns, one week at a time, over the Primus stove in the tiny nine-by-five fo'castle.

A LOST ISLAND

Steve, who was at the tiller, looked vaguely troubled, but offered no comment; neither did we. "Leave a man to his job" had become our watchword through many vicissitudes. But when night followed day with customary inexorableness, but without producing anything more tangible than the same empty expanse of ocean, Steve was constrained to mutter, a sure preliminary to coherent speech.

"One of three things has happened," he announced: "the chronometer's got the jim-jams, the chart's wrong, or the blinking island has foundered."

As skipper of the *Dream Ship*, it devolved upon me to verify these surprising statements, which after a superhuman struggle (being probably the worst mathematician on earth) I did.

By our respective observations and subsequent calculations, the ship's position proved identical. According to instruments, we were at that moment plumb in the middle of Tower Island. It was thoughtless of it to have evaporated at the very moment when we so sorely needed it as a landmark. We said so in

AN OLD-FASHIONED CANE MILL IN BARBADOS

One-third of the area of Barbados is planted in sugar-cane, yielding as high as 50,000 tons
of sugar and 10,000,000 gallons of molasses annually.

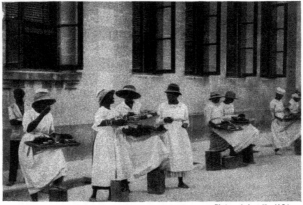

Photographs by J. Harold Stieg

SALES GIRLS IN BRIDGETOWN, BARBADOS: THEIR COUNTERS ARE TRAYS AND THEIR
WARES ARE SWEETMEATS

strong terms. We were still saying something of the sort when a small, high-pitched voice came from aloft:

"Land O!"

Peter, in striped white and green pajamas, was astride the jaws of the gaff. Steve and I exchanged relieved glances and, with a lashed tiller, we all went below for a rum swizzle, the inevitable accompaniment to a landfall. We had reached the Galapagos Islands.

The southeast "trade" was blowing as steadily as a "trade" knows how, and there was nothing between us and San Cristobal (Chatham) Island, the most populous of the group; consequently I slept the sleep of a mind at peace until awakened by a well-known pressure on the arm.

FACING AN UNKNOWN DANGER

"Come and take a look at this," whispered Steve, so as not to wake Peter in the opposite bunk. .

"This" proved to be a solid wall of mist, towering over the ship like a precipice. The trade wind had fallen to a stark calm, and the *Dream Ship* lay wallowing on an oily swell. A young moon rode clear overhead, and myriads of stars glared down at us; yet still this ominous gray wall lay fair in our path.

"It ought not to be land," said Steve, "but I don't like the look of it."

Neither did I.

We stood side by side, straining our eyes into the murk. A soft barking, for all the world like that of a very old dog, sounded somewhere to port. Splashes, as of giant bodies striking the water, accompanied by flashes of phosphorescent light, came at intervals from all sides and presently the faint lap of water reached our ears.

"Mother of Mike!" breathed Steve, "we're *alongside* something."

At that moment, and as though impelled by some silent mechanism, the pall of mist lifted, revealing an inky black wall of rock not fifty yards distant!

My frenzied efforts at the flywheel of the motor auxiliary were as futile as I had more than half expected. Who has ever heard of these atrocities answering in an emergency? We had no sweeps. To anchor was a physical impossibility. The lead-line vanished as probably twenty other lead-lines would have vanished after it, in those fathomless waters. So we stood, watching the *Dream Ship* drift to her doom.

"CLAWING OUR WAY ALONG A ROCKY WALL"

What happened during the next hour is as hard to describe as I have no doubt it will be to believe. The Galapagos Islands are threaded with uncertain currents, and one was setting us now onto the rocky face of an islet cut as clean and sheer to the sea as a slice of cheese. We should have touched but for our fending off. There is no other way of describing our antics than to say that we clawed our way along that rocky wall until, at the end of it, a faint air caught the jib, the foresail, the mainsail, and we stood away without so much as a scratch.

Sunrise that morning was the weirdest I have ever seen. There are over two thousand volcano cones in the Galapagos Islands, and apparently we were in the midst of them. On all hands and at all distances were rugged peaks one hundred to two thousand feet high, rising sheer from a rose-pink sea into a crimson sky.

Sleek-headed seals broke water alongside, peered at us for a space with their fawn-like eyes, barked softly, and were gone. Pelicans soared above our truck and fell like a stone on their prey. Tiny birds, yellow and red, flitted about the deck or flew through the skylights and settled on the cabin fittings with the utmost unconcern.

Down in the crystal-clear depths vague shapes hovered constantly—sharks, dolphin, turtle, and ghastly devil-fish.

All life seemed confined to water and air; never was dry land so desolate and sinister as those myriads of volcanic cones. Yet one of them was peopled with human beings. Which? We were lost, if ever a ship was lost, in the labyrinths of an ash-heap.

BECALMED IN THE WATERS OF THE ASH-HEAP

All we knew was that Cristobal was the easternmost of the group. We sailed east, only to be becalmed inside of an hour and to lose by current what we had gained by wind.

Photograph by Ralph Stock

AT ANCHOR OFF ONE OF THE WEST INDIES

Close to this same group a sailing vessel has been known to have her insurance paid before she reached port. The calms run in belts of varying widths, and unless a ship can be towed or kedged to one side or the other, there is nothing to prevent her remaining in the same spot for six months.

Our water would not last half that time, and there is little on any of the islands except Cristobal.

We began to think. We continued to think for four mortal days, until the fitful southeast "trade" revived, as by a miracle, and we were bowling along at a seven-knot clip. What a relief was the blessed motion of air! We hardly dared breathe lest it should drop.

It held, and we made what we took to be Cristobal. The dinghy was lowered, the ship cleared up for port, and we began to discuss the possibilities of fresh milk, eggs, and bread.

But it was not Cristobal Island. Neither were three others that we visited, all as alike as peas—a chain of ash-heaps; an iron bound coast of volcanic rock, broken here and there by a dazzling, powdered coral beach.

I admit that to the professional seafarer our inability to find Cristobal must appear ridiculous. For his benefit I would point out that we were not professional seafarers, but a party of inconsequent and no doubt over-optimistic land-lubbers engaged in the materialization of a dream—to cruise through the South Sea Islands in our own ship; that what navigation we knew had been learnt in three weeks, and that I would invite any one who fancies his bump of locality to test it in the Galapagos Islands.

LANDING AT CRISTOBAL

We had more than half decided to cut out Cristobal and its five hundred inhabitants and shape a course for the Society Islands, 3,500 miles to the southwest'ard, when Steve gave a yelp like a wounded pup.

"I see Dalrymple Rock," he chanted, as one in a trance, with the binoculars to his eyes. "I see Wreck Point, and a bay between 'em, with houses on the beach. What more do you want?"

How supremely simple it was to recognize each feature by the chart—when there was an unmistakable landmark to go by. What fools we had been not to—

But we left further recriminations till a later date. At the present it was necessary to enter Wreck Bay through a channel three hundred yards wide, without a mark on either side, in the teeth of a snorting "trade," and with a lee tide.

At one time during the series of short tacks that were necessary to get a slant for the anchorage we were not more than fifty yards from the giant, emerald-green rollers breaking on Lido Point to port with the roar of thunder. To starboard one could see the fangs of the coral reef waiting for us to miss stays to rip the bottom out of us.

But the *Dream Ship* did not miss stays, and finally we shot through the channel into Wreck Bay, and anchored in three fathoms of water off a rickety landing stage.

THE OWNER OF CRISTOBAL BOARDS THE DREAM SHIP

While the agony of removing a three-weeks' beard was in progress a crowd had assembled on the beach, and presently a boatload of three put off to us. Steve, who had picked up some Spanish

during three misspent years in Mexico, received them at the companion with a new-born elegance that matched their own.

They proved to be the owner of the island, a good-looking youth of about twenty-five; the chief of police (presumably "chief," because there is only one representative of the law in the Galapagos), a swarthy Ecuadorean in a becoming poncho, and a little, wrinkled old man with a finely chiseled face and delicate hands.

The owner of Cristobal informed us in excellent French (he had been four years in Paris previous to marooning himself on his equatorial possession) that the island was ours and the fullness thereof; that he also was ours to command, and would we dine with him that evening at the hacienda, it being New Year's eve?

The "chief" of police demanded our ship's papers, which, when placed in his hands, he gracefully returned without attempting to read, and gave his undivided attention to a rum swizzle and a cigar.

DAD HAD HEARD RUMORS OF THE WAR

The little old man, whom we soon learned to call "Dad," sat mum, with a dazed expression on his face and his head at an angle, after the fashion of the deaf. When he spoke, which he presently did with an unexpectedness that was startling, it was in a low, cultured voice and in English! "What about this Dutch war he had heard rumors of during the last year or two? With Germany, was it? Well, now, and who was winning? Over, eh, and with the Allies on top? That was good, that was good."

He rubbed his wrinkled hands and glared round on the assembled company with an air of triumph, but without making any appreciable impression on the owner of Cristobal or the "chief" of police.

"Dad" was a type, if ever there was one, of the educated ne'er-do-well, hidden away in the farthest corner of the earth to avoid those things that most of us deem so desirable. He had a split bamboo house on the beach, a wife who could cook, freedom, and God's sunlight. What more did man desire?

He had run away to sea at the age of seventeen, run away *from* sea two years

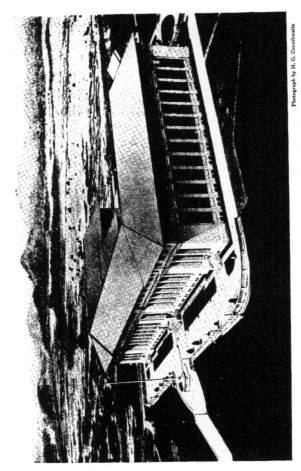

AN AIRPLANE VIEW OF THE PANAMA CANAL ADMINISTRATION BUILDING, BALBOA HEIGHTS

According to the testimony of the adventure-seekers aboard the *Dream Ship*, "things are done in a few seconds and utter silence" on the Panama Canal "that would take hours and pandemonium elsewhere." The miracle of passing a ten-thousand-ton liner from the Atlantic to the Pacific through six locks and forty miles of tortuous channels has been performed in six, and a half hours (see text, page 10).

later at the Galapagos Islands, and remained there ever since. This was the second time he had spoken English in fifty years; so we must excuse his halting diction. But the tales he could tell, the tales! He was here when the pirates of the South American coast murdered for money, even as they have a knack of doing to this day, and hid the loot at their headquarters in the Galapagos Islands, silver and gold—boatloads of it.

He had built a cutter with his own hands and sailed in search of this same loot, only to encounter the then owner, still guarding his ill-gotten gains, though reduced to nakedness and hair. At a distance "Dad" had seen him first and, mistaking him for a mountain goat, had shot him through the heart.

It was the first man he had killed, and he could not stay on the island after that, especially at night.

Afterward I asked the owner of Cristobal if one might believe half the old man said, and he nodded gravely.

"There is much, also, that he does not say," he added with a smile.

TREASURE STILL LIES HIDDEN IN THE GALAPAGOS ISLANDS

There is undoubtedly treasure still lying hidden in the Galapagos Islands. Two caches have been unearthed, silver ingots and pieces of eight respectively. The finder of one built himself a handsome hotel in Ecuador, the other drank himself to death in short order. But there is definite proof that there is more.

As a field for the treasure-hunter, it is doubtful if any place in the world offers better chances of success today than the Galapagos Islands; but—and there is always a "but"—the uncertainty of wind and current among the islands make it impossible for a sailing ship to undertake the search, a motor auxiliary is too unreliable, and even a small steamer is too large for the creeks and reef channels it would be necessary to negotiate.

With a full-powered launch and diving apparatus, and a parent ship in attendance, and unlimited time, money, and patience—but these be dreams beyond the reach of a penhiless world-wanderer—dreams, nevertheless, that will assuredly one day be realized.

No one thinks of the Galapagos Islands. Situated a bare six hundred miles from the American coast-line, in the direct trade route between the South Pacific Islands and the United States of America, this group is seldom visited more than twice a year, and then for the most part by Ecuadorean schooners.

The veriest atoll in the South Pacific receives more attention, and with not a tithe of the cause. The cause? Well, come with us to the hacienda of the owner of Cristobal and you shall see.

PRECISELY ON THE EQUATOR

For this purpose it is necessary to transfer one's activities from the heaving deck of the *Dream Ship* to the equally heaving back of a mountain pony and lope for an hour up a winding, boulder-strewn track through a wilderness of low scrub and volcanic rock.

"Still an ash-heap," you think; "nothing but an ash-heap." Then you surmount a ridge, the last of half a dozen, and rein in to breathe your pony and incidentally to marvel.

You remind yourself that you are precisely on the Equator; yet it is positively chilly up here. A green, gently undulating country, dotted with grazing cattle and horses, patches of sugar-cane, coffee bushes, and lime trees, stretches away to a cloud-capped range of mountains.

The soil is a rich, red loam, almost stoneless, and scarcely touched by the plow. There are 3,500 head of cattle at present on Cristobal Island, and it could support 50,000 with ease. There is no disease and no adverse climatic condition with which to contend, and at three years' old a steer brings $100 (gold), live weight, at Guayaquil—when a steamer can be induced to call and take it there.

There are a few hundred acres under cultivation when there might be thousands, and two hundred bone-lazy peons do the work of fifty ordinary farm hands.

Looking down on this fertile valley, it is hard to realize that one is standing on the lip of a long-extinct crater, that in reality Cristobal is a series of these, dour and uninviting to a degree, viewed from outside, but veritable gardens within. And there are four other islands in the Galapagos group—some smaller, some

Photograph by H. G. Cornthwaite

PEDRO MIGUEL LOCKS ILLUMINATED AT NIGHT: PANAMA CANAL

"Something happened to us that night at Pedro Miguel. Looking back on it all, I can hardly persuade myself that it is not a dream" (see text, page 15).

larger, than Cristobal—uninhabited and exactly similar in character. Nominally, they belong to Ecuador. Here, surely, is a new field for enterprise.

In the midst of the valley, situated on a hillock and surrounded by the peons' grass houses, is the owner's house. Here we met, at a dinner of strange but appetizing dishes, the accountant and the *comisario,* the former a rotund little gentleman with very long thumb nails (the insignia of the brain worker), which he clicked together with gusto, when excited or amused; the latter a tall, handsome youth and something of an exquisite, if one may judge by cream-colored silk socks and an esthetic tie.

DINNER, A CHEERFUL OCCASION

It was a cheerful occasion, followed by the best coffee I have ever tasted and songs to a guitar accompaniment.

Out in the compound, under the stars, the peons also indulged in a New Year *fiesta;* so that by midnight the place was a blur of tobacco smoke, oil flares, thrumming guitars, gyrating, bright-hued ponchos, with their owners somewhere inside them, dogs, chickens, and children.

Every one seemed thoroughly happy and contented. And, after all, what else matters? That is the Ecuadorean point of view, and who shall say it is a bad one?

ONE OF THE MANIFOLD JOYS OF ONE'S OWN SHIP

A star-lit ride to the beach, a few strokes of the oars, that carve deep caverns of phosphorescent light in the inky waters, and we are again aboard. And herein lies one of the manifold joys of one's own ship. One may travel at will over the highway of the earth, carrying his home and his banal, but treasured, belongings with him. Like the hermit-crab, he may emerge where and when he wills, take a glimpse at life thereabouts, and return to the comfort of accustomed surroundings—a pipe-rack ready to hand, a favorite book or picture placed just so.

Sheltered by a coral reef that broke the force of the Pacific rollers and with holding-ground of firm, white sand, we made up arrears of sleep that night, and scattered after breakfast to explore the beach.

There was a lagoon swarming with duck, not half a mile inland, that attracted Steve and his new twelve-bore

THE "DREAM SHIP" FLOATS OUT OF THE MIRAFLORES LOCKS (PANAMA CANAL)
INTO THE PACIFIC OCEAN

gun like a magnet. Peter interviewed the lighthouse-keeper's wife anent cooking for us during our stay, and I—I lazed; it gives one time to notice things that escape the attention of the industrious.

A steam-engine was chugging somewhere behind the belt of stunted trees that fringed the beach, and I found it to be a coffee-grinder fueled, if you please, with sawed lengths of lignum-vitæ—a furnace of wood at something like $5 a stick in most countries! I should like to have seen the face of a block-maker of my acquaintance at such vandalism. But here it is nothing of the sort. Little else in the way of indigenous scrub grows on Cristobal.

WHY DAD DIDN'T FINISH HIS SHIP

Mechanically gravitating t o w a r d "Dad's" split-bamboo abode, I came upon him seated on a log, staring meditatively at the crumbling skeleton of what had been, or was at one time going to be, a ship.

"Why didn't you finish her?" I shouted into his "best" ear.

He stared at me in a daze; then burst forth in Spanish, until I succeeded in convincing him that he might as well talk double Dutch.

"Of course, of course," he muttered. "I forget; Lord, how I forget. It's queer to me that I can speak English at all after all these years; but I can; that's something, isn't it?"

"Sure thing," I yelled: "keep it up. Tell me why you didn't finish your ship."

He pondered the matter; then spoke slowly:

"I told you of the other I built—and why. Well, I ran her on a reef—splinters in five minutes. Took the heart out of me for a bit, that did.

"Then I began to think of that loot again. I do still, for that matter; can't help it. You see, I think I know where it is. So I started on this one." He nodded toward the hulk, silhouetted against the crimsoning sky.

"I'd got to the planking when it occurred to me I'd want a partner for the

Photograph by Earle Harrison

THE PUBLIC SQUARE, PANAMA CITY

The cities of Panama and Colon, at the Pacific and Atlantic ends of the Panama Canal, are under the authority of the Republic of Panama, but complete jurisdiction is given the United States in both cities and their harbors in all that relates to sanitation and quarantine.

job, at my age; and who could I trust? They'd slit your throat for $10 in those days. They murdered the present owner's father in cold blood. I wouldn't put it beyond 'em to do the same to this one if it wasn't that he's a smart lad and carries the only firearms on the island.

"No one's come here since, no one that I'd trust. . . . Then, too, what if I found the stuff? What good would it do me—now?" He spread out his delicately shaped hands in a deprecating gesture. "I should die in a month if I left here. Finest climate on earth, this is. . . ." Suddenly he laughed — a low, reminiscent cackle of mirth.

"But that wasn't all that decided me. I'd got to the planking, Guayaquil oak it was, and I was steaming it on when a nail drew, and that plank caught me in the chest, knocked me six yards, and broke a rib. It's broken yet, I guess; there was no one to mend it. Well, that finished it. I wasn't meant to build that ship."

He stopped abruptly and stared down at his battered, rawhide shoes.

SHALL THE DREAM SHIP BECOME A TREASURE SHIP?

The inference was obvious.

"Well, what about it?" I suggested.

He looked up at that.

"I've been thinking about it ever since you came here," he confessed. "I'll go with you; but mind this, you musn't curse me if nothing comes of it. I don't promise anything. All I say is I think I know where the stuff is, if some one hasn't gotten it."

"I'll let you know tomorrow," said I, and left him sitting there.

Was the man senile? There was nothing to make one think so. Was he a liar? There was equally nothing to prove it. At least half his story was a matter of island history.

We of the *Dream Ship* held a committee meeting on the subject of loot that evening. We discussed it from every angle, and came to the conclusion that, with the present atrocity called a motor auxiliary and the weather conditions of the group, we might take three days over the business and we might take three months; that the chances of finding something were outweighed by the risk of losing the ship, and that we were in pursuit of something sufficiently visionary, anyway, so we had better get on with it.

The voting went two to one against, and I leave you to guess who's was the deciding voice.

I give this interview with "Dad" for what it is worth, and simply because I see no prospect of undertaking the search as it should be undertaken. I am aware that it reads like the purest romance, but it is true in every particular, as any one will soon discover on visiting Wreck Bay, San Cristobal, Galapagos Islands.

The old man still waits there on the beach for a ship and some one he can trust; but judging by his frail appearance (he is seventy-seven), he will not wait much longer.

Often during the days that followed I found myself standing at the *Dream Ship's* rail, looking seaward to a dim outline of mountains against the blue, and wondering. . . . But only the ashheap knows.

*　*　*　*　*　*

Tomorrow the owner of Cristobal, the accountant of the elongated thumb nails, and the exquisite comisario are to dine with us (on Heaven and our cook alone know what), the next day we hunt the wily duck among the lagoons and marshes of the island, and the day after that, D. V., we continue the pursuit of our dream across three thousand miles of South Pacific Ocean, west-sou'west, to be exact.

Chapter III

More nonsense has probably been written about the South Sea Islands than any other part of the world. The library novelist, the globe-trotting journalist, and a reading public athirst for exotic romance have all contributed to this end; so that at the very outset of this paper I find myself at a loss. In short, "these few remarks" may be taken as an apology and a warning.

I have nothing to offer on a par with the standard article, such as struggles with sharks, conflicts with cannibals, or philandering with princesses. My line, I fear, consists of facts as I find them.

THE COMISARIO PLEADS FOR PASSAGE ON THE DREAM SHIP

The *Dream Ship* lay in Wreck Bay, Galapagos Islands. Her crew had just finished watering, or rather transferring 300 gallons of a doubtful-looking fluid from the beach reservoir to the ship's tanks by means of kerosene tins, a rickety landing stage swarming with sand flies, and an equally rickety dinghy.

We were, in fact, enjoying a spell to the accompaniment of vast quantities of coconut milk before setting sail for the Marquesas Islands, 3,000 miles distant, and were in no mood for an interruption, which is probably why it came. A pigmy figure on the landing was apparently dancing a hornpipe and emitting strange cries.

"Who is it, and what the — does he want?" I queried with customary amiability.

"It's the *comisario*," said Steve, with binoculars upheld in one hand and a brimming coconut shell in the other, "and he's probably found that we need a bill of health or clearance or something."

I believe I sighed. I have a notion that Steve swore, and I am quite sure that we rowed ashore and interviewed the *comisario*, the handsome youth whose silk socks and passionate tie contrasted strangely with his surroundings. He still danced.

"He says that it is necessary that he should accompany us," Steve translated.

"To the Marquesas?"

"To anywhere."

"Really, and where does the necessity come in?"

After still further variations of the hornpipe and a prodigious outflow of Ecuadorian Spanish, the following was evolved:

They were after him—a trifling indiscretion in the matter of issuing grog licenses to the peons. The Ecuadorian Government was to blame. They expected an official to live on $20 a month and nothing else! How was it possible? Moreover, the President himself, elected on a wage basis of $40 a month and bring

THE RICKETY LANDING STAGE IN WRECK BAY, SAN CRISTOBAL, OF THE GALAPAGOS
GROUP, THE "ASH-HEAP" OF THE PACIFIC

One of the crew of the *Dream Ship* in a dinghy is completing the job of transferring 300
gallons of doubtful-looking fluid from the beach reservoir to the vessel by means of kerosene
tins. This water was transformed into an "aquarium" of energetic animalculæ before the
ship reached the Marquesas Islands (see text, page 30).

"DAD," WHO HOLDS THE SECRET OF HIDDEN TREASURE AMONG THE
GALAPAGOS ISLANDS (SEE TEXT, PAGE 27)

THE OWNER OF SAN CRISTOBAL ISLAND (HORSEMAN IN THE CENTER) AND TWO OF
HIS ADMINISTRATION OFFICIALS

The owner is a young man of twenty-five, who spent four years in Paris before marooning himself upon his equatorial possession. His hospitality to the *Dream Ship's* crew took the form of a New Year's celebration.

Photographs by Ralph Stock

THE PEON QUARTER ON SAN CRISTOBAL, THE MOST POPULOUS ISLAND OF THE
GALAPAGOS ARCHIPELAGO

Photograph by Ralph Stock

PETER INTERVIEWS THE LIGHT-KEEPER'S WIFE ON SAN
CRISTOBAL (CHATHAM) ISLAND

The name of the Galapagos Archipelago is derived from *galapago*,
a tortoise, on account of the giant species peculiar to the islands.
Owing to the isolation of the Galapagos group, these tortoises were
of special value to Darwin in his studies relating to the "Origin of
Species."

be mentioned, exceed-
ingly tired of cooking.
The *comisario* seized
on our silence.

SEÑOR BILL JOINS
THE CREW

"Maybe we thought
he could not work!"
With a dramatic
gesture he tore from
his neck the passion-
ate tie, from his feet
the silk socks, from
his back a violently
striped shirt, a n d
stood revealed in a
natty line of under-
vests.
"Poor devil," said
I, thinking of the
Dream Ship's fo'cas-
tle in a seaway.
"P o o r nothing,"
s a i d Steve. "H e
wants work; let him
have it."
And that was how
S e ñ o r ———, hence-
forth known as "Bill,"
came to join t h e
Dream Ship.
We sailed and sailed
before a steady south-
east trade wind for
twenty-two days, dur-
ing which the *comi-
sario* suffered alter-
nately from seasick-
n e s s, homesickness,
and sheer inability to
to do anything but
smoke cigarettes and
sleep. Our water tanks, under the magic
wand of the Galapagos beach reservoir,
transformed themselves into aquariums
of energetic animalculæ, and our entire
biscuit supply crumbled to dust under the
onslaughts of a particularly virulent red
ant.
But these be incidentals to life aboard
dream ships, and at the first sight of
Nukuhiva faded to little more than amusing memories.
A fine island this—as fine a volcanic
island as one will find anywhere. Sheer

your own blankets, would be getting the
boot in a short three months, and with
him went every one—every one!
What was then to happen to the officials he had placed in power? More
important still, what was to happen to
this particular official? He must accompany us. It was the only possible solution. He would work. *Carramba*, how
he would work! and for nothing but his
passage to anywhere—anywhere!
Steve and I exchanged glances. The entire crew of the *Dream Ship* was, it may

walls of c l o u d - capped rock 6,000 feet high, some liter- ally overhanging the crystal - clear water, and all embossed and e n g r a v e d w i t h strangely patterned basalt. There are pil- lars, battlements, and turrets; so that with half - closed eyes it seems one is a p - proaching a temple, a medieval castle, a mosque of the East. A n d the valleys— deep, river-threaded, verdure - choked val- leys — fading away into mysterious pur- ple mists! But it is little better than an impertinence to at- tempt a description o f Nukuhiva after Melville's "Typee."

GUESTS OF FRENCH IN TAI O HAE

For once the mon- strosity in our engine- room was induced to exert three of its four cylinders, and we en- tered the harbor of Tai o Hae in style. It was as well, for a trim trading schooner flying the French flag was at anchor close inshore, and her en- tire crew lined the rail to see what man- ner of insect had invaded her privacy.

"Where are you from?" hailed a sur- prisingly English voice as soon as our anchor chain had ceased its clamor.

"London," we chorused.

"Well, I'm damned!" came a response, evidently not intended for our ears, but audible nevertheless.

In rather less than three minutes a whaleboat-load of visitors was aboard the *Dream Ship*, and the silent bay echoed to a fusillade of question and counter- question.

Photograph from Ralph Stock

ALLEGED MUSIC ON THE CLARINET: CABIN OF THE "DREAM SHIP"

Owing to the fair winds encountered throughout most of the voyage, it was seldom necessary for more than one of the three voyageurs to remain on deck at a time. The watches were appor- tioned four hours on and eight off.

There followed a dinner at the trading station, on a wide, cool veranda, where, under the influence of oysters, California asparagus, fowl, bush-pig, taro root, and French champagne, we became better ac- quainted with our hosts, two as amiable Frenchmen as ever I met. They repre- sented a trading company of Papeete and lived as only Frenchmen appear to know how to live.

The Marquesans, we gathered over coffee and cigars, were dying rapidly of consumption, introduced in the form of Panama fever by laborers returning from

"LAND O!"

This is *not* Peter discovering lost Tower Island of the Galapagos group, however (see text, page 19).

canal construction. The fever afterward developed into the white plague by reason of the natives' unresisting, if not acquiescent, nature. And when all were gone, what then? Chinese.

The Chinese appear to be the answer to most questions in the South Pacific today. They come; it costs them but $50 to land; and after that they grow—*mon Dieu,* how they grow!

And can nothing be done? A shrug of the shoulders and the offer of a refilled glass are the answers of the Frenchman. But a short time now and he personally will be in a position to return to his beloved Paris, or Marseilles, or Brittany.

But we had lately returned from dealing with the Boche; so had our hosts. We drank respectively to the Royal Field Artillery, the Mitrailleuse, the Machinegun Corps, and the incomparable French Infantry. What of it, if we continue the sport on the morrow, among the wild cattle and goats of Nukuhiva? Tomorrow, then, at 5 o'clock.

OFF ON A HUNT FOR WILD CATTLE AND GOATS

The schooner, scheduled at daylight to load copra worth $500 a ton, was cheerfully detained for the trip and loaded to capacity with bottled beer, coughing Marquesans, and a variegated armory of firearms.

We sailed down a coast that it is a sore temptation to describe and landed by whaleboat on a surf-pounded beach. Thereafter we plodded, crawled, and stumbled over as vicious a country as it is possible to imagine—crumbling shale, razor-edged ledges, and deceptive tablelands of knee-high grass that only served to hide the carpet of keen-edged volcanic rocks beneath.

And the heat! But a representative of the incomparable Infantry led the way; and who would not follow to the death, out of very shame? At each halting place the elan of this same representative seemed to increase. Sitting cross-legged on a rock in the meagre shade of a scrub tree, he would discourse on any subject under the sun, while his audience gasped, emptied the perspiration out of their boots, and cursed the *cantene* (a gigantic

RECIPROCAL ABLUTIONS ABOARD THE "DREAM SHIP"

native bearing an almost as gigantic sack of bottled beer) for lagging.

I was under the impression that the game was to have been wild; hence my surprise when a herd of something like 150 goats of all ages, from the bearded and maned veteran, or "stinker," down to the daintiest kid, cavorted up to our resting place and sniffed at us inquisitively. It was necessary to fling stones to keep some of the more daring at bay.

So much for goat-hunting in the Marquesas. It is evident that these beasts are so "wild" that they know nothing of man; and who shall say they have missed much in consequence?

FOUR CATTLE BAGGED FROM A HERD OF 50

The cattle are a different matter. Shy as deer, they must be warily stalked and shot mostly on the run, at anything from 100 to 150 yards; also, they have an engaging habit of turning when wounded and giving the huntsman the worst possible time in their power, which in the case of a hefty bull or cow with calf is not inconsiderable.

There must have been a herd of some-thing like 50 grazing on the precipitous hillside, and the first shot, fired by an over-anxious Marquesan, against strict orders, sent them scuttling like antelope out of the valley and over a ridge. One fine bull received his medicine from my trusty little Winchester on the very brink, collapsed, and rolled like an avalanche of meat to the bottom.

We bagged four of this herd, and the Marquesans fell on them, quartering and selecting with extraordinary skill, and finally carrying 100 pounds each of solid meat to the beach five miles below. How this last feat was accomplished by a band of ramping consumptives I have no notion, though I saw it done. I only know that after carrying two rifles and a gun over the same country I literally tumbled onto the beach, bruised and bleeding and trembling from sheer fatigue. Even the representative of incomparable Infantry admitted to being tired, and, thank heavens, he looked it!

A NATIVE DANCE, A PAGEANT OF HISTORY

It had been a successful day, I was given to understand, and there followed

Photograph from Dr. Hugh M. Smith

PORT TAI O HAE, NUKUHIVA, MARQUESAS ISLANDS

"A fine island this—as fine a volcanic island as one will find anywhere. There are pillars, battlements, and turrets; so that with half-closed eyes it seems one is approaching a temple, a medieval castle, a mosque of the East" (see text, page 30).

34

in consequence song and dance aboard the *Dream Ship* until dawn touched the peaks of Tai o Hae.

A native dance is a dreary and monotonous affair to the average white man, because he does not take the trouble to understand. He sees before him an assembly of posturing, howling natives, and seldom realizes that he is witnessing a pageant of history that has never been written or read.

The performance opened with a pantomimic representation of the cruise of the *Dream Ship*. According to the actors' ideas, all aboard suffered acutely from seasickness, were utterly unable to stand upright, and continually looked for land under the shade of an upraised hand. Our vigor in battling with storms was extraordinary; we stumbled over rope ends, clung to the rigging, nearly capsized, and one of us fell overboard, to be rescued, amid shrieks of laughter, by means of a boat-hook and the seat of his pants.

We were a joke, there was no doubt about that, and any one who takes a 10,000-mile journey in a 23-ton yacht to the Marquesas and wants to be taken seriously had better go elsewhere.

From such trivialities the performers passed on to what was evidently their stock repertoire—the history of the Marquesas as handed down from father to son. It was all there in gesture and chant—mighty battles with their neighbors the Paumotans, cannibalism, peace, the advent of the white man with his rum, the plague that still consumes them, and all enacted without resentment.

That is the most astounding thing, that these people who were living their own lives, and surely as happy lives as ours, bear no ill will for the incredible sufferings our civilization has brought among them. Perhaps they do not think, and if so it is as well.

Conceive yourself, if you can, oh denizen of Park Lane, Fifth Avenue, or Champs Élysées, a healthy, upstanding, unclad savage of the South Seas, and living your own life.

You may be a cannibal; and are there no cannibals, and worse, west of Suez? You will be a warrior and fight for your country and your women folk. Is there anything wrong about that?

You will have a stricter moral code than most white folk, but that cannot be helped. You will hunt and fish and gather fruit for your family—in fact, you will live in the only way you know how to live, in contentment.

One day an extraordinary-looking object called a white man presents himself and informs you that you are not living in the right way at all. A much better way, according to this gentleman, is to exchange a ton of your coconuts for a bottle of rum or a death-dealing instrument made of rusty iron.

You are a tolerant sort of person, and you listen and drink his rum. The next day you have an insufferable headache, and, logically concluding that he has poisoned you, you kill him.

But that is not the end. Replicas of him keep arriving, and you find you need his rum and his rusty iron, the one for its elevating properties, the other for its dispatch in dealing with enemies.

OFF TO HUNT PEARL SHELLS

To revert to safer topics, there is pearl shell in the Marquesas. The representative of incomparable Infantry told us so while we sat on his incomparable veranda one morning, consuming large quantities of papia, rolls, honey, and coffee, each in his particular brand of pajamas.

The information brought upon our serene lives at Tai o Hae the white man's blight of avariciousness. Was this thing possible, with shell at $1,000 a ton, delivered at Philadelphia? Yes; he, the incomparable, had seen it through a waterglass, in anything from five to fifteen fathoms, between the islands of Hivaoa and Tahuata.

Why had it not been prospected? It was doubtful if any but he and the natives knew of its existence. Undoubtedly it was worth looking into. He made us a present of the information, to do with as we willed. His cook was an old Paumotan diver, who would no doubt accompany us—Pascal!—accompany us to the island, a bare 90 miles distant. We could take samples of shell to the company in Papeete, and no doubt make arrangements — Pascal!! — arrangements with them to advance working capital in return for a lien on the shell—Pascal!!!

Photograph by Ralph Stock

PASCAL, A PAUMOTAN NATIVE WHO PROVED A WIZARD AS A COOK, BUT A SNARE
AND A DELUSION AS A PEARL FISHER

"He could produce savory messes from a kerosene tin, remain under water three minutes, discourse entertainingly in pidgin-English, French, German, Marquesan, and Paumotan, and secure a ship's provision without the annoying triviality of paying for them."

"Monsieur." An enormous Paumotan native stood in the doorway smiling benignly.

He would accompany us. He would cook and he would dive.

PASCAL PROVES A REMARKABLE FORAGER

We sailed that evening, the deck being littered with green bananas, live chickens tied by a leg to bulwark stanchions, a rabbit, firewood, a stove composed of a kerosene tin half filled with earth, and— Pascal.

There was apparently nothing that this extraordinary man could not do. He knew every island of the Marquesas like the palm of his hand. He could produce savory messes from a kerosene tin, remain under water three minutes, discourse entertainingly in pidgin-English, French, German, Marquesan, and Paumotan, and secure a ship's provisions without the annoying triviality of paying for them.

"But whom do we owe for all this?" I asked him, eying the menagerie that surrounded us.

Pascal smiled and waved a hand.

"Rabbit no money," he informed us; "chickens, bananas, all no money. I get um."

Here surely is a solution of the "high-cost-of-living" problem. Take Pascal to the profiteering areas and the thing is done.

Dawn revealed to us Tahuata close abeam. Each island of this group seems more lovely than the last: waterfalls pouring 3,000 feet to the sea, blow-holes at the base of rocky cliffs that spray the air with spindrift and miniature rainbows, deep bays with coral beaches at their head.

But the beauties of nature were not for us on this occasion; we were prospecting. It was a serious business. There might be money in it. After this I can scarce believe that in Paradise itself the white man will not be dogged by the curse of opportunism.

Photograph by Ralph Stock

SOUTH SEA ISLANDERS PAY A VISIT TO THE "DREAM SHIP"

Leaving the *Dream Ship* at anchor a cable's length from shore, we took to the dinghy and explored the floor of the ocean thereabouts through water-glasses. This was the place, Pascal informed us, and, sure enough, there was shell, old barnacle-encrusted shell, but widely scattered.

What of a few samples? Pascal grinned and shook his head. "Shark," he muttered apologetically; which, being interpreted, meant that he refused to dive. He pointed out that in the Paumotus it was different. In the Paumotus there was always a reef-surrounded lagoon, where few sharks found entrance. In the Paumotus men dived in couples as a safeguard. In the Paumotus—

In vain we pointed out that we happened to be in the Marquesas and not the Paumotus; that he had been hired to dive in the Marquesas; that we were really very angry—in the Marquesas. He grinned.

HO FOR TAHITI

In rather less than half an hour, and to Pascal's utter amazement, we had put him and his belongings ashore, paid him

his wages, and were under way for Tahiti.

Ah, Monsieur of the incomparable, I rather suspect you of "pulling our legs." Or was it that your innate enthusiasm ran away with you? Or that we should have been less hasty? I do not know. All I know is that you spoke truth; there is shell in the Marquesas—and it is likely to remain there.

CHAPTER IV

From the moment I first set eyes on an atoll it fascinated me, and its lure has not departed with the years.

Think of any place in the world that you have seen, and an atoll is different. It is the fairy ring of the sea. Out of the depths it comes, rearing a vegetation and a people of its own, and often into the depths it goes, leaving no trace. How? Why? Scientists murmur something about the vagaries of the corallitic polyp; but, not being a scientist, I prefer my theory of the fairy ring. That was how it looked to me several years ago, and that was how it looked again from the masthead of the *Dream Ship*.

NATIVES SPEARING FISH IN THE SOUTH SEA ATOLLS

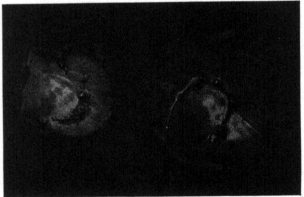

Photographs by Ralph Stock

PEARL BLISTERS ARTIFICIALLY PRODUCED BY "MR. MUMPUS," OF THE SOUTH SEAS
(SEE TEXT, PAGE 43)

The pearl is a disease of the oyster; introduce the disease and you will get a pearl, according to this authority, who hopes some day to be able to produce whole pearls as readily as he now produces half-pearls.

We had left the Marquesas seven days previously, and were now becalmed in that maze of atolls known as the Paumotu, or Low Archipelago.

Imagine a circular beach of glistening coral sand and green vegetation from five to fifty yards wide, thrust up through the sea for all the world like a hedge, and inclosing a garden of coral fronds submerged under water so still and clear as to be hardly visible, and you have an atoll as I saw it from the masthead.

There were myriads of them — big atolls, little atolls, fat and thin atolls—fading away into the shimmering heat haze of the horizon. The fairies must have been mighty busy down this way.

I descended to the deck and things mundane. "What to do when becalmed in a network of coral reefs and seven-knot currents" was the problem that confronted us.

A RACE OF MERMEN

I had no text-book on the subject, but by some miracle the thing we called an engine was persuaded to fire on two of its four cylinders, and the *Dream Ship* tottered through the narrow gateway in the hedge—I should say, pass in the reef—and came to anchor in the garden—I mean lagoon.

It was sunrise, and already the pearling canoes were putting out from the village and scurrying to the fishing grounds over the glassy surface of the lagoon.

A fine people, these of the atolls—upstanding, deep of chest, a race of mermen if ever there was one. From birth up, if they are not in the water they are on it or as close to it as they can get. Take them inland and they die. So they squat on their canoe outriggers, smoking, chatting, laughing, until the spirit moves them (nothing else will), and one of their number drops from sight, feet first, with hardly a ripple.

You look down and you see him, as though through green-tinted glass, crouched on the sloping floor of the lagoon. He is plucking oysters as one would gather flowers in a garden. There is no haste in his movements, nothing to indicate that there is any time limit to his remaining down there, under any-

thing from five to fifteen fathoms of water.

A minute passes, two minutes; still he pursues his leisurely way, plucking to right and left and thrusting the shells into a network bag about his neck.

UNDER WATER THREE MINUTES

The man of the atolls is in a world of his own, where none but his kind can follow, and they still squat on their outriggers, chatting and laughing like a crowd of boys at a swimming pool.

One alone seems interested in the diver's movements; his mate, a fair-skinned woman, with streaming blue-black hair, leans over the gunwale of the canoe, looking down through a kerosene tin water-glass.

The diver's dark figure against the pale-green coral becomes more blurred; a stream of silver air bubbles floats upward. Three minutes by the watch have come and gone. To the landsman it seems incredible; and even then there is no haste, no shooting to the surface and gasps for breath.

The dark body becomes clearer in outline as it emerges from the depths, and slowly, quite slowly, floats upward until a jet-black head breaks water and the diver clings to the gunwale of the canoe, inhaling deep but unhurried breaths and exhaling with a long-drawn whistle peculiarly his own.

In what way this whistle helps matters, it is impossible to say, but whether a habit, a pose, or an aid in the regaining of breath, it is universal throughout the Paumotus; so much so that a busy afternoon with the pearlers sounds more like a tin-whistle band than anything else.

With the people of the atolls the ability to remain under water for long periods is more than an art; it is second nature. Instinctively, they do just those things that make one breath suffice for three minutes and sometimes four.

Preparatory to a descent they do not take a deep breath and hold it until the surface is reached again. They fill their lungs with a normal amount of air, which lasts them about a minute and a half; the other minute and a half is occupied in its exhalation. Then, too, every movement below water is made with the utmost conservation of energy; yet a good

Photograph by Ralph Stock

A MAN OF THE ATOLLS POLING HIS CANOE TO THE PEARLING GROUNDS

Upstanding, deep of chest, the natives of this part of the Pacific are veritable mermen,
living in and on the water almost from birth.

diver can bring up 150 kilos of shell in a day, which means in the neighborhood of 600 francs.

DEAFNESS, PARALYSIS, AND IMPAIRED VISION THE DIVER'S LOT

And it is just these same nimble dollars that tempt the Paumotan to abuse his talents, even as others are tempted the world over. For the sake of a few more shells, another cluster a little farther down, he remains below just that trifle longer than is good for him, and in time it tells. The eyes become bloodshot and start from the head, he goes deaf, or paralysis seizes him.

"But the women are the worst," a sun-baked trader informed me; "the worst or the best, as you like to put it," he added grinning. "They'll go on till they burst, or pretty near it. Bargain-counter instinct, I guess. We call it the 'bends.'"

"The 'bends'?"

"Yes, one of 'em goes down, and down; sees some more shell a little lower, and some more a little lower than that. Then she's reaching out for one last flutter at something like twenty fathoms when they get her—the 'bends,' I mean. You can see her fighting against them, but it's no good; they bring her knees to her chin, and she can't straighten up, and she drops the last lot of shell she's gathered, and hates that worse than the 'bends.'"

"What does she do?"

"Nothing, except lie there crumpled up until her mate fetches her up and massages her back to life. Then she's no sooner conscious than she's down again."

HE SWIMS WHEN HE CAN NO LONGER WALK

"Water never kills this crowd; it takes dry land to do that. Why, there's a diver close on fifty years old here, paralyzed clean down one side. He can't walk, but he can swim. He gets them to carry him down to the reef and heave him in; says it's the only place he can get any comfort."

"How about sharks?" I asked.

"Oh, there are sharks all right, but the diver's mate looks after that; gives the signal and they're all in after him double quick."

"Finish him off with knives, eh?"

The sun-baked trader smiled reminiscently.

"Well, hardly," he said. "A dead shark makes a square meal for the others, and that's all. What they need is an example, and they get it. They're cruising about sometimes when they come on one of their number with no tail, one fin, and sundry other decorations that wouldn't exactly please the S. P. C. A. He is not nice to look at, and they clear out of a place where such things are possible.

"When an island's thrown open for pearling, we spend weeks mutilating sharks before the divers'll go down, and small blame to them, I say. Sharks are—well, sharks."

A PEARL RUSH IN THE PAUMOTUS

The casual reader picks up a good deal of information about "gold rushes" and such like romantic undertakings from the plethora of novels on the subject; but who has ever heard of a "pearl rush"? Yet they occur every year in the Paumotus.

The group belongs to the French, and is administered from the local seat of government at Papeete, Tahiti. Here a heterogeneous collection of humanity awaits the opening of the pearling season like a hovering cloud of mosquitos.

There are pearl buyers from Paris and London, representatives of shell-buying concerns from Europe and America; British, Chinese, and Indian traders, speculative schooner skippers and supercargoes, not to mention the riffraff of the beaches, all intent on pickings from the most prolific pearling islands in the South Pacific.

And this is the law of the group—infringed, circumvented, broken, but still the law—that although under French Government, the Paumotus and all they produce belong to the Paumotans.

FRANCE PROTECTS THE NATIVE DIVERS

Still further to protect the native, diving apparatus is banned throughout the group. The oyster, as he brings it from the water, is the diver's property. He must open the shell aboard his canoe before touching land, remove the flesh, and, after testing it for pearls (usually by kneading it so thoroughly between finger

Photograph by Ralph Stock

A PEARL DIVER PREPARING FOR A DESCENT

during the closed season.

With shell at $1,000 a ton in Philadelphia (the largest consumer at the present time) and pearls soaring to apparently limitless heights, all will be well when work starts.

And the diver? From long experience of mosquitos, he is by no means slow. Shortly before the season opens he is presented with a bill that would cause most of us to register apoplexy. He looks at it, grins, and proceeds to dive. He also proceeds to make caches of shell on the floor of the lagoon, only bringing up half of what he collects in payment of his debts. At night he retrieves his cache and sells for cash to the smaller mosquitos who infest the beach.

As for pearls, from the moment the diver's finger and thumb encounter foreign matter in the flesh of the oyster, he becomes

and thumb as to crush the life out of it), throw it back into the lagoon to propagate its species. Should he find a pearl, it is his also.

It is then up to the cloud of mosquitos before mentioned to get both shell and pearl out of him as best it can. One can imagine the buzzing and biting that ensues.

From the buyer's point of view, the sooner and the deeper he gets a good diver into his debt the better. He then has some hold. Consequently he spoonfeeds his selected divers like the infants that they are. Tinned delicacies of all sorts, Prince Albert suits of unbelievable thickness and cut, silk socks, and stockings are a good diver's for the asking,

about as communicative on the subject as his catch. Should the truth leak out, his find will promptly be confiscated in payment of his everlasting debts, or the wily pearl-buyer will use threats of exposure to reduce the price.

No, the diver, if he is up to snuff, will work his passage to Papeete on a schooner, sell to a Chinaman, who neither asks questions nor tells tales, and proceed to enjoy himself according to his lights.

Blossoming into a Prince Albert suit, a red tie, and silk socks, he will hire a car, load it up with lady friends and execrable rum, and vanish into thin air for a fortnight, at the end of which time he has somehow contrived to get rid of

Photograph by L. Gauthier
FISHING MEN OF TRITI, NEAR PAPEETE, TAHITI

all he possessed and is perfectly prepared to return to his atolls and his debts. He has lived like a white man and cheated the mosquitos; what more can Paumotan heart desire?

PROGRESS HAS NOT DESTROYED PICTURESQUENESS OF PEARL FISHING

The thing we call progress has slain the picturesque in most industries of this world, but not so with pearling in the Paumotus. During the season, the beach of one of these atolls resembles an Old World fair more than anything I can call to mind.

A crazy merry-go-round brays and rocks in the shade of the palms, luring the adventurous to invest three pearl shells in a ride on a broken-necked camel. The ubiquitous movie "palace" has reared its unlovely head, and for more shell or five coconuts one may witness on the shores of a South Sea lagoon the battered remnants of a love affair enacted not far from Los Angeles. I have often wondered what happens to all the worn-out films in the world. Now I know.

This season, and for the first time, the people of the atolls are to be initiated into the mysteries of ice-cream. Truly, the mosquito stops at nothing.

It was down in this part of the world that I met Mr. Mumpus, though that is not his name. To reach him you must pick your way in the motor auxiliary through a maze of reefs, lie off and on, because there is no pass into his lagoon, and plod through blazing sand in a temperature of ninety in the shade, which there is not. But it is worth it.

THE PEARL-MAKER OF THE SOUTH SEAS

You will probably find him in the pearl orchard, a green-lined umbrella in one hand and a dripping oyster shell in the other. He will stare fixedly for upward of half a minute and then say: "How the devil did you get here?" with a brusqueness that is alarming until you get used to it.

In my own case I indicated the *Dream Ship*, looking particularly smart in her recent coat of white paint.

Photograph by Ralph Stock

TAHITIAN BEAUTIES IN A TARO PATCH

It is from the rootstock of this plant that one of the staple foods of the Pacific Islands is made. The fermented paste is known in Hawaii as poi. In temperate regions taro is our ornamental caladium or "elephant's ear."

"What! in that thing?" remarked Mr. Mumpus.

I was smitten to silence for a space.

"I heard you were making pearls," I told him on regaining something of my equanimity, "and thought you might be so good as to tell me about it."

For some reason, probably the appearance of myself or my ship, Mr. Mumpus took pity on me.

"A PEARL IS A DISEASE OF THE OYSTER"

"Come up to the house," he barked, and led the way to a rambling erection of corrugated iron and palm leaves containing, as far as I could make out, a gaping "boy" of uncertain origin, some empty soap boxes, and a microscope.

"There's nothing new in what I'm doing here," he told me over two brimming shells of coconut milk, "nothing that the Chinese have not been doing for centuries. The pearl is a disease of the oyster; introduce the disease and you will get a pearl."

"Quite," said I.

"No one has succeeded up to the present," continued Mr. Mumpus, "but there is no reason why it should not be done in time, no reason at all. I am appreciably nearer than I was a year ago, for instance. In the meantime I am producing

THE "UPA UPA," ONE OF THE FAMOUS NATIVE DANCES OF THE SOCIETY ISLANDS

Papeete, the principal settlement of the Society Islands, is the metropolis of the southeastern Pacific, just as Honolulu is of the northeastern. It attracts as varied an assortment
of humanity as any port in the world.

the ordinary blisters, or half pearls, with various foundations. You see, the cestode—"

But I cannot hope to set down here all that this amazing man told me in scientific jargon, as he strode back and forth across his mat-strewn floor.

He was a doctor by profession, had tired of it, and had come to the islands to pursue his hobby of pearl culture. He takes an oyster from the lagoon, opens it very carefully by the slow insertion of a wooden wedge, and places a pilule of beeswax against the main muscle. The mantle of the oyster then covers it with mother-of-pearl, and in the course of a few months our friend cuts from the shell a very fair imitation of a half pearl.

But, as most people are aware, the real pearl comes from the flesh of the oyster, and it is on the production of the genuine article that Mr. Mumpus centers his efforts. He breeds oysters in the lagoon and dissects them under the microscope for signs of the parasite that undoubtedly causes the pearl. He injects into the flesh of others all manner of foreign matter.

Down there on his speck of an atoll he treats the oyster as a surgeon treats an interesting case and, who knows? some day there may burst upon an astonished

THE HARBOR OF PAPEETE, TAHITI

It was in this port that the *Dream Ship* lost its cook, the young exquisite of Cristobal. He has renounced the life of a wanderer and may be seen today presiding over the soft-goods counter of a French establishment in Papeete.

world the name of a man who can make pearls, and that name will not be Mr. Mumpus.

CHAPTER V

DREAM'S END

Regretfully leaving the people of the atolls, the *Dream Ship* set sail for Papeete, Tahiti, and arrived there without mishap.

A greater contrast in two groups of islands a bare day's journey apart can hardly be imagined than that between the coral reefs of the Paumotus and the cloud-capped volcanic peaks of the Societies. It is like approaching another world.

At the pass in the barrier reef a genial French pilot took charge and secured us the best berth in the harbor. Here the coral wall that forms the beach is so sheer that it is possible to make fast to the trunk of a flamboyant, as though to a bollard on a quay, and walk ashore on a gangplank, which we of the *Dream Ship* promptly did, and dined in splendor at the best hotel.

With unaccustomed collars chafing our leathern necks, and perspiring freely under the burden of clothes after a régime of towel and sola topee, we consumed iced vin rouge, poulet roti with salad, and omelette au maitre d'hôtel. Papeete was a pleasant place in that hour.

Indeed, Papeete is a pleasant place at any hour. It is the metropolis of the southeastern Pacific islands, just as Honolulu is of the northeastern, attracting as varied an assortment of humanity as any in the world.

Here we have the planter of vanilla and coconuts, the trader in anything from copra to silk stockings, the pearl-buyer, the schooner skipper, and the ubiquitous adventurer on their native heath and under conditions that make it possible for each to live and prosper.

THE DREAM SHIP LOSES ITS COOK

The French may be wrong from our iron-bound, Anglo-Saxon point of view, but they certainly have the knack of making life a more enjoyable affair under their administration than under any other at the present time.

It was at Papeete that we of the *Dream Ship* lost our cook. It may be remembered that in the Galapagos Islands, 5,000 miles back on our tracks, we rescued an exquisite Ecuadorean Government official from a delicate position by christening him "Bill" and installing him in our culinary department, where he was expected to work his passage to Australia.

He proved to be an expert cigarette smoker and little more, so that when he approached us after the first night in Papeete and intimated that he found it necessary to leave, we were neither surprised nor pained.

And so you may see to this day "Bill" of the biscuit-colored silk socks and esthetic tie, leaning gracefully over the soft-goods counter of a French store, extolling the virtues of a new line in underwear or gallantly escorting a bevy of Tahitian beauties to the movies of an evening.

"Bill" has found his niche in the scheme of things, and who can say more?

THE BENEFICIAL STREAM THREATENS TO BECOME A FLOOD

The main trouble in the Societies, as elsewhere in the Pacific islands, is scarcity of labor. Each group in this mighty ocean is struggling with the problem at the present time, and has not yet succeeded in reaching a solution.

The native will not work. He does not believe in toiling for others when he is a self-supporting land-owner himself; and, when you come to think of it, why should he?

The Pacific islands, ambitious for development, are consequently forced to turn for help to the more congested quarters of the globe, such as India and China, and herein lies the danger. The influx has already begun, and there is not a doubt that in time it will swell from a beneficial stream into an overwhelming flood, unless ultimately returned to its source by a conduit of stringent legislation.

Between the Societies and Australia, there is a regular line of steamships calling at Rarotonga, Samoa, and New Zealand, and it was in order to avoid this cut-and-dried route that we of the *Dream*

A PEARL DIVER ON THURSDAY ISLAND, OFF THE NORTHERN COAST OF AUSTRALIA

Here the undersea worker is seen incased in a diver's suit, but in the pearl fisheries of the atolls to the east the natives pluck the oysters at depths of from five to fifteen fathoms without any equipment save a network basket slung about the neck in which to deposit the shells. They remain under water as long as three minutes. The Japanese control the pearl fisheries of Thursday Island (see text, page 52).

Photograph Courtesy of Charles J. Glidden

Ship headed for Palmerston Island, a mere speck on the chart, 600 miles distant.

LIZARD MEN LIVED ON MUREA

On the way, we called in for water at Murea, a fairy isle of fantastic volcanic peaks and fertile valleys, where legends still live. There were lizard men on Murea in the old days, it appears—an agile race of dwarfs, who lived on the inaccessible ledges of the mountain range and descended periodically on the coast dwellers, bearing off their wives and other valuables. They carried a short staff in either hand, giving them the appearance of lizards, as they scrambled back to their fastnesses where none could follow.

To prove his words, the Murean native of today will point out uniform rows of banana plants growing in clefts of rock among the clouds, the crops of the lizard men! How otherwise came they to be there? He would be a wise man who could find the answer.

Leaving Murea, the *Dream Ship* passed close to the wreck of a French gunboat piled high on the reef (page 50) as a warning to others not to tamper with coral, and stood away for Palmerston.

December to April is the hurricane season in this part of the Pacific, when the schooner skippers from Rarotonga and other places in the direct path of the cyclonic disturbance flee to the comparative safety of Papeete, and the *Dream Ship* left in April.

Luckily we escaped hurricanes, but for three days violent wind and rain squalls burst upon us, with no warning from the barometer, and we experienced the first real discomforts of the voyage.

A DELEGATION FROM "MISTER MASTERS"

Palmerston Island was a welcome sight, as welcome as it was unique. It is doubtful if such another gem adorns the earth. Neither atoll nor island, it is a perfect combination of both, a natural necklace of surf-pounded coral strung with six, equidistant, verdant islets, the whole inclosing a shallow lagoon slashed with unbelievable color.

Such was Palmerston as we approached it before a stiff southeast "trade," to be welcomed by a fleet of amazingly fast luggers and their astonished crews. "Who were we? Where had we sprung from? Had we any matches?"

To our own astonishment, the questions were fired at us in English and, what was more, English of a vaguely familiar pattern. It is a strange thing to hear one's own tongue fluently bandied among a brown-skinned people on an isolated speck of earth in mid-Pacific. But there was no opportunity of solving the riddle just then.

"Let go!" "She's set." "Lower the peak; lower the main!"

The *Dream Ship* had come to anchor on the northwest side of the reef, well sheltered from the almost eternal southeast "trades" of these latitudes, and the pilot, a six-foot figure of bronze sketchily attired in a converted flour sack, was addressing us with a courtesy as unusual as it was refreshing.

"With our permission, he would take us ashore at once. Mister Masters himself had given instructions."

PALMERSTON ISLAND, THE PLACE TO WAIT FOR THE END

The "Mister Masters himself" settled it. We tumbled into one of the luggers, tumbled out at the reef, and stood knee-deep in swirling waters, while the pilot and his crew towed their craft against a ten-knot current through a tortuous boat passage. Then aboard once more, and away at an eight-knot clip through a maze of coral mushrooms, bumping, grazing, ricochetting, until finally sliding to rest on a glistening coral beach.

"Mister Masters himself," a dignified old gentleman with a flowing white beard and the general air of a patriarch, met us at the veranda steps of his spacious home, and inside of ten minutes we were sitting down to a meal of meals.

I have Palmerston Island securely pigeonholed in my own mind as the spot of all others in which, when the time comes, to sit down and wait for the end. The outside world, in the shape of a schooner from the Cook group, intrudes itself but once a year. The lagoon and the neighborhood islets are a mine of interest to the naturalist or sportsman, and the people have a simple charm that is all their own.

Photograph by Ralph Stock

THE WRECK OF A FRENCH GUNBOAT ON THE REEFS OF MUREA, LAND OF THE
LIZARD MEN

The lizard men are supposed to have been agile dwarfs who lived on the almost inaccessible ledges of a mountain range, from which point of vantage they were accustomed to make periodical raids on the coast settlements.

Many years ago one John Masters leased Palmerston Island from the British Government and, not believing in half measures, took unto himself three native wives. By each he had a large and healthy family, which he reared in strict accordance with his own standards of social usage. That they were sound standards is evidenced in the people of Palmerston today. They read, write, and speak English, this last with an accent vaguely reminiscent of the southwest of England. They are courteous, hospitable, and honest to a degree nothing short of startling, these days, and although naturally inbred, they do not show it, either mentally or physically.

One thing alone troubles the John Masters of today. To whom do he and his island belong? The war has changed all things. The Cook group, of which Palmerston has been declared a far-flung unit, is administered by New Zealand. Is "Mister Masters himself" to be taxed, governed, and generally harried by a people who hardly existed when his father took over Palmerston. It looks like it.

Au revoir, little island. Some day in the not-very-distant future a decrepit, irritable old man will return to your hospitable shores in search of peace, and if you are then as you are now, which Heaven send! he will assuredly find it.

We of the *Dream Ship* had no large-scale chart of the Tonga group, our next port of call; so that when we sighted the island of Tonga Tabu at dusk, two weeks later, we hove to and waited for dawn. Even then two more days and nights elapsed before we had found the Eastern pass through the maze of reefs that surround the island of Tonga Tabu, and hove to in the passage awaiting a pilot.

We could see his station and flagstaff on a sandy islet, but no flag in answer to our own. We waited and continued to wait, while a four-knot current carried us up the ever-narrowing channel to within fifty feet of the coral bar at its end. And then it was that the motor auxiliary that I have so consistently reviled throughout these pages vindicated itself by saving the *Dream Ship* from certain destruction. It went! Literally inch by inch it fought the current, and continued to fight it for the hour or more we were obliged to wait on the pilot's pleasure.

Something ought to be done about those passes into Tonga Tabu. The pilot does not know the difference between an international code signal and a burgee, so he admitted when he had clambered aboard and allowed his canoe to float away on the tide.

"I didn't know who you were," he apologized. "Thought you might be a local trading cutter."

There are moments too full for words, and this was one. In stony silence he steered us through the most fearsome network of reefs we had yet encountered, and the *Dream Ship* was soon made fast to a buoy not twenty yards from Nukulofa wharf.

A SINISTER FIGURE BOARDS THE DREAM SHIP

It was here that the port doctor boarded us, in company with a genial gentleman, who, if I had known then what I know now, would never have set foot aboard. He said very little until later in the day, when I met him at the cosy Nukulofa Club.

"Do you want to sell that boat of yours?" he asked me.

"No," said I.

"Will you sell her?" he corrected himself.

"Not for what any sane man would be inclined to pay," I told him.

"And what is that, may I ask?"

More as a jest than anything else, I named a figure sufficiently preposterous to raise a laugh from most people. But the genial gentleman did not laugh.

"You would take no less?" he suggested gravely.

"Not a cent," said I. "As a matter of fact—"

"I suppose a draft on —— will satisfy you?"

"What's that?" I stammered.

"I'll take her," said the genial gentleman. "I was saying that—"

But I heard no more. I had sold the *Dream Ship!*

Confession is said to be good for the soul, but I have not noticed much improvement in the state of my own since making the above statement. Imagine parting for pelf with a home that has conveyed you across twelve thousand miles of ocean; or, better, try to imagine selling your best friend, and you have some idea of my feelings since the transaction.

THE DREAM SHIP VANISHES IN A COMMERCIAL TRANSACTION

There was no going back on it. I have not the moral courage for such deeds. The draft lay on the table before me. I had a pocketful of money and no ship. I have never been more miserable in my life.

It took me the best part of an hour's aimless wandering over the powdered coral roads of Nukulofa to summon the necessary courage to break the news to the crew of the *Dream Ship*, but by the end of that time I had some sort of scheme evolved.

Between Tonga and Australia there were no islands of particular interest, anyway. We would continue our journey by steamer—it would be a pleasant change—and in Australia I would invest my ill-gotten gains in a far more magnificent vessel than the *Dream Ship*.

On this "more magnificent" craft we would carry out our original program of cruising up the Queensland coast to the islands of the northwest Pacific, and so home via Java, Colombo, and the Suez, thereby avoiding the monotonous passage between Tonga and Australia.

Rather clever, I thought.

Nevertheless I prefer to draw a veil over the communication of this brilliant scheme to the rest of the crew. It is enough that we took our departure by steamer according to schedule and without daring to look back on the good ship we had left behind.

We then proceeded to rub shoulders with a horde of fellow-passengers, who no doubt regarded us as unattractive as we regarded them; to consume beef tea at 11 o'clock, and push lumps of wood about the deck with a stick for want of something better to do.

A VAIN SEARCH FOR ANOTHER SHIP

In Australia I went in search of the "far more magnificent ship," yet one

HOUSE-BUILDING IN TONGA

It was here that the master of the *Dream Ship* said farewell to his craft.

which must be small enough to be handled by a crew of three and sufficiently staunch to withstand anything.

I found a country struggling with the same problems that vex the rest of the world at the present time, yet possessed of potential resources and a cheery assurance; but I found no ship.

I journeyed to New Zealand, and there beheld a prosperous, immensely earnest people, encumbered about much sheep-raising, dairying, and over - legislation both at home and in their newly acquired territory in the South Pacific; but I beheld no ship.

I scoured the Queensland coast all the way to Thursday Island, where the Japanese are permitted to carry off the major part of the profits from bêche de mer (a sea slug which makes the most nutritious soup in the world), pearls, pearl shell, and trocas shell (second only to mother-of-pearl for the manufacture of buttons, etc.) for the simple reason that the majority of Australians cannot be induced to leave their pet city and tackle the industry; but still I found no ship.

And so the heart is gone out of things. The dream is ended.

TREASURE-HOUSE OF THE GULF STREAM

The Completion and Opening of the New Aquarium and Biological Laboratory at Miami, Florida

By John Oliver La Gorce

Author of "Devil-fishing in the Gulf Stream," "A Battle-ground of Nature: the Atlantic Seaboard, "Pennsylvania, the Industrial Titan of America," etc.

NO LONGER can the land animal kingdom of the earth and its peculiar relation to mankind be called a mystery, for painstaking scientists and intrepid hunter-explorers through the centuries have penetrated to the remote places of the world and brought back to civilization minute accounts of the habits and characteristics, the skeletons and skins, as well as living specimens of wild animal life. As a result, we find that today only at rare intervals is a new and distinct species of quadruped or biped made known to us.

Our knowledge of the denizens of the deep is another story. In this department of zoölogical research, however, though the recognized species have increased from fewer than 300 to more than 12,000 within less than two centuries, there are numerous varieties yet to be recorded, many more to be studied, and large areas rich in marine fauna still to be explored scientifically for the common good of mankind.

Since the dawn of human history, man has studied land animal and bird life—in fact, he now knows much of prehistoric creatures long since extinct.* But the "waters under the earth" still hold countless fascinating secrets which challenge the ichthyologist, who pursues a branch of science pertaining to the study of fish life only a few hundred years old, with a world of sub-sea life to conquer, especially among the warm waters of the semi-tropic regions.

THE PART THE POOR FISH WILL PLAY IN A WORLD PEACE

This challenge now has a mighty urge, in that a mounting population faces a dwindling pro rata food supply, and must

* See "Hunting Big Game of Other Days," by Barnum Brown, in the May, 1919, NATIONAL GEOGRAPHIC MAGAZINE.

turn to the sea, as its primitive ancestors once did for an entirely different reason, if it would assuage its hunger and avert the national "land hunger" which is a potent stimulus to war. In the light of a better realization of the economic causes of wars, it is not stretching the imagination to say that he who discovers a new food-fish supply is an apostle of future peace.

Once more, as in its other natural resources, the United States of America is favored among nations. Paralleling our eastern coast for hundreds of miles, the Gulf Stream, that mightiest river of the ocean, which sweeps northeastward with such giant force to turn back the icy waters of the Arctic from our shores, performs another and less widely recognized service in depositing upon America's southeastern threshold a gift of fishes which some day may be regarded as providential, if not miraculous. Indeed the map-minded person might even picture a peninsular hand, in the shape of Florida, reaching out to receive this boon, nourished in the warm waters of the kindly current.

The Gulf Stream is, in truth, a happy hunting grounds for scientist, amateur angler, and professional fisherman. In its waters there have been found some six hundred varieties of fishes, composing practically one-fifth of the entire fauna of the American continent north of Panama.

The most southerly city on the Florida mainland is Miami, nestling beside the limpid waters of Biscayne Bay, separated from the ocean by a peninsula which completely protects the city from the lashing of an angry ocean during seasonal storms. At Miami Beach has been constructed an aquarium and biological laboratory (latitude 25 degrees 46 minutes north and longitude 80 degrees 7 min-

53

THE AQUARIUM BUILDINGS FROM BISCAYNE BAY: MIAMI BEACH, FLORIDA

The Biological Laboratory is located in the right wing of the building. The Aquarium grounds and gardens are being rapidly developed, and already contain numerous varieties of beautiful palms and sub-tropical flora. One of the Aquarium collecting boats is moored to the dock.

Photograph from Walter A. RaKeyser

THE WINDMILL AND CASINO, MIAMI BEACH, FLORIDA

The waters of the Gulf Stream itself lave the shores of Miami Beach and afford delightful bathing the year round. The Aquarium and Biological Laboratory is situated near by.

ONE OF THE EXHIBITION CORRIDORS: MIAMI AQUARIUM

The interior of the Aquarium building is especially designed for the best arrangement and grouping of the fifty large tanks in which the hundreds of unusual and gorgeously colored fish can be seen and studied by the visitors. During the day the only illumination within the corridors is the sunlight, which enters from skylights directly above each tank, and the light thus diffused through the sea water within the tanks creates a very realistic atmosphere of the ocean's depths.

utes west), which, because of its ideal location and equipment, will take rank with the great aquariums of the world.

HUMAN INTEREST IN THE QUICK

Humankind takes a deep interest in animate things, and fish seem to have a peculiar and potent appeal to man. The child turns from toy and pet to gaze upon goldfish in a tiny bowl; the adult will sit by stream or in a boat by the hour in the hope of landing a "string." Angling, in fact, makes the whole world related. It is one of the few sports that knows no flag nor race.

A striking proof of this interest is manifested in the fact that each year the visitors to the New York aquarium, located on the tip of Manhattan Island, are twice as many as those who go to the more conspicuous and accessible Metropolitan Museum of Art on upper Fifth Avenue.

May the reason of 'this fascination not be the racial memory of that far-gone time when our remote ancestors, still too primitive to invent weapons to give them sure advantage in hunting wild animals, turned to stream and ocean inlet for a palatable, abundant, and ever-ready food supply?

The wonder is that science, which has been defined as "intelligent curiosity," should have waited so long to turn to that field which offers a vast, unexplored content of animal creation. That Protean observer, Aristotle, studied fish life, but from his day nearly twenty centuries intervened before the Swedish savant, Peter Artedi, "Father of Ichthyology," met an untimely death by drowning in a Holland canal, but left enough notes of his observations to enable Linnæus to publish them (in 1738), and thus establish a starting point for modern study of genus and species.

AN OCTOPUS IN ONE OF THE MIAMI AQUARIUM TANKS

The octopus is a source of fascination to most people in spite of its repulsive appearance. It has a large, ugly head, a fierce-looking mouth armed with a pair of powerful, horny jaws shaped much like the beak of a parrot, and topped with two diabolical eyes set close together that can send forth a demoniac glare when angry. The grotesque head is mounted on a somewhat oval body from which radiate eight arms usually united at the base by a membrane. The arms, or tentacles, are provided with rows of suckers with which to clasp and cling to its prey with uncanny strength and quickness. The octopus has the faculty of instantly changing color before the very eyes, and is constantly doing strange and weird things, which always attract the attention of the passer-by.

Twice fish figured importantly in our national life. The inland stranger who visits Boston may smile at the "sacred codfish," which is so conspicuous in the decoration of the State House; but a study of the Bay State's early history will impress every American with the major part fishing played in the industrial history of his country. Moreover, the prominence of fish food in the conservation program that helped toward a glorious victory in the World War is a matter of recent memory.

THE ECONOMIC SIDE OF THE STUDY OF MARINE FAUNA

Now there is not only the food problem urge to impel scientific study of fish, but many other fish products, such as cod-liver oil, menhaden oil as a linseed-oil substitute in paint manufacture, seal oil for miners' lamps, and the possibilities of fish guano as fertilizer, fish meal as cattle food, shark skin for leather, and fish oil for glue, to warrant a closer scrutiny of the industrial uses of fish.

Popular interest and industrial possibilities are two reasons why humanized geography is such a compelling subject. The Miami Station not only will afford visitors an opportunity of getting a bird's-eye view of the little-known life forms of ocean depths, but it will offer unique opportunity for scientific observation and study of these sub-sea citizens.

It is difficult to transplant and keep alive the denizens of the warm seas, for they do not take kindly to the colder waters of the north; therefore, to exhibit them successfully, not only must clear

Undersea Photograph by Dr. W. H. Longley

UNDERSEA STUDY OF A FAMILY GROUP OF YELLOW GRUNTS

To realize the full value of this amazing photograph, one must remember that these multihued fish are at home among the coral and sea-fans of their natural habitat, many feet beneath the surface of the Gulf Stream. The yellow grunt is one of the species of fishes which makes a croaking or grunting sound, a fact from which it derives its name. A distinguishing feature of this fish is its bright red or orange color at the base of the jaws and inside the mouth. The color patch is revealed to its fullest extent when the mouth is opened wide in the presence of an enemy, or when it invites the services of the butterfly fish to enter between its jaws and extract certain parasites attached to the walls of its mouth (see Color Plate VII).

and uncontaminated salt water itself be transported from miles out in the ocean to the tanks of the city aquariums in the north, but the water must be kept heated the year round to the proper temperature of their southern habitat.

More fortunate is the Miami Aquarium, which is located within a few hundred yards of the outlet of Biscayne Bay into the old Atlantic; for it has salt water from the Gulf Stream itself available for changing in the tanks at every turn of tide, if necessary, and there is no necessity for artificial heating all year around, as the water is never below 63° F. in winter nor above 85° F. in summer.

THE MIAMI AQUARIUM HAS EXCEPTIONAL EQUIPMENT

The Miami Aquarium is equipped with fifty exhibition tanks, each with a visible area of 4 x 6 feet. One of the glass-front tanks is 36 feet long, 15 feet in width, and 10 feet deep—probably the largest display tank in the world. In it may be shown fish up to 12 feet in length. The exhibition tanks are arranged along corridors, in the general form of a Maltese cross, with a central rotunda.

The only light is that admitted from skylight openings directly over each exhibition chamber, so that the sun's rays filtering through the waters of the tanks give the interior of the aquarium the atmosphere of the ocean bottom itself, and the multihued and wonderfully beautiful fish citizens of the tropics stand out in their regal colors and without the optical distortion which arises from artificial illumination against glass. To further create the atmosphere of the natural habitat of these fish, the tanks are lined with coral rock and festooned with living specimens of the wondrous flora of the ocean bed.

This plant life also is needful to make the captured specimens feel at home in their new environment, and, with such peaceful and customary surroundings, most of them soon become domesticated and seemingly unaffected by their transplanting. Indeed, they are relieved of the burden of the high cost of living and are even willing to give up their pursuit of prey, since their natural food is supplied at regular intervals.

Most people who live far from the subtropic seas, especially those in inland America, have little conception of the wondrous beauty of the colored fish of our southern waters.

FISH TINTS THAT CHALLENGE THE RAINBOW

Elsewhere in this number will be found a series of four-color reproductions of life portraits of some of the more common of these richly colored specimens. These studies (see Plates I to VIII, pages 61 to 68) were made by a noted artist, who watched the fish within the tanks of the aquarium day in and day out, studied their color phases, and the ability of many of them to change their tints and hues, as does the chameleon, until he was able to transfer a suggestion of their rainbow coloring to the canvas.

To the student of ichthyology, the completion and opening of the Miami Aquarium early in January, 1921, will be an occasion of moment, for this station is the only one of any size on the entire South Atlantic seaboard, and is located but twelve miles from the axis of the Gulf Stream.* The Biological Laboratory, equipped with tables for individual or class use, offers opportunity for the scientist and student to pursue these engrossing studies with every convenience of supply and equipment and with their study subjects ever available under most favorable conditions. The institution will specialize in the investigation of the migration of food-fish and the artificial cultivation of the spiny lobsters, stone crabs, et cetera.

Instead of having to go to the great Italian station at Naples, or the Museum of Oceanography at Monaco, students of fish life will be offered the facilities outlined, in their own country, for our own subtropic waters have all that the Mediterranean affords and much besides.

THE PERSONNEL OF THE AQUARIUM

The director of the Miami Aquarium, Mr. L. L. Mowbray, has acquired an extensive knowledge of warm-sea fish, in studies extending over many years. He

*See "The Grandest and Most Mighty Terrestrial Phenomenon: The Gulf Stream," by Rear Admiral John E. Pillsbury, in the NATIONAL GEOGRAPHIC MAGAZINE for August, 1912.

built and had charge of the aquarium at Bermuda and, after developing that to its full possibilities, was associated later with the work of the Boston and New York aquariums.

Mr. Mowbray has had charge of the installation of the complicated tanks and interior equipment of the Miami Aquarium and, with his assistants, already has obtained from every available nook and hiding place among the Florida Keys and the Bahamas more than 2,500 fish specimens for exhibition purposes. These range from the lordly tarpon to the gentle angel-fish. In the aquarium grounds are open tanks in which are sea-cows, otters, and alligators.

FOOD VALUE OF WARM-WATER FISHES TO BE ESPECIALLY STUDIED

The president of the Miami Aquarium Association is Mr. James Asbury Allison, whose great interest in sport fishing brought about a desire to make available a laboratory where investigations might be carried on concerning the food value of warm-sea fish, and thus enlarge the food supply of the country.

One of Mr. Allison's desires is to develop practical data concerning the food worth of certain fishes at different periods of the year. For example, it will be valuable to housewives to know that a mullet at six cents a pound may be, during certain months, because of what it eats during that time, as valuable in food content as the halibut or sea bass, which cost four times as much, and can be prepared for the table in an equally appetizing manner. Not only will the aquarium seek information of this character through scientific study, but, having ascertained the facts, it will place them at the disposal of the public in popular, understandable form.

FIRST OF THE AQUARIUM EXPEDITIONS FINDS A FLAMINGO COLONY

Already the Miami Aquarium has achieved a success in sending an expedition to Andros, the largest, but least known, of the Bahama Islands, to relocate the most beautiful of the larger birds of the world, the glorious flamingo, once indigenous to Florida, but which no longer exists on the American continent—indeed, it is making its last stand in the New World on this island in the Bahama group.

The party of naturalists, ornithologists, and artists, after weeks of effort in the tidal swamps and uncharted bayous, finally located the flamingo colony and collected valuable data.

Upon the return of the expedition to Nassau, permission was given by the colonial government to bring back to Miami a sufficient number of the birds for propagation purposes, and they will be located in a giant aviary on the beautiful shores of Flamingo Bay, only three miles from the aquarium buildings. It is hoped that in this natural habitat the birds will reacclimate themselves and multiply in large numbers, so that they may once more take their place in the natural history of the United States.

A method by which the aquarium intends to popularize the study of fish life will be by making motion pictures of the peculiar habits of fish, of their movements in the water, and their ability to take on a protective coloration when frightened or otherwise disturbed. Motion pictures also will portray the hatching of eggs, the development of the spawn by its natural instincts, showing its efforts toward self-preservation and desire to escape the fate that constant warfare in the seas portends.

EMINENT AUTHORITIES ON NATURAL-HISTORY SUBJECTS AMONG ADVISERS

Carl G. Fisher is Vice-President of the Association, John Oliver La Gorce, Secretary and Treasurer.

The advisory committee is composed of Alexander Graham Bell; Gilbert Grosvenor, President of the National Geographic Society; Dr. Barton W. Evermann, President of the California Museum of Science; Henry Fairfield Osborn, President of the New York Zoölogical Society; Dr. Hugh M. Smith, U. S. Commissioner of Fisheries; Thomas R. Shipp; Dr. David Fairchild, agricultural explorer; Dr. Charles H. Townsend, Director of New York Aquarium; Dr. Charles D. Walcott, Secretary of the Smithsonian Institution; Dr. Carl H. Eigenmann, of the Indiana University; Dr. E. Lester Jones, Director, Coast and Geodetic Survey, and other well-known naturalists.

THE SQUIRREL FISH OR SOLDATO (*Holocentrus ascensionis*)

These bright hued habitants of the tropical seas are to be found in the waters surrounding the Bermudas, Florida, the West Indies, St. Helena and Ascension Island. They reach a length of two feet, and are considered a good food fish.

THE PORK FISH (*Anisotremus virginicus*)

This important food fish, found from Florida to Brazil, reaches a length of fifteen inches, and lives in large numbers about coral heads and reefs. It is easily trapped by market fishermen.

FOUR RESPLENDENT TYPES OF ANGEL-FISH

The Blue Angel-Fish (*Angelichthys isabelita*), shown at the lower left, feeds chiefly on crustaceans, and lives among the coral reefs of the Florida Keys and the Bermudas. The Black Angel-Fish (*Pomacanthus arcuatus*), shown at the upper right, is found from New Jersey, through the waters of the West Indies and as far south as Bahia, Brazil. It is one of the most beautiful of reef dwellers. The French Angel-Fish (*Pomacanthus paru*), shown at the lower right, is found from Florida to Bahia, and reaches a foot or more in length, but is not considered a good food fish. The Rock Beauty (*Holacanthus tricolor*), upper figure, is rarely found in Florida waters, but swims as far south as Bahia. It lives in the deeper parts of coral reefs, and is most difficult to trap.

THE SPADE FISH IS ALSO KNOWN AS THE WHITE ANGEL (*Chaetodipterus faber*)

This excellent food fish, which attains a length of from two to three feet, is caught by hook from Cape Cod to Rio de Janeiro. It is especially abundant on our South Atlantic Coast.

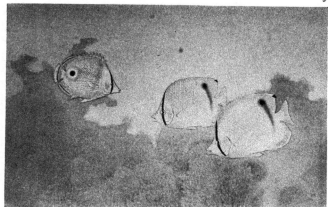

THE FOUR-EYED FISH (*Chaetodon capistratus*) **AND BUTTERFLY FISH** (*Chaetodon ocellatus*)

The Four-Eyed Fish, shown in the left top corner, is a parasite hunter. It even goes into the mouths of larger fishes which remain perfectly still while the little fellow hunts for its prey. The Butterfly Fish is one of the most conspicuous of reef dwellers. Both species are found in Florida and West Indian waters.

THE GREEN MORAY (*Lycodontis funebris*)

This largest of eels, which sometimes reaches a length of eleven feet, is an excellent food fish. It is found in tropical seas from Bermuda and the Florida Keys to Rio de Janeiro, and from the Gulf of California to Panama and the East Indies.

THE SEA HORSE (*Hippocampus*)

This is the only fish which possesses a prehensile tail. With its curious appendage, it holds to seaweed while feeding on small crustaceans. The female deposits her eggs in an external abdominal pouch of the male, where they are hatched. The Sea Horse is found in all warm seas, including the Caribbean, the Black Sea, and the waters south of Japan. One species is found from South Carolina to Cape Cod.

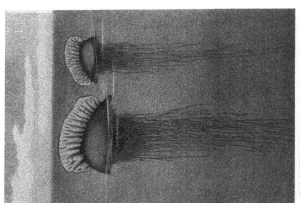

THE PORTUGUESE MAN-OF-WAR (*Physalia arethusa*)

Floating on the surface with the tide and currents, in search of food, this curious sea creature trails its tentacles behind it for forty feet. The tiny fishes upon which it preys become helpless after coming in contact with the stinging cells of the tentacles. The Portuguese Man-of-war is found in tropical seas, but sometimes strays as far north as Cape Cod. Among the tentacles of this creature the little Portuguese Man-of-war Fish hides from its enemies.

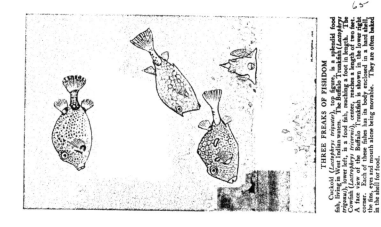

THREE FREAKS OF FISHDOM

Cuckold (*Lactophrys trigueter*), top figure, is a splendid food fish, living in West Indian waters. The Buffalo Trunkfish (*Lactophrys trigonus*), lower left, is a food fish, reaching a foot in length. The Cowfish (*Lactrophrys tricornis*), center, reaches a length of two feet. A face view of the Buffalo Trunkfish is shown in the lower right corner. Each of these fishes has its body enclosed in a hard shell, the fins, eyes and mouth alone being movable. They are often baked in the shell for food.

THE QUEEN TRIGGER-FISH (*Balistes vetula*)

Found in Florida, Bermuda, West Indian and Bahama waters, and in the Indian Ocean, the Queen Trigger-fish lives on rocky and grassy bottoms. It has a variety of nicknames, such as "Oldwife," "Oldwench" and "Cochina.". It takes the hook readily and is esteemed as seafood. This fish gets its name from the fact that the first dorsal fin is composed of a short, stout, rough spine, with a smaller one behind it and usually a third so placed that by touching it the first spine may be set or released.

THE ROCK HIND (*Epinephelus adscensionis*)

This spotted beauty inhabits tropical American waters from Bermuda to Brazil, and is often encountered on the east coast of Florida. It lives in rocky places, and is highly esteemed as a food fish, reaching two feet in length.

THE SHARK SUCKER (*Echeneis naucrates*)

This curious inhabitant of warm seas attaches itself by means of a suction disk to sharks, turtles, and other large denizens of the deep. On the African coast it is used by natives to capture turtles. The fisherman attaches a cord to the shark sucker's tail, and allows it to swim among the turtles. When it has attached itself to one, the turtle is quickly hauled in.

VI

THE YELLOW TAIL (*Ocyurus chrysurus*)

This excellent food fish, reaching a length of three feet, is one of the gamiest of the snapper tribe. It is found in the waters off the coast of Bermuda, Florida, and the West Indies, as far south as Brazil.

THE BLUE STRIPED GRUNT (*Heamulon sciurus*)

This food fish reaches a foot in length and lives about rocky shores from the Bermudas as far south as Brazil. It feeds on worms and crustaceans.

VII

THE SERGEANT MAJOR OR COW PILOT (*Abudefduf saxatilis*)

As its name, saxatilis, implies, this inhabitant of tropical American waters lives among the rocks. It attains a length of six inches and is not used for food.

THE RAINBOW PARROT-FISH (*Pseudoscarus guacamaia*)

Weighing as much as sixty pounds, the Rainbow Parrot-fish is the largest of its family. The flesh is soft but of very good flavor. It is found from the Florida Keys to Rio de Janeiro, and lives on mollusks, worms, and several species of algae.

INTERESTING CITIZENS OF THE GULF STREAM

By Dr. John T. Nichols, Curator of Recent Fishes

WE THINK of tropical seas as the home of a gaudily colored assemblage of fishes. In a sense, this first impression is correct. Active, short-bodied, elastic-scaled, spiny-finned, bright-colored species here occupy the center of the stage.

As a matter of fact, tropical shore-lines are the great metropolis of the world's fish life. The evil-visaged snake-like Moray (Plate III), one of the most degenerate of true fishes, threads the hidden passages among the coral over which Blue Angel (Plate II) and red, green, or parti-colored Parrot-fish (Plate VIII) are swimming.

Out on the open sand, spotted floun-ders lie, matching their background so as to be well nigh invisible, or little gray gobies move about like shadows, eager to escape detection.

Countless varieties of fishes are hiding in every patch of weed. Schools of sil-versides, anchovies, and herring dart through the stretches of open water.

It is their function, in the scheme of things, to feed on the minute organisms so abundant in sea water, to multiply prodigiously, and in turn form a basic food supply for a great variety of larger fishes.

To do this and at the same time con-tribute something to the forces of evolu-tion, their numbers must be conserved, however. Their silvery sides render them difficult of observation by hungry eyes below, and they are available only to the quick and the keen.

ENORMOUS QUANTITY AND DIVERSITY OF LIFE IN THE GULF STREAM

Over the heat equator warm air is con-stantly rising. Heavier cooler air from higher latitudes flows steadily in to take its place, and, deflected by the earth's ro-tation, becomes the easterly trade winds, before which millions of waves, reflecting the clear deep blue of the ocean depths under their white crests, go dancing to the westward.

The whole surface of the tropical At-lantic moves, drifting toward the coast of America, is caught and turned about in the Gulf of Mexico, and shoots out past the Keys and the east coast of Flor-ida as the Gulf Stream.

Inasmuch as many young marine fishes and other animals regularly drift in ocean currents, it is easy to understand what an enormous quantity and diversity of life the Gulf Stream must carry.

Furthermore, such waters, when they enter the Gulf, have already flowed under a tropical sun for many, many miles. The Gulf of Mexico is not a place for them to lose calories, and Gulf Stream water has a considerably higher temper-ature than the 79 degrees found, in gen-eral, at the surface of the open ocean on the Equator.

TRULY TROPICAL FISHES IN FLORIDA WATERS

It follows that shores bathed by such water have as truly tropical fishes as if they were situated much farther south.

Essentially the same fishes extend from Florida to Brazil. Scattered representa-tives of this great tropical fish fauna of the western Atlantic are drifted to the capes of the Carolinas and, to a less ex-tent, in summer, even to New England. We have seen a stray Spade Fish (*Chæ-todipterus faber*) (Plate II) on the New Jersey coast and a little Butterfly Fish (*Chætodon ocellatus*) (Plate III) washed ashore on the south side of Long Island, New York.

It is a little over ten years ago that the writer made a first trip to Florida. After a prolonged period of more or less dis-tasteful, though necessary, indoor activ-ities during a northern winter, he found himself suddenly foot-loose on the Miami water-front.

The yachting party that he was to join here on a collecting trip among the Keys was somewhere up the coast, stuck on a sand-bank. Meanwhile, there was noth-ing to do but sit and swing one's heels.

Photograph by John Oliver La Gorce

THE ETERNAL STRUGGLE BENEATH THE SEA

A school of giant tuna fish feeding on myriads of sardines. The tuna were evidently blood-mad and the white patches of water were occasioned by their great bodies breaking above the surface as they hurled themselves among their prey. This is a graphic illustration of the never-ending struggle beneath the wave where the big ones eat the small ones, and only the fittest survive. It is also an evidence of Nature's safeguard against overproduction of species. The swiftly striking tuna charging with wide-open mouth, causes the little sardine to jump for his silvery life, but, alas, the instant he shows as much as a fin above the surface low-swinging gulls, attracted from miles around by the disturbed waters, seize him from above. This picture was taken in mid-Gulf Stream and the area covered by the huge school of sardines was several acres.

The first objects of interest were the brown pelicans flapping by. Why they did not break their necks on the bottom when they dove precipitously from a height into water not more than two or three inches deep, was something of a problem.

FISH THAT WEAR VIVID REDS, GREENS, YELLOWS, AND BLUES

But the pelicans were not alone in their ability to see fish. It was soon discovered that a number of interesting species could be observed swimming along the shore. None were more beautiful or as easily identified as the little schools of Pork Fish (Plate I), with their bright yellow markings set off by the bold black pattern on head and shoulders. This fish scarcely belongs with the true, gaudy reef fishes, but rather with those less dependent on the protection of the reef, the golds and blues and rose colors of whose livery are often extremely beautiful, yet seldom striking enough to make the fish conspicuous in the water.

By no means all fishes whose haunts are on and among tropical reefs are brightly colored, but there are a great number of active species found there which wear vivid red, green, yellow, blue, orange, etc., and which, furthermore, are marked in the boldest patterns, frequently with black.

Good examples are the Rock Beauty and the Blue Angel-fish (Plate II). Various parrot-fishes, butterfly - fishes, etc., belong to this class.

Naturalists have offered in explanation that the reef itself was as full of color as a garden of varied flowers, wherein the very brightness of the fishes rendered them inconspicuous. To most observers, however, a coral reef as a whole appears rather monotonous in tone, the many varied fishes swimming about giving it the principal note of high color, and these not only easily seen but readily identified.

SOME FISH CAN AFFORD TO BE CONSPICUOUS

How many northern fishes can one see and recognize as easily, swimming in the water, as the black and yellow Sergeant Major (Plate VIII), for instance?

Granted that, in general, these colors render the fish conspicuous, can they be classed as warning colors, like the black-and-yellow striping of wasps? Apparently not, for there are plenty of predaceous fish which eat some of them and would doubtless be pleased to consume more.

Immunity colors, they have been called most appropriately. The idea is that a wide-awake, active fish on a coral reef has so many avenues of escape from its enemies, so many projections to dodge behind and holes to hide in, as to be practically immune from attack. It can afford to be as conspicuous as it likes.

Be this as it may, the striking patterns are a great convenience to the ichthyologist, who has to separate one species from another, for nowhere else does one find so many different, but closely related, species living side by side, each doubtless differing from the others in habits in some way, be it ever so slightly.

THE NUMEROUS FAMILY OF SEA BASSES

One of the principal families of fishes in our southern fauna is the sea basses, to which the gigantic Jewfish, the rockfishes, groupers, hinds, and so forth, belong. These are all fishes which resemble our northern Sea Bass. They are big-mouthed and voracious species, living for the most part about rocky or uneven bottom, though also swimming out over open stretches of sand.

Many are food-fishes of importance. They have leathery mouths, so that when once hooked they are not easily lost. Though well formed and by no means sluggish, they are solitary and sedentary, as contrasted with the equally abundant predaceous family of snappers, for instance.

Always lurking on the lookout for smaller fishes to come within striking distance, and sometimes associated in considerable numbers at favorable localities, they do not range about, hunting in schools, like the snappers.

The colors of this group are varied and sometimes extremely beautiful, in none more so than in the small Rock Hind (Plate VI), whose home is in the bright lights of the coral reef. But the plan of coloring is such as to lower, not raise, the

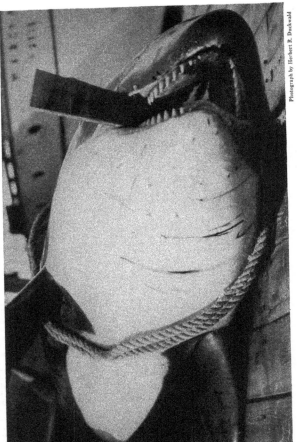

Photograph by Herbert R. Duckwald

KILLER WHALE, THE GREAT WOLF OF THE SEA

The ferocity of the "killers" strikes terror to the other warm-blooded animals of the deep. They are known to swallow small seals and porpoises entire, and they attack large whales by tearing away their lips and tongues. When attacking large prey they work in packs. This specimen was captured in the Gulf Stream between Miami and the Bahama Islands.

MIAMI BEACH AT EBB TIDE

Now and then "on-shore" winds bring strange sea visitors to this beautiful beach from the far-flung reaches of the Gulf Stream. However, a change of wind and tide once again sweeps the sands as clear as a ball-room floor. The near-by aquarium frequently profits by unusual specimens that are brought to shore in the sargasso weed and other forms of sea flora, which afford a hiding place for minute fish of many kinds.

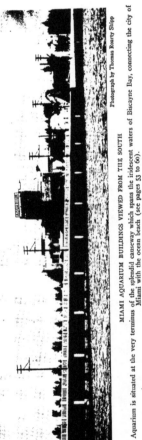

Photograph by Thomas Roarty Shipp

MIAMI AQUARIUM BUILDINGS VIEWED FROM THE SOUTH

The Aquarium is situated at the very terminus of the splendid causeway which spans the iridescent waters of Biscayne Bay, connecting the city of Miami with the ocean beach (see pages 53 to 60).

visibility of the fish. Contrast, for instance, the color plans of the Rock Hind and the bizarre Rock Beauty (Plate II).

CHAMELEONS AMONG THE FISHES

These groupers, rock fishes, and hinds, furthermore, have the power of undergoing complete color changes almost instantaneously. The color tone becomes lighter or darker and the markings become bold or fade and disappear. Such color changes can be seen to advantage in individuals kept in an aquarium. There can be no doubt that in the fishes' natural environment they adapt it to the bottom it is swimming over, and, further, that inconspicuousness may aid in its getting a full meal at the expense of its smaller associates.

There is a related fish which has a color pattern almost exactly like that of the Rock Hind, namely, the Spotted Hind. The principal technical difference between the two is that one has minute scales on its maxillary and the other has not—a characteristic about as obvious to the layman as what the fish is thinking about. The Spotted Hind's squarish tail fin, with a broad, blackish border, affords an amateurish, but simpler, way of telling it.

The fish life of warm shores is one of contrasts. In contrast to the big-mouthed sea basses, there are species, usually sluggish, which have very small mouths, depending for their subsistence on the great abundance of small sea animals found about tropical reefs and ledges, or seaweeds. To capture such small creatures does not require great agility.

THE MALE SEA-HORSE HAS AN INCUBATOR POUCH

The sort of life they lead has probably been taken up gradually, through long periods of time, and many of them have meanwhile acquired remarkable and sometimes quite unfishlike characters of form and structure. None is stranger than the little Sea-horses (Plate IV), with body encased in rings of bony mail, horse-shaped head set at right angles, and prehensile tail to grasp the seaweed where they are hiding, body floating upward erect in the water.

The male sea-horse carries the eggs in a pouch situated under his tail, until they are hatched and the young large enough to fend for themselves.

Sluggish small-mouthed species frequently have hard nipper-like teeth, as the small animals which they eat are many of them shelly.

As it is difficult for them to get out of the way of larger predaceous fish, they are variously protected against attack, mostly being colored more or less in resemblance to their surroundings. The trigger-fishes have a stout dorsal spine which locks erect, as well as a very thick leathery hide which must be of some protection. The gaudy colors of the Queen Trigger-fish (Plate V) are an exception among such forms.

A somewhat related flat-sided filefish scarcely swims about at all, but drifts with the tides, more or less head downward, and can be easily captured in the hand. It is so striped as to be readily overlooked, however, among the eel-grass which is drifting with it.

HOW THE SWELL-FISH FRIGHTENS ITS ENEMIES

The swell-fishes have the power of suddenly inflating the body with water or air until they assume an approximately globular form several times the normal diameter, which must be disconcerting to any enemy about to seize one. The porcupine-fish, in addition to doing this, has the body everywhere covered with long, sharp spines which project in every direction like the quills of a hedgehog. Many persons who are familiar with the inflated skins of swell-fishes and porcupine-fish used by the Japanese as picturesque lanterns will be surprised to learn that both are common in local waters.

The trunk-fishes, instead of being protected in this way, have the body encased in a bony shell, like a turtle. In the East Indies there are rectangular species, but ours are all three-cornered, beechnut-shaped. They go by various names—cuckold, shellfish, and so forth, the Cowfish (Plate V) being a species with two hornlike spines projecting from its forehead. They are excellent eating, cooked in the shell like a lobster.

The back muscles of the swell-fishes are sometimes eaten, but make a risky delicacy, as there are well authenticated instances of severe poisoning from eating these fishes. The poison seems to be localized in the viscera and to permeate the rest of the fish after death.

SOME FISH ARE RISKY DELICACIES

In some quarters of Japan swell-fish is highly esteemed when prepared for the table with care, but there is a Japanese proverb to the effect that before eating swell-fish one should have one's last will and testament in good order.

Poisoning resultant from eating certain species of tropical fishes is a subject which will repay further study. In Cuba several kinds are reputed dangerous and their sale prohibited in the larger markets. Among them are the Great Barracuda (see illustration, page 80), Green Moray (Plate III), and certain species of the Carangiidæ, or crevally family. On the other hand, this same Barracuda is particularly favored as a food-fish in Porto Rico, as it is known to subsist entirely on clean, live food.

It is said in Cuba that by no means all the fishes of these species are poisonous, and that the smaller ones are safer. The symptoms of poison are sometimes alimentary disorders, sometimes skin troubles. The cause is not known, but Mowbray, writing in the New York Zoölogical Society Bulletin, November, 1916, presents a strong case in favor of the hypothesis that such tropical fish poisoning is in most cases due to improper marketing. He says: "It is probable that if, when caught, the fish were eviscerated and bled, a case of poisoning would be a rarity."

Bulletin No. 1 of the Madras (India) Fisheries Bureau, 1915, thus emphasizes the importance of properly marketing fish in a tropical climate: "Of all general food, fish is most liable to taint and most poisonous when tainted. . . . Fish not kept alive *must be cleaned and washed at sea* and properly stowed. This brings them to shore with a much decreased chance of taint, even if several hours intervene."

SNAPPERS ARE THE MOST IMPORTANT SOUTHERN FOOD-FISH

As food-fishes, the snappers are perhaps the most important southern family.

Photograph by L. L. Mowbray

THE WHITE ARMED ANEMONE

Sea-anemones, closely resembling beautiful and many-hued chrysanthemums, are found among the rocks in quiet waters along the Gulf shores. This low form of animal life feeds by arresting with its outspread petal-like tentacles small particles of food floating by, which it then draws toward the central mouth. From a muscular base the anemone can move very slowly from place to place, one observation in the New York Aquarium showing a travel of forty-eight inches in the course of twenty-four hours. They have no food value for man, but are sometimes eaten by fish.

A snapper is an all-around, up-to-date fish, an evolutionary product of the keenest of all competition in the fish world, that at the tropical shore-line.

There is nothing peculiar or freakish about the snapper. He is just thoroughly successful and modern, active, adaptable, and clever—trim-formed, spiny-finned, keen-eyed, smooth-scaled, and strong-toothed.

Almost anywhere one goes one can see little schools of the Gray Snapper through the clear tropical water, skirting the shore or the edge of the mangroves, on the lookout for small fry to satisfy their appetites, and at the same time with a weather eye out for possible danger. It would seem a simple matter to catch one on hook and line, but no fish is warier about being thus ensnared.

Several species of snappers are almost equally abundant, the Muttonfish and the Red Snapper, which is taken in comparatively deep water, being perhaps the most important commercially.

The excellence of the Red Snapper is

Undersea Photograph by Dr. W. H. Longley

PORTRAIT TAKEN BENEATH THE SURFACE OF THE SOUTH ATLANTIC

Porkfishes and tang against a background of live coral six feet or more under water among the Florida Keys. The shadowy object, suggesting an irritated porcupine, near the lower right corner, is a purple sea-urchin with spines erect.

widely known, and quantities of this fish are shipped to distant northern markets. For baking, a fine large one has few equals. Bright red color in fishes has often a peculiar significance, which will be spoken of later.

Though not exactly a snapper, the excellent table-fish known as the Yellow Tail (Plate VII) belongs to the snapper family. It is somewhat more elongated than the true snappers, with lines more graceful, and its tail-fin is more deeply forked. One sees immediately that it is a freer, swifter swimmer, navigating wider stretches of more open water.

WHY SWIFT SWIMMING FISH HAVE FORKED TAILS

Most marine animals which swim, especially swiftly and continuously, have a forked tail-fin. This shape of tail avoids the space immediately behind the axis of the body where the stream-lines following the sides (of a moving fish) converge. A rounded or pointed tail which would occupy such area would be a drag.

Whales and porpoises, though they move the tail up and down instead of from side to side, have a forked tail-fin, only it lies in a horizontal instead of a vertical plane. The wide ranging members of the mackerel family and other more or less related marine fishes have a forked tail-fin set on a firm, narrow base; and the freest swimming sharks (mackerel sharks and the Man-eater) have acquired a tail of the same shape, though the ordinary shark tail is weak and unsymmetrical.

Fresh-water minnows almost invariably have a forked tail-fin, waters which they have to traverse being considerable in relation to the small size of the fishes themselves.

In the blues and greens of the waters

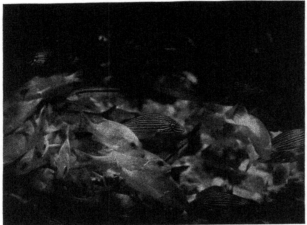

Photograph by Dr. W. H. Longley

PANIC UNDERSEAS

This wonderful photograph was taken, not in an aquarium tank, but about eight feet under water in the Gulf Stream, with an especially designed camera. Posing for their portraits are gray snappers, yellow goatfish, grunts, a parrot-fish, and a schoolmaster, nocturnal fish, which, as a rule, rest quietly all day. The seeming confusion is due, however, to the presence in their neighborhood of a barracuda, that veritable tiger of the warm seas and the natural enemy of all small fish.

through which it swims, the Yellow Tail's bright yellow tail probably makes a shining mark, though its colors otherwise are well calculated to give it a low visibility. Are we to conclude from this that there are no larger fishes which prey on it? No; there pretty surely are such fishes, though it may well be so swift as to escape many which would otherwise do so.

DEEP SWIMMING FISH ARE OFTEN RED IN COLOR

As regards concealment, having a yellow tail must be a disadvantage to it, and is a character which would doubtless have been lost in the keen competition of the tropical waters where it lives, were there not, on the other hand, some compensating advantage. It may be a badge of identification, useful to a school in keeping together.

It has been previously mentioned that the Red Snapper comes from deeper water than other common snappers. There is a tendency for fishes which swim deep down under the blue or green sea and yet within the range of surface light penetration to be red in color. A great many are not, to be sure, but a larger proportion are red here than elsewhere, frequently a clear bright striking red all over.

It seems almost a pity that the light in which they live is so green that the color, red, must appear an intangible neutral gray! Perhaps it gives them a useful inconspicuousness down there, or perhaps it absorbs a maximum amount of the dim, strongly blue-green sunlight, which is in some way beneficial.

One of the commonest species of the surface reefs, the Squirrel Fish (Plate I),

MIAMI AQUARIUM COLLECTING CRUISER "L'APACHE"

L'Apache, Captain C. W. Peterson commanding, is one of a fleet of three power cruisers used for investigating fish habitat in southern Florida waters and among the Bahama Islands. The *Allisoni* and *Chub,* sister ships, were built for the purpose of gathering and bringing in live specimens in their especially constructed live wells (see pages 53 to 60).

has a regular, bright, "deep-water" red color. But the mystery of how it comes to such a color is easily explained, for it has similar relatives living deeper down. Evidently the Squirrel Fish has recently come up in the fish world, and its big eyes indicate that it has not yet adjusted itself to the bright light of the surface sun, but is more or less nocturnal.

The Gulf Stream runs so close to the coast of Florida that, when the wind is right, quantities of the drifting yellow gulf-weed it carries are washed ashore and into the bays. A variety of fishes hide in and about this weed.

One of the commonest and perhaps the most interesting, namely, the Mouse Fish, spends its entire life in the drifting sargassum. Colored in wonderful mimicry of this habitat, its shape also, grotesquely irregular, covered with leaf-like processes or flakes, heightens the resemblance, so as to make it well nigh invisible. This protection against larger fish which might disturb it probably also serves the purpose of camouflage to enable it to approach and capture smaller fish, crabs, and shrimps.

The Mouse Fish, for its size, has a large mouth and appetite in proportion. Many other species hide in the weed when young and, as a rule, have colors to match at that time of life, though later these may be quite different.

THE PORTUGUESE MAN-OF-WAR HAS A FAITHFUL COMPANION FISH

The rainbow-tinted pink, blue, or purple bubble-like floats of the Portuguese Man-of-war (Plate IV) drift at the surface over all tropical oceans and are sometimes washed in close to the shore in numbers. With them comes an interesting little fish, *Nomeus,* the sting of which is exceedingly severe, which never strays far from the tentacles which stream below the Man-of-war.

When traveling by steamer along the Florida coast the writer has watched for *Nomeus,* and from where he stood on deck has seen one and sometimes more individuals lying suspended in the clear water, their blackish ventral fins conspicuously spread, always within a short distance of a Man-of-war, floating above.

Comparatively few kinds of fishes are abundant "off-sounding," away from the influence of the shore-line, and these may

THE SEA TIGER—A BARRACUDA

Because of the clarity of the waters of the Gulf Stream and with the ever-occurring carpet of white sand on the bottom to be found along Floridian shores, this unusual photograph of a five-foot barracuda was obtained by simply holding a kodak over the side of the boat and snapshotting the big fish swimming along six feet or more below the surface. Because of the splendid illumination afforded by the sun on the white sand, even the shadow of the fish, as well as the little tufts of sea flora, was recorded.

be divided rather sharply into the hunters and the hunted. Mouse Fish and *Nomeus,* belonging to the latter class—the one hides, the other lives under the protection of a powerful companion.

WHEN THE FLYING-FISHES PLAY

Flying-fishes, which are abundant, have an even more interesting method of escaping their enemies, leaping above the surface and, with favorable wind conditions, shooting through the air for perhaps as much as an eighth of a mile, supported by their long, stiff breast-fins, widely spread at right angles to the body. When there is a whole-sail breeze blowing, they seem to fly also for sport.

A flock of little flying-fishes no bigger than herring, all in the air at once, gleaming blue and white silver in the sun, is one of the most beautiful sights of a tropical sea. The very thought of it takes one back to the broad blue expanse of trade-wind ocean, warm decks lurching under foot, spray singing through the shrouds, squawking tropic birds and bellying square-sails which swing against a background of fleecy cloud and sky.

In spite of their agility, flying-fishes form the chief food of the little schools of Oceanic Bonitos, and of the Dolphins, swiftest, most graceful, and most highly colored of marine fishes, which prowl over the high seas.

THE PRIMEVAL SHARK IS STILL WITH US

Ages before modern fishes, of which we now find such countless variety in tropical seas, had been evolved in the slow process of evolution, there were sharks which differed comparatively little from those of the present day. Intermediate forms have become antiquated and dropped out, but the primeval shark (Plate VI) is still with us. Especially in the tropics they occur in great abundance.

Prowling singly along the edges of the reefs, over the shallow flats, or through offshore stretches of open water, they hunt largely by sense of smell, and congregate in numbers wherever food is abundant.

When a whale is being cut up at sea it is astonishing how quickly the slender offshore Blue Sharks gather to the feast; it would almost seem from nowhere.

By far the most abundant sharks numerically are the ground sharks (*Carcharhinus*). There is probably no tropical or temperate coast-line where one or more species of this genus do not enter the bays and inshore water at the proper season to give birth to their young.

SHARKS PROPAGATE UNLIKE MOST OTHER FISHES

Though relics of a bygone age, as far as bodily structure is concerned, sharks, of all fishes, have the most highly developed reproductive system. Some lay a few large eggs, each one protected by a horny shell, but for the most part the egg stage is passed through within the body of the parent fish, and the young are born well grown and able to fend for themselves.

The Black-tip Shark (*Carcharhinus limbatus*) is a small species of ground shark, females of which are taken with young in the Bay of Florida in April. They are frequently hooked by tarpon fishermen, who erroneously call them "mackerel shark," and put up a spirited fight when so hooked. They are usually between five and five and a half feet in length, and the young, about three to six in number, are two feet long, or a little less, when born.

We have data concerning another ground shark, *Carcharhinus milberti,* the Brown Shark, which gives birth to its young in Great South Bay, New York, in midsummer. The mother sharks are a little larger—six or seven feet—the young, however, of about the same size, but more of them, eight to eleven having been recorded for this species. Some kinds of sharks which grow much larger have a proportionately larger number of young.

While evolution has been molding other more modern fishes into a great variety of forms to fit every niche in the infinitely varied but unchanging environment of tropical seas, the shark has always been much as we find him today.

A FISH THAT UTILIZES A SHARK AS A TAXI

It is not surprising, therefore, that there is a fish which owes its very re-

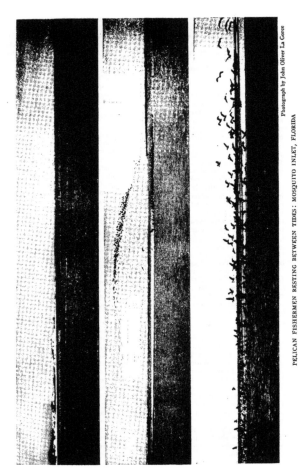

PELICAN FISHERMEN RESTING BETWEEN TIDES: MOSQUITO INLET, FLORIDA

A dignified policeman-like bird is the pelican as he sails along with a wingspread of seven or eight feet. All dignity is cast aside, however, when his keen eye sights a fish, for he flops down from a considerable height and strikes the water with a crash and splash that should by all natural laws break every bone in his body, yet apparently never disturbs him. Recent investigations by the Biological Survey prove that instead of a wholesale consumer of valuable food fish, the pelican of our South Atlantic coast lives almost entirely on menhaden, a fish of comparatively little value.

markable structure and habits to the presence of sharks. This is the slender Shark Sucker (Plate VI), which has the anterior portion of its body horizontally flattened, and a remarkable oval structure, with movable slats like those of a blind, on the top of its head. With this apparatus it attaches itself firmly at will to the shark's broad side and thus as a "dead head" passenger is transported through long stretches of ocean without any effort on its own part.

The Shark Sucker is boldly and very beautifully striped with black and white, but can change its color almost instantly to a dull, uniform gray matching the side of the shark to which it is clinging. It sometimes attaches itself also to other large fishes, such as the Tarpon, or to turtles.

A related species, the true Remora, is found clinging to those sharks which swim through the high seas far from shore. A third is found clinging about the gills of spearfish or marlin swordfish, as they are called by California anglers. A fourth, with very large and strong sucking disk, has been found attached to whales.

All of these may, loosely speaking, be called Remoras. They are sometimes erroneously spoken of as "Pilot-fish," for the Pilot-fish is an entirely different small species related to the Amber Jack, which swims in front of or alongside of sea-going sharks and is vertically banded with black.

THE REMORAS ARE ONE OF OCEAN'S MYSTERIES

Among the fishes of the world the Remoras occupy the position of a genius with unknown ancestry. There is nothing else like them, and to what manner of fishes they may be related is one of the mysteries of old ocean.

Fish life of the shallow pools so often found along a rocky shore at low tide will repay careful study. Such a pool may be a few yards long, with a very irregular outline, full of nooks and crannies, and a few square feet of sand covering its lowest point.

Here the young of several types of fishes act out in miniature the drama which their elders are playing on the reef. Only the villains of the play, the larger predaceous fishes, are absent, at least for the present, until the returning flood inundates the isolated pool to make it once more a part of the big salt water, and we retreat up the beach.

The stage setting is extremely simple: the jagged blackish bottom of the pool, small area of gray-white sand, a little patch of brownish seaweed in one place, either growing there or drifted in at the last high water. From a distance half a dozen small fishes are visible, swimming actively about.

Nearer view shows them to consist of two or three Sergeant Majors (Plate VIII), instantly recognized by the black and yellow uniform in vertical stripes; a couple of Beau Gregorys, with bright blue heads and yellow tails separated by a slanting line of demarcation, and a young Wrasse striped lengthwise with black on a pale ground.

THE WRASSE CHANGES ITS COLOR INSTANTLY

If one attempt to catch a fish of either of the former species, it displays great alertness and agility, dodging about the many projections and irregularities of rock. But now we have the Wrasse cornered and believe we have it in an instant, when suddenly it has disappeared.

Surely it did not dodge past and make good its escape in that way. Where can it be? Two or three minutes of careful scrutiny are rewarded. There it is, motionless, squeezed into a crevice of the side of the pool just large enough to hold it.

Swimming actively about, it was scarcely less conspicuous than the Sergeant Majors, but it has now, furthermore, changed color, so as to have a very low visibility in its sheltered nook. Here we have an illustration in detail of how various theoretical types of coloring work out. While swimming about with them the Wrasse had a conspicuous *immunity* pattern like the Sergeant Majors; now, in the twinkling of an eye, it is a concealingly colored fish.

THE SAND FLOUNDER DEFIES DETECTION

We have been speaking of fishes which no one will hesitate to admit are conceal-

Photograph by John Oliver La Gorce

PROMINENT MEMBER OF THE NUMEROUS RAY FAMILY

The whip ray, or spotted sting ray, as he is also known, is now and then seen in the shallow waters adjacent to Miami. The ray uses its broad cephalic fins much as a bird its wings and seems to fly rather than swim through the water. It is beautifully marked with many golden-brown rings. It is not edible.

ingly colored; but, lying in plain view on the sand, there is a little pale-colored Sand Flounder so exceedingly inconspicuous that it is unlikely that we shall see it unless the water is drawn out of the pool and its inhabitants raked into our collecting bottles.

Even now the possibilities of such a pool have not been exhausted.

NOISY FISHES OF THE DEEP

One thinks of fishes as leading a life of perpetual silence down there under the waters. This generalization is not in all cases true, however. Lying anchored in a small boat at night in Florida waters, one may sometimes hear a school of sea-drum go swimming by below. "Wop, wop, wop," they seem to say. Then there is the little Trumpet-fish, so called, whose identity is open to question, technically speaking, that will at times lurk under the boat and intrigue you with its elfin tooting.

Many species utter croaking or grunting sounds when caught, the various species of grunts owing their name to this habit.

Grunts are fish somewhat resembling snappers in appearance and to a certain extent in habit, but smaller and less vigorous. They are variously and artistically colored in grays, blues, and yellows. The Blue-striped or Yellow Grunt (Plate VII) is yellow, with blue length-wise stripes. The Common Grunt has many narrow stripes of deep, clear blue on the head, the scales of the shoulder region enlarged and conspicuous, bronze in color, with grayish borders. The French Grunt is light bluish gray, with broad, undulating, irregular stripes of yellow; and there are many other varieties.

Grunts have bright red or orange color at the base of the jaws and inside the large mouth. The color is not visible when the mouth is closed.

So wonderful and varied are the fishes of our warm seas that one could write on and on about them did time permit; however, in a later issue of THE GEOGRAPHIC will appear another and more extensive color series of the brilliant fish of the Gulf Stream.

EVERY-DAY LIFE IN AFGHANISTAN

By Frederick Simpich and "Haji Mirza Hussein"

The following article is based upon the observations of "Haji Mirza Hussein" during his stay in Kabul as the guest of the Amir of Afghanistan and upon information gathered during his caravan travels throughout the country. Haji Mirza Hussein is a pseudonym adopted by this European observer, whose mission was of both political and military significance; hence the compulsion to travel in the disguise of a Persian pilgrim. Mr. Simpich, who has translated and edited these notes, was formerly United States Consul at Bagdad and has traveled extensively through south Persia and India.

THE BUFFER State of Afghanistan, historic shock-absorber between Great Britain and Russia in middle Asia, years ago put up a "Keep Out" sign, a "This Means You" warning, to all white men and Christians. The land is "posted,"—to use a poacher's phrase—posted against trade and concession hunters, against missionaries, and against all military and political hunters in particular.

Time and again the British have pushed up from India to invade this high, rough region hard by the "roof of the world." More than once their envoys have been massacred or driven back, or imprisoned, with their wives and children, in the frowning, gloomy citadel of Kabul; and once a retreating white army "shot it out" almost to a man, scattering its bones all the way from Kabul back to the Indian frontier.

THE "KEEP OUT" SIGN IS STILL UP

In sheer drama, in swift, startling action, in amazing, smashing climax, no chapter in all the tales of the romantic East is more absorbing than this story of Britain's wars with the Afghans. And Russians, too, in the splendid glittering days of the Tsars, waged their fierce campaigns from the North, over the steppes of Turkestan, with wild Cossack pitted against wary Afghan.

But the "Keep Out" sign is still up. Today the foreigner is no more welcome in Afghanistan than he was a hundred years ago. Forbidden Lhasa itself is no more exclusive than brooding, suspicious Kabul, the capital of this isolate, unfriendly realm of fanatic tribes, of rocks, deserts, irrigated valleys, and towering unsurveyed ranges (see pages 86-87).

No railways or telegraph lines cross this hermit country or run into it, and its six or seven million people are hardly on speaking terms with any other nation.

Night and day, from stone watchtowers and hidden nooks along the ancient caravan trails that lead in from India, from Persia and Russia—trails used long ago by Alexander and Jenghiz Khan—squads of bearded, turbaned Afghans, with imported field-glasses and long rifles, are keeping watch against trespassers from without.

For reasons of foreign policy, the Amir has long felt the necessity of secluding his little-known land to the greatest possible extent from the outside world. Only a few Europeans, mostly British, but occasionally also an American and now and then a few Russians or Germans, have had permission to come into this country and to sojourn for a while in its curious capital. But even on such rare occasions as when a foreign engineer, or a doctor whose services are badly needed, is admitted by the grace of the Amir, the visitor is subject to a surveillance that amounts almost to imprisonment.

No ambassadors or ministers, not even missionaries, are permitted to reside in this forbidden Moslem land. "Splendid isolation" is a sort of Afghan tradition, a conviction that the coming of the foreigner will spell the end of the Amir and his unique, absolute rule.

THE AMIR NEVER WALKS

Today no other monarch anywhere wields such undisputed authority or is in closer touch with the every-day life of his subjects. He personally runs his country's religion, its foreign affairs, and

THE GREAT CITADEL OF KABUL, CAPITAL OF AFGHANISTAN

Photograph from Frederick Simpich

This is one of the most famous old fortresses in all Asia. The wall, which is in ruins, runs back over the mountain. There is a fairly good hospital in Kabul, over which presides a Turkish physician. The Amir has also invited several Indian physicians to make their homes in his capital. In the summer many residents of Kabul live in tents.

A GENERAL VIEW OF KABUL AND THE BEMARU HEIGHTS, WHERE A BRITISH ARMY OF OCCUPATION WAS STATIONED IN 1840

Kabul is situated on the river of the same name, at an elevation of nearly 7,000 feet above sea-level. Its population is variously estimated at from 60,000 to 180,000. It commands all the passes from the north through the Hindu Kush, and through it have passed the successive armies of invasion led by Alexander the Great, Jenghiz Khan, and other conquerors on their way to India.

Photograph by P. O. Crawford

THE FORT OF BALA HISSAR, WHICH CROWNS A HEIGHT OF 150 FEET, COMMANDING
THE PLAIN ON WHICH IS BUILT THE CITY OF KABUL

The Bala Hissar was partially destroyed forty years ago by the Amir Abdur Rahman. It
has never been restored.

he even supervises much of its commerce. He also owns and censors the only newspaper printed in all Afghanistan. Incidentally, he keeps 58 automobiles, and he *never* walks. Even from one palace to another, he goes by motor over short pieces of road built especially for his pleasure.

From the World War, though he took no active part in it, the Amir emerged with singular profits. His old and once rival neighbors, Great Britain and Russia, drawn together as allies in the world conflict, left him a free hand, and in 1919 Great Britain officially recognized the political independence of this much-buffeted buffer State, to whose rulers she had so long paid a fat annuity.

With an area of 245,000 square miles, Afghanistan is, next to Tibet, the largest country in the world that is practically closed to the citizens of other nations. But political life at wary, alert Kabul is in sharp contrast to the meditative seclu-

Photograph courtesy Air Commodore L. E. O. Charlton

AN AIRPLANE VIEW OF THE AMIR'S PALACE AT KABUL: AFGHANISTAN

The Amir has neither airplanes nor radio stations in his military establishment, but he knows a flying-machine when he sees it, for the British flew up from India during the Anglo-Afghan "unpleasantness" in 1919 and dropped a few persuasive bombs in the vicinity of the city.

sion and classic aloofness of the pious lamas at Lhasa. Amir Amanullah Khan, through his agents in India and elsewhere, is in close touch with the world's current events; and, as the last remaining independent ruler of a Moslem country, now that the power of the Caliph at Stamboul is broken, he wields a far-reaching influence throughout the Mohammedan world; also, because his land happens to lie just as it does on the map of the world, it is plain that for a long time to come he will be an active force in the political destinies of middle Asia. Like Menelik of Abyssinia, Queen Lil of the Hawaiian Islands, or the last of the Fiji kings, this Amir, remote and obscure as his kingdom is, stands out in his time as a picturesque world figure.

The Amir's word, his veriest whim, is law to his millions of subjects. He is, in truth, the last of the despots, a sort of modern Oriental patriarch on a grand scale. His judgments are, of course, based primarily on the Koran, or on the common law of the land; for there is no statute book, no penal code, and no court.

HIS WORD MEANS LIFE OR DEATH

To keep the wires of politics, of military and economic control, in his own hands, the Amir vests subordinate authority only in his relatives and close friends; and woe betide the incautious underling who dares think for himself or act contrary to the Amir's wishes; for in this primitive, secluded region there still survive many unique and startling methods of "rendering a culprit innocuous."

Drawn by A. H. Bumstead

A MAP OF AFGHANISTAN AND ITS BORDER LANDS

Afghanistan has an area equal to twice that of the State of New Mexico, and has a population variously estimated at from 6,000,000 to 7,000,000. The inaccessibility of the country is due to its distance from the sea, its inhospitable sands, and the lofty mountain fastnesses which almost encircle it.

The Amir reserves to himself the right of passing death sentences. The cruel Afghan forms of punishment, such as shooting a prisoner from the cannon's muzzle, sabering off his head, stoning him to death, burying him alive, cutting off his hands and feet or putting out his eyes, are seldom employed nowadays; yet often the criminal himself will choose a quick, though violent, exodus to paradise rather than suffer long imprisonment in a filthy iron cage, perhaps to die eventually of starvation.

The way of the transgressor in Af-

ghanistan continues to be uncommonly hard, however. Time and again, in the recorded history of this land, deposed amirs, troublesome relatives, and political enemies have been deliberately blinded, there being a tradition here that no man with any physical affliction may hold a public office of honor or profit.

CARAVANS OF 120,000 ANIMALS ENTER AFGHANISTAN

Politically, Afghanistan is divided into four provinces: Afghan Turkestan, Kabul, Kandahar, and Herat. Topographi-

THE AMIR'S CLOCK TOWER IN KABUL, CONSTRUCTED BY EUROPEANS IN 1913

Clocks of all kinds, particularly those with alarm bells, cuckoos, and musical attachments, are popular in Afghanistan, but time itself is no object here. An audience with the Amir often lasts from 9 o'clock in the morning till late at night, interrupted by a long, ceremonious repast.

cally, its most conspicuous features are the high peaks in the northeast; where it touches the great Hindu Kush, the Tirach Mir attains a height of over 23,000 feet.

Through these mountains of northeast Afghanistan wind some of the most picturesque and historic trails of the whole world. For centuries the trade between Turkestan and India has flowed over these high passes, and the story goes that often these annual caravans number as many as 120,000 loaded animals, including camels, mules, and horses.

Alexander the Great founded Herat and Kandahar, and here and there are ruins and monuments that mark the marches of the ancient Greeks through the valley of Kabul, of Loghar and Bactra.

At Aibag and elsewhere in Afghanistan are also found the crumbling ruins of Zoroastrian fire temples, the best preserved of which is probably the "Tup-i-Rustam" ruin at Balkh. Near Tacht-i-

Rustam, also, several prehistoric caves have been found, their walls decorated with carvings of giant sunflowers.

The city of Balkh, like Babylon, apparently lived through three or four different civilizations; its ruins show that one city after another has occupied this site, each one being built above the ruins of its predecessor.

Among the oldest ruins in existence are these fire temples in Afghanistan.

A BABEL OF TONGUES AND RACES

Afghanistan is a Babel of races and tongues; more than half its population are not Afghans at all. The majority group embraces the Iranian-Aryan Tadjiks, who inhabit the settlements and large towns; the Mongolian Hazarahs, who roam the mountainous central regions of the country, and the Turkomans and Uzbegs of northern Afghanistan. The real Afghans, or "Pahtos" (Pathans), as they call themselves, live in the high ranges stretching from the Solimans past

FORT JAMRUD, WHICH COMMANDS THE INDIA TERMINUS OF THE FAMOUS KHYBER PASS (SEE PAGE 108)

Photograph © W. D. Holmes

Beginning at Jamrud, 10½ miles west of Peshawar, the pass twists through the hills for 33 miles in a northwesterly direction, terminating at Dakka, in Afghanistan.

Ghazni and Kandahar to the west, toward Herat.

Authorities differ as to the exact origin of the Afghans, but the old theory that they are of Semitic extraction is now discredited; it seems more probable that they are merely a mixture of Turanian tribes, developed here through many centuries of raids, migrations, and tribal changes.

In physical appearance the Afghan is a sort of Turco-Iranian type, the minor tribal divisions in the east of the country showing also a mixture of Indian blood. (The name "Afghan," or "Agwan," is of comparatively recent usage.)

The tribes are divided into minor clans, called "Khel," and they live almost entirely off their herds of cattle, camels, and sheep. Here, as in India, deaths from snake-bites are numerous; scorpions and tarantulas also enliven the nomad's life, and in winter the felt-floored tents are alive with vermin. Few real Afghans are found in the settlements or towns; they instinctively cling to the wild, free life of the open ranges.

War is the chief occupation of all these tribes; they constantly quarrel among themselves and seldom intermarry.

PERSIAN CULTURE HAS MOLDED AFGHAN SOCIAL LIFE

Though the language of the Afghan originated from the old Iran idiom, it shows now the mark of Indian influence. In writing, the Afghan uses a sort of Arab character—that is, one of those alphabets which as children we used to call "fishworm letters." His meager literature, modeled after the poetry of Persia, is also influenced by Islam.

Persian culture has molded the social life in Afghanistan through centuries; notwithstanding the religious hatred between the Sunnis and the Shias, Persian customs have been more or less adopted in the upper ranks of all middle Asiatic Moslem society.

From the Persians the Afghans got the idea of marrying more than one wife; but, like the Persians, too, they have found, to their dismay, that polygamy is nowadays more expensive than exciting.

Sometimes, when the Amir wants to favor his faithful officials with presents,

A VIEW OF THE KHYBER PASS FROM ALI MASJID, LOOKING TOWARD AFGHANISTAN

Photograph © W. D. Holmes

The trade between India and Afghanistan is borne on camels and mules, which travel in caravans, but during the extremely hot summer months, when the temperature often rises to 118° in the shade, it is found that mules endure the heat and hard work much better than camels.

93

A CARAVAN IN THE KHYBER PASS

In the morning this pass is open to caravans coming into Afghanistan, while in the afternoon those bound for India have the right of way.

Photographs by Haji Mirza Hussein

TYPES OF INDIAN SOLDIERS RECRUITED BY THE BRITISH TO GUARD THE AFGHAN FRONTIER

"The Khyber Rifles" is one of the most famous military organizations of the British service in India. The two battalions, commanded by British officers, are recruited from members of the Afridi tribe.

Photograph by P. O. Crawford

AN AFGHAN POST-OFFICE ALONG THE ROAD FROM JALALABAD TO KABUL

If an American wishes to write to a friend in Afghanistan, he must address his letter in care of the Afghan postmaster, Peshawar, India, who will forward it to its destination in the Closed Kingdom. The amount of postage must either be deposited with the Afghan postmaster at Peshawar or paid by the recipient. Letters are dispatched by runners twice a week and require three days for delivery between Peshawar and Kabul. Newspapers, books, and bulky packages are held in Peshawar until they can be dispatched conveniently in batches on horseback.

or perhaps to play practical jokes in certain cases, he distributes women among them; but these "gifts" often prove so troublesome that no great degree of gratitude is apparent among the recipients.

BOYS OF FOURTEEN MARRY GIRLS OF TEN

Family life, however, seems to be rather more intimate and private in Afghanistan than in Persia. Usually the young Afghan does not see his bride before the day of the wedding. Female relatives conduct the preliminary skirmishes, a sort of courtship by proxy which is later followed by negotiations between the bridegroom and his future father-in-law.

Marriage is celebrated at a very early

Photograph by P. O. Crawford

AFGHAN GUARDS ON THE ROAD BETWEEN JABUL-US-SIRAJ AND JALALABAD

Although the influence of the Amir of Afghanistan is far-reaching and his authority is absolute, many of the outlying tribes enjoy a measure of independence. Such tribes and their villages are still presided over by their chiefs, called Khans, Malaks, or White Beards. These chiefs are chosen for life. Second-hand uniforms are among the principal articles of import into the land of the Amir.

age, especially in the northern parts of the country, where boys of fourteen marry girls of not more than ten or twelve years of age.

Amir Habibullah Khan (who was assassinated in 1919) had a harem of over 100 women, and among these, strangely enough, were a few Europeans. The present Amir, Amanullah Khan, has but one wife.

The women of Afghanistan are kept in more rigid seclusion and are more closely veiled than the women of any other Moslem land. The Afghan is notoriously jealous of his harem, and few indeed are the men of the outside world who have ever looked on the face of an Afghan woman of the towns. With the desert women, wives and daughters of the nomads, it is different; the Koran permits them to go unveiled.

AFGHAN WOMEN ARE NOT TAUGHT TO READ OR WRITE

Like the Arab, the Afghan considers it unnecessary and even unwise that women should learn to read or write. No girls are admitted to the bazaar schools and no mullahs are employed to teach them, and Afghanistan knows nothing of women teachers.

In spite of their illiteracy, however, many individual Afghan women wield no little influence in tribal affairs, and, as a rule, the wives of the upper classes lead a comfortable and apparently happy life. They are lavishly provided with every luxury of food and dress which Afghan means can afford, and they visit constantly from one harem to another to gossip, sing, and play games. To be left childless is counted life's saddest misfortune.

About the time the little girls of the family put on their veils, the boys of the same age must begin their studies. First of all, a boy is taught to ride; then to hunt and shoot. The horse is the Afghan's constant companion.

The education of middle and lower class boys is in charge of the mullahs, or teachers. Usually a shabby house or convenient nook in the bazaar is utilized as a school-room, the boys sitting on the floor and studying aloud. The pupils are often surrounded by an interested group

of long-haired, wild-looking camel-drivers or visiting nomads.

The government contributes nothing to maintain public schools. Often the better families send their sons to be educated at universities in India.

Few Afghans have acquired any considerable knowledge by travel in other countries.* The late Habibullah Khan probably surpassed all his subjects in intellectual attainments, for he had specialized in history and the sciences. Next to him, the most educated Afghan of today is the editor of the only Afghan newspaper, the *Saradj-ul-Akhbar*. This editor, who has traveled much in India and Turkey, is at the present time also holding the position of Minister of Foreign Affairs.

The longest journey any Afghan has ever undertaken was made by Nasrullah Khan, the brother of the murdered Amir, who traveled to England in 1895.

The present Amir has never left his country; his brother, however, has been in India several times. Yet, on the whole, an eager desire for learning is innate in every Afghan, and of late years not only Indian, but also British, culture and customs have begun to influence the better classes of the people.

The Afghans call their language "Pushtoo." For official matters, however, the Persian idiom is used and understood over most of the country. The Turkish and Mongolian tribes in western and central Afghanistan speak their own tongues. The ruling Amir knows Persian, some Pushtoo, and Turkish.

THE AMIR LOVES PICTURES AND IS A GOOD AMATEUR PHOTOGRAPHER

Foreign newspapers, most of them coming from India, are most carefully read at the Amir's court, where they are translated by hired students trained in India. The Amir delights in illustrated newspapers and is himself a fairly good photographer.

The Afghan works no more than is absolutely necessary to make his living. The upper classes consider it their privi-

* The only considerable group of Afghans who seem ever to have gotten far from home is a colony of men taken to Australia some years ago for handling camel caravans on the Australian deserts.

A GROUP OF AFRIDIS WITH THEIR PICTURESQUE RIFLES AND SHIELDS

The Afridis are said to have Israelitish blood in their veins. In fact, the Afghans claim to be descended from King Saul, and declare that they were among the people carried away captive from Palestine by Nebuchadnezzar. The whole of the Khyber Pass lies within the country of the Afridis.

A COMPANY OF AFGHAN INFANTRY UPON ITS RETURN TO KABUL AFTER A PUNITIVE EXPEDITION AGAINST SOME OF THE AMIR'S REBEL SUBJECTS

The real military strength of Afghanistan rests not so much upon the shoulders of its poorly organized and indifferently equipped regular army, but upon the inhospitable character of the country itself, the scarcity of roads, and the aptitude of the inhabitants for guerrilla warfare.

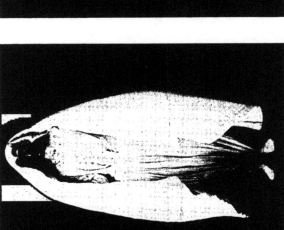

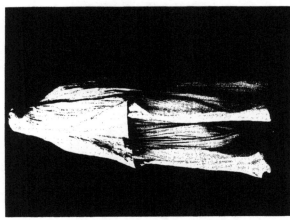

Photographs by Haji Mirza Hussein

FRONT AND REAR VIEWS OF THE COSTUME OF AN AFGHAN WOMAN OF THE UPPER CLASSES

As no Afghan woman would ever consent to have her photograph made without the veil, these costume pictures were posed for the photographer by an English woman at one of the Indian frontier posts. The women of the poorer classes wear cotton bloomers, slippers, and a cotton tunic.

The wealthy Afghan woman wears a round cap embroidered with gold thread. The hair, parted in the middle, is arranged in tiny braids caught in a black silk embroidered bag, worn under the gold cap. Married women wear a fringe of hair, often curled, on each side of the face.

The
r in
d'n
peer
graph in a black silk embroidered bag worn under t c
hair women wear a fringe of hair, often curled, in c
fave.

AN AFGHAN CHIEF WEARING HIS WHITE FELT CLOAK

Photograph by Maynard Owen Williams

The turban of the Afghan serves many purposes; for, in addition to covering the head, one end is allowed to hang down on the shoulder and is used in turn as a pocket handkerchief, a purse, and a dust veil. The turban is worn over a *kullah* (cap).

AN AFGHAN MERCHANT OF KABUL

Photograph by Haji Mirza Hussein

Note the white ivory ring worn on the big toe. The fancy beaded slippers are worn without socks. The average man of the middle class in Afghanistan wears tomboons, or pajamas, gathered in at the waist and falling in folds from hips to ankles.

THE MAT AND SKIN HUTS OF AFGHAN TURKESTAN ARE TIED DOWN WITH GUY-
ROPES, LIKE TENTS

Huts are divided into small rooms by curtains, and several families are often found in one
hut. The floors are covered with felt.

Photographs by Haji Mirza Hussein

AN AFGHAN SHEEP HERDER'S TENT

Little girls help with the sheep, goats, and camels. The heaviest work, around the wells,
is done by the men; water is drawn by camel-power, the animals walking away from a well,
lifting the water by means of a rope drawn over a wheel. Thus, the length of the beaten
path leading out from the well shows its depth to water.

MOST OF THE HOUSES OF AFGHANISTAN ARE CONSTRUCTED OF SUN-BAKED
MUD BRICKS

The roofs are made by spreading long rush mats over poles placed as rafters. Upon the mats is smoothly laid about six inches of mud. Bits of hollowed wood are set in the mud to serve as rain-spouts.

Photographs by Haji Mirza Hussein

MANY HOUSES OF AFGHANISTAN RESEMBLE BEEHIVES

To secure strength against the weight of the winter's snow and to shed water, the roofs of these structures are dome-shaped. Although the climate is healthy, due to the hot, dry air, in which bacilli do not thrive, the country is often visited by epidemics, owing to the unclean and insanitary dwellings of the natives.

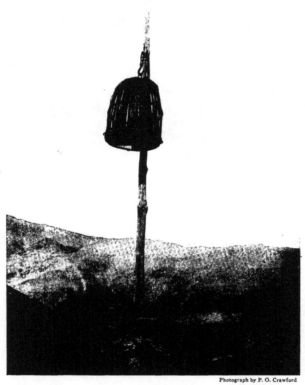

Photograph by P. O. Crawford

AN IRON MAN-CAGE NEAR THE SUMMIT OF LATABUND PASS, AFGHANISTAN

This pass has always been infested with thieves, and during the reign of Abdur Rahman one of the felons was caught, put in this iron cage, raised to the top of the pole, so that his friends could not pass food or poison to him, and here he was left to die (see pp. 105 and 106).

lege to exploit the poor, and the burden of taxation is very heavy.

As for entertainment, the people, especially the wealthy, are fond of games and of sports. Hunting, horse-racing, wrestling matches, and gymnastic games are popular. Recently, football and tennis have been adopted by the upper-class youngsters of Kabul. Ram fights, cock-fighting, and even fights between male quail are favorite diversions, and throughout all Afghanistan dancing is indulged in and the public declamation of ballads is warmly applauded.

ODD COMBINATIONS IN EUROPEAN COSTUMES

Every better-class Afghan owns a piano, imported from Bombay, which he plays with one finger, keeping his foot on the loud pedal constantly. When Haji Mirza Hussein played for them, using ten fingers at a time, they were overcome with amazement and admiration. A tale is told of one man at Kabul who sawed the legs off his grand piano, so that he might play it while sitting on the floor, Afghan fashion.

Costumes vary in different parts of the country. In the East the garments approach the Indian style, and of late years a few natives have even appeared in European dress. Lately, the Amir has introduced European uniforms and suits for himself and his whole staff of officials.

European hats and uniforms of all styles, imported in quantities from India, are often worn in the most singular combinations. One servant in Herat was seen wearing a tile hat, even when sitting in the house. It was held in place with an elastic band, which was passed under his chin. His body was wrapped in an old blue coat with brass buttons, which had strayed from the wardrobe of a railroad conductor in Germany. He had on baggy Afghan pants, with his bare feet sticking out from under the big blue coat.

THREE KINDS OF HEADGEAR ARE WORN

Often the Amir is accompanied by a sort of court jester, who wears a gray tile hat of extra height (like the Boers formerly wore), and colored tights. Instead of the harlequin's sword, he carries a fly-swatter!

The typical national dress of the Af-

Photograph by Haji Mirza Hussein

THE "RUBBISH" IN THE BOTTOM OF THIS MAN-CAGE WAS ONCE A SUBJECT OF THE AMIR

ghan consists of a longtailed calico shirt, white pants, leather shoes or boots, and a tanned sheepskin coat elaborately embroidered with yellow silk; this coat is sometimes replaced by a long toga of red cloth.

Three kinds of headgear are customary. Some wear a low, many-colored cap; others a blue or white turban, which is frequently gold-embroidered with a flap hanging down behind to protect the neck from the sun. In some provinces men wear the *kullah*, a colored cap that looks like a Turkish fez, but which widens toward the top.

In the house and at work women wear long calico shirts, wide, colored pants like the men and head-cloths above gold-embroidered caps. Their street dress consists of long, wide pants and a blue or black overdress, the costume being completed by a loose garment that covers the head and upper part of the body, just allowing the eyes to look through a latticed insert like a strip of mosquito bar. The feet are stuck in large red slippers.

AN EXECUTION-POST AMONG THE BARREN MOUNTAINS OF AFGHANISTAN

Such atrocious forms of punishment as shooting a prisoner from a cannon's muzzle, stoning him to death, burying him alive, and cutting off his hands and feet or putting out his eyes are seldom employed nowadays in Afghanistan, but the iron man-cage is still a favorite method of making the wicked cease from troubling (see pages 104 and 105).

Photograph by P. O. Crawford

Photograph by P. O. Crawford

THE RAM'S HORNS AT THE HEAD OF THE GRAVE INDICATE THAT THIS IS THE BURIAL PLACE OF A HOLY MAN OF AFGHANISTAN

Instances are reported where pious men have been deliberately assassina ed by certain tribes and then buried in the tribal village, in order that the community might wax affluent on the toll collected from pilgrims coming to visit the shrine.

AMANULLAH KHAN, THE AMIR OF
AFGHANISTAN

His word is absolute law to his people. He
alone has the power to pronounce death sentences. He runs his country's trade, religion,
and politics and censors its only newspaper.

The bill of fare of the Afghan is very
simple and reflects the poverty of the
country. Bread, fruits, vegetables, tea,
sweet milk, sour milk, and cheese are the
main foods. Rice, mutton, fowl, and
sweets cooked in various ways are found
on the tables of the well-to-do. The
average Afghan has no particular fond-
ness for wine or spirits.

Tobacco raised in the land is of inferior quality; the better sorts are imported from Persia, Russia, India, and
Egypt. The Amir Habibullah Khan always had a good private stock of Havana
cigars. Both young and old people take
snuff.

Tea, sweetened and unsweetened, is
the favorite drink and is consumed in
prodigious quantities. When you go to
see an Afghan, you can hardly escape before swallowing four or five cups of tea;
it is, therefore, no trifling gastronomic
feat to pay several visits in one afternoon, the more so if the polite host (with
a view of honoring the European guests)
has the tea served in big Russian glasses.

The right hand is always used in eating and drinking, the left hand being considered unclean.

Dogs, though numerous and useful,
are looked upon as unclean, and pious
people never touch them.

THE AFGHAN WILL NOT KILL FLEAS

Animals that go badly lame on the
march or camels that get snow-bound in
the mountain passes are abandoned to
their fate. Afghans never kill such animals, as we might do, to put them out of
their misery. They believe that the lives
of all living things are in the hands of
Allah, and that man sins if he presumes
to interfere with the Supreme Will.
Afghans will not even kill fleas or other
vermin; they merely pick them off and
throw them away!

The trade of Afghanistan is moved
entirely by caravans and is largely in the
hands of Hindus and Tadjiks. The chief
route lies through the famous Khyber
Pass, the great gateway from India,
which has been fortified by the British
Government (see pages 92-94).

This pass is open every week, on Tues-
days and Fridays, except in very hot
weather, when it is available to trade
only on Fridays. A most rigid scrutiny
is exercised by the Amir's agents on all
who come and go. As soon as caravans
from India enter the country, their Indian leaders are turned back and heavily
armed Afghan guides take their places.

Some of these Afghan caravans, organized with military precision, number
thousands of camels and a proportionate
number of guides and camel-drivers. In
the morning the Khyber Pass is open for
caravans coming into Afghanistan, and

in the afternoon for those routed in the opposite direction. The pass is absolutely closed between sundown and sun-up.

Camels leaving the country are usually loaded with wool, skins, dried fruits and vegetables, assorted gums, and spices. Thousands of horses are also driven along for sale in India as cavalry and polo mounts.

THE AMIR STUDIES MAIL-ORDER CATALOGUES

Supplying the wants of the Amir and his court is an interesting undertaking and is usually accomplished by his own agents, who reside in the cities of India. All goods consigned to him come in duty free; he buys anything that strikes his fancy, and often amuses himself by studying the pictures in mail-order catalogues.

In his various palaces and government offices the Amir has installed a few American desks, typewriters, sewing-machines, and clocks.

The Yankee fountain pen and cheap watch are popular in Kabul. Most imports, however, come from India and China. Of late much Japanese merchandise is finding its way into the country. Either directly or through reshipping, India supplies Afghanistan with cotton goods, hardware, sugar and tea, dye materials, and silver bars for the coining of money.

Gun running and the smuggling of ammunition, which flourished for many years, have recently been restricted by British supervision of the Indian frontiers.

Though camels and pack-horses (yabus) are mostly used for transport, it is not at all uncommon to see elephants, and even wheelbarrows, on the Afghan trails.

The main road between Kabul and Peshawar has been improved by the Amir, and a few American trucks belonging to him are used on this stretch of road for hauling freight. These trucks are operated by Hindu chauffeurs.

Along all the caravan trails in the country are good, solid caravansaries, built of stone and clay, situated about a day's march from each other.

The most important caravan roads leading out of Afghanistan are: In the west, from Herat to Meshed; in the north, from Maimene and Aktcha to Kerki; in the east, from Kabul to Peshawar and in the south from Kandahar to Quetta.

Important cities like Kabul, Kandahar, Herat, Maimene, and Mazar-i-Sharif are connected by fairly good caravan roads, which, over various long stretches, can be used by motor cars. The Amir has good motor roads built in and around Kabul to link up his palaces.

Owing to the aggressive pursuit and harsh punishment meted out by the Amir's troops, the once famous robbers of the Afghan hills have almost disappeared, so that caravans, even in the desert districts, can now travel in safety; but in some provinces near the borders constant quarrels and raids are going on among hostile tribes.

AFRAID OF RAILROADS AND TELEGRAPHS

Afghanistan maintains a postal service with horsemen and couriers on foot, but it is not yet linked up with the International Postal Union. For fear of opening his empire to foreigners, the Amir has so far objected to the building of railroads and telegraphs.

Much remains to be done in the direction of developing the trades and arts. Like the Persians, the Afghans have abandoned many good old national home industries and now buy mostly cheap European goods. Apart from a few xylographs, some crude adornments for women, a little silk and felt, and a few simple woven tissues, no products of native skilled labor are on the market. And even much of what is produced in these few lines is merely an imitation of Western and Eastern art. Small industries supply only the most urgent needs of the lower classes. The rich people buy their luxuries from abroad, and the poor make shabby shift with the cheaper fabrics.

In Kandahar a small colony of native artists supports itself by carving prayer beads, many of which are sent by the pilgrims for sale in Mecca.

In military matters Turkish influence is noticeable, and Turkish officers are used as instructors. In all Asia no fighting force is more picturesque or presents a more astonishing mixture of ancient and modern fighting methods than does the army of the Amir. Most of his troops are mounted, either on horses or camels, and a few of his better regiments of cavalry are organized somewhat after the Anglo-Indian style. The regulars are recruited mostly from among the town-dwelling Tadjiks.

The Malkis, or territorials, are organized and used in the various provinces as a sort of home guard. Some of them use flintlocks, and many depend on the spear and the long, curved sword for dispatching an enemy at close quarters.

This army is about 70,000 strong. Save a few field howitzers and mountain guns it has no artillery.

The real Afghans belong to the Sunni sect of Moslems. Here, as elsewhere in Moslem countries, the Sunnis have no close relations with the Shia Persians or with the Hazarah Shiite tribe in the high central region. Nor do the Afghans feel at all kindly toward the non-Mohammedan Indians who venture in for barter and trade. The Turks, however, being Sunnis, are popular with the Afghans.

Every year companies of pious Afghans make the pilgrimage to Mecca. Shia Hazarahs, as well as some of the Sunni Afghans, journey to the sanctuary of Iman Rizas at Meshed, in northeast Persia, to say their prayers, and some few even venture on the long and perilous journey across Persia, to the shrines of Kerbela and Nedjef, in Mesopotamia; also, since the days of the Zoroastrians, a tomb at Mazar-i-Sharif, in northern Afghanistan, has been a shrine that has drawn pilgrims from all over the country.

Smaller shrines and sacred tombs are found in various villages throughout Afghanistan. The rumor is current that aged persons of great piety have been deliberately assassinated by certain tribes and then buried in the tribal villages, in order that steady profits might be reaped from pilgrims coming later to pray at the good men's tombs.

Afghanistan's willful isolation of herself has, of course, affected the life of her people. Even among the different tribes within the country, jealousies and ethnological differences are conspicuous. The high mountains and frequent deserts so separate the cultivated and inhabited districts that tribal customs and habits, tongues, and religious differences are found here in sharper contrast than in most other countries of the East.

As a race, the Afghans are more observant of the Koran's prohibition law than some of their fellow-Moslems farther west. Only now and then, when a caravan comes up from India, the less orthodox element in Kabul enjoys a brief period of alcoholic relaxation.

The Amir keeps at Peshawar a political agent, who occasionally pays a visit to the Viceroy of India; and, since Afghanistan's formal independence of 1919, envoys have been sent to Persia and one is perhaps now in Soviet Russia.

But because of the Afghan's chronic aversion to all foreigners, and the clever exclusion policy of the Amir, aided by nature's own barriers of sand wastes and almost inaccessible mountain ranges, it is likely that for a long time to come foreign influence will spread but slowly in this isolated land.

AFGHANS FOLLOW WORLD EVENTS

Yet the Amir and his military aristocracy follow intently all big events in the turbulent outside world. America is spoken of with sympathy and admiration, and, despite the prevailing illiteracy, many Afghans display an amazing knowledge of geography and current history. During the World War even the nomads on the steppes had fairly accurate news of great battles, and they had heard of air raids and submarines.

Today all Islam is in ominous ferment. Though the World War is officially ended, fights and disputes are still sweeping over Asia. Eventually and inevitably Afghanistan must again become the object of rivalry among big powers that rub shoulders in the East.

Anticipating many requests from members for copies, suitable for framing, of the frontispiece to this number of THE GEOGRAPHIC, "The Argosy of Geography," a limited de luxe edition has been printed on heavy art mat paper, postpaid in the United States, $1.00. The February GEOGRAPHIC will contain as a supplement, a map of the New Europe, in colors, size 30 x 32 inches.

VOL. XXXIX, No. 2   WASHINGTON   FEBRUARY, 1921

CZECHOSLOVAKIA, KEY-LAND TO CENTRAL EUROPE

By Maynard Owen Williams

Author of "Russia's Orphan Races," "The Descendants of Confucius," "Syria, the Land Link of History's Chain," "Between Massacres in Van," etc.

CZECHOSLOVAKIA is an excellent example of a cultured nation which, owing to the overthrow of the old order in Europe, is now a free land.

It was on American soil that the plans of freedom of this nation were developed; its Declaration of Independence was written in an American city and shortened to meet the space limitations of an American newspaper. In success or failure, this key-land to central Europe cannot but be of interest to America and to the world.

Prague, the capital of the new republic, is one of the most interesting of the world's cities, and to one who comes to know its charms it has a peculiar appeal. The view of the ancient palace of Hardčany from the opposite end of the old Charles Bridge is one long to be remembered, and, although I have seen it by many varying lights. I think the most memorable picture of it was at night, during a recent river festival (see p. 118).

A RIVER FESTIVAL AT PRAGUE

On a platform on one of the islands that dot the Moldau (Vltava), a spirited performance was being given. The ink-black waters of the stream were gashed with blinding beams from the searchlights on shore, and the steamers and smaller boats that crowded the river were shadowy hulks vastly magnified by their reflections.

To the waving of white handkerchiefs and the flutter of a hundred snowy skirts whirling in gay dances on the platform, the bright scene ended and the lifted searchlights rolled up a curtain of darkness before the open stage; then, sweeping their aluminum shafts into the air until they pointed to a spot just above the Royal Palace, the lights were quickly lowered, so that the sharp spires of St. Vitus leaped upward to spear the haze, and the fairy mass of palace wall, surmounted by the delicate tracery of the cathedral, stood out like a silver casket against a leaden sky.

The favorite view of Prague is from a hideous view-tower on the Petřín. From its top one can see the Bohemian forest on the Bavarian frontier and the other low ranges that inclose the great plain of Bohemia; but as a vantage point for viewing Prague, it is distinctly disappointing. Even the high spires of the St. Vitus Cathedral cut the hillside instead of the skyline, and the rolling city, caught in the boomerang curve of the river, seems much flatter than it really is.

A better point of vantage is the view-tower in the grounds of the early eighteenth century Schönborn Palace, now the residence of the American Minister to Czechoslovakia. From that lower level

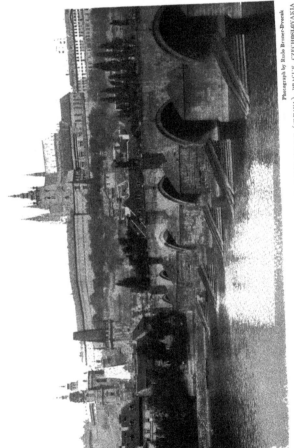

Photograph by Rudo Bruner-Dvorak

A VIEW OF THE ANCIENT PALACE OF HRADČANY AND MALA STRANA FROM THE MOLDAU (VLTAVA), PRAGUE, CZECHOSLOVAKIA

On the crest of Hradčany is the great palace where Bohemia's kings lived and the Estates of Bohemia met. Latterly it has been the home of the Crown Prince of Austria, and there the Emperor of Austria was entertained. The Bohemian kings were crowned in the Cathedral of St. Vitus, which may be seen overtopping the palace. Herein is the Chapel of St. Wenceslaus, where the insignia of Bohemian kings is deposited. The Prince of Wales now has an identical crest, with three feathers and the motto "Ich Dien," and tradition has it that the blind Bohemian King John dropped it during the Battle of Crécy, where it was found and later adopted by Edward, the Black Prince.

the ragged skyline, which is Prague's chief charm, is visible and one gets a more intimate view of the spots where history has been made in many a hard-fought fight. The numerous spires remind one of the pikes and spears of a warrior host and the domes of the churches suggest the round-topped helmets of the Swede invaders.

When the sun begins to set behind the Petřín, the saw-toothed towers of the cathedral, to which one's gaze so frequently returns, stand out dull brown and edged with darker tones against a hazy sky; but the dun brown buildings beyond the romantic towers of the historic bridge of Prague respond to the farewell kiss of the setting sun as do the towering columns of Baalbek, glowing with a mellow light. Then one suddenly realizes why the Bohemians call their beloved capital Zlata Praha, Golden Prague.

"THE GATHERING OF THE FALCONS"

No rainbow following storm was more colorful than Golden Prague during the Seventh Sokol Congress. Six former festivals had attracted the attention of the world and had, in increasing measure, alarmed the rulers of the crazy-quilt empire of the Hapsburgs.

The Congress was no longer a thinly veiled challenge to the tyrant's power, but a testimony of triumph to those Czechoslovak volunteers who had fought on many a far-flung battle line and to those others, forced to stay at home, who had tirelessly plotted the overthrow of the hated Hapsburgs.

One witnessing the gathering of the Sokols, or Falcons, from all parts of the new republic was in danger of thinking that once again he had fallen a willing victim to histrionic craftsmanship, backed by all the persuasive power of mass and color, of movement and music. But the formerly oppressed lands that now constitute Czechoslovakia have long been the stage for the preliminary scenes of this stirring drama, and the actors in the moving climax have been life members of a countless cast.

So when the day of liberation came, and fresh-faced maids wearing the gay costumes of their race grouped themselves about the base of the ugly, rugged

monument to Huss which stands in the old town square of Prague, one almost expected those stern features to lighten with triumphant warmth at the joy of the delivery from Teuton tyranny.

City and program were alike full to overflowing. Participants and spectators came by the hundreds of thousands, most of them bringing their own food with them. Sokols came from Chicago and other parts of America with food drafts, for, without this help from outside, jubilation would have brought starvation in its train.

Miles of crowded cars dumped their polychrome loads in Prague and then rushed away to bring more joyous souls. Most of the visitors had never before seen their beautiful capital, whose bridges, streets, and palace walls are steeped in history. Eager patriots of all ages were massed about every museum, public building, and point of vantage, and diversity of color was furnished by innumerable costumes that would make one think that when the rainbow was designed only the softer colors were employed.

The streets were moving bands of brightest hues, their flag-trimmed flanks a sea of flaming red and dazzling white, with many Stars and Stripes to thrill the visitor from transatlantic scenes. The great avenue, Václavské Náměstí, was a sea of surging shapes, moving in formless masses till the grand parade arranged in regular patterns the chaos of color whose brightness so intrigued the eye (see pages 120 and 121).

FUNERAL FLAMES FOR THE VALIANT

This noble street, two hundred feet in width and nearly half a mile from end to end, has lately echoed to the tread of many glad parades in which the Czechoslovak volunteers, now home from war, were greeted by their womenfolk and marched in light-foot ranks between long lines of girls in holiday attire. There also burned the funeral flames, lighted in honor of those who had given their lives to their country's cause, and under the evening sky vast shadowy masses knelt in thankful tribute to those valiant souls who died on foreign soil. But on this day the troubled past was pushed aside

PRAGUE AND A BANNER OR TWO

Parades and other street gatherings are ingrained habits with Prague citizens.

WOMEN'S SOKOLS IN PRAGUE: A FIELD OF HEADS PLANTED IN TRENCHES OF HUMAN ARMS (SEE PAGE 116)

The sokols were fertile soil for maintaining race solidarity and national consciousness during a period when Hapsburg domination prevented any gathering which might be construed as military or political. Women played a major part in fostering this patriotism, and they had their reward when the Czechoslovak Declaration of Independence said, "Our democracy shall rest on universal suffrage; women shall be placed on an equal footing with men politically, socially, and culturally."

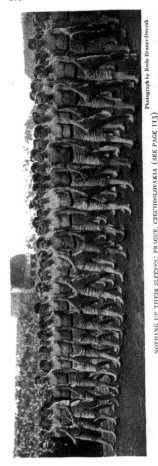

NOTHING UP THEIR SLEEVES: PRAGUE, CZECHOSLOVAKIA (SEE PAGE 113)

Photograph by Rudo Bruner-Dvorak

with reverent touch to frame the passage of the living thousands around which the hopes of the future cling.

The Prague parades cannot be reduced to black and white. Their rosy joy, which rolled up cheers in countless waves of sound, defies description.

Nor can one convey the tremendous impression made by twelve thousand men, or an equal number of women, moving like some orchestra whose music is attuned to eye instead of ear. High above the tribunal a hidden leader plays upon that orderly array and produces eye-music through the exercise of fifty thousand legs and arms.

From both ends of the great stadium there pour broad streams of womanhood, their red caps and white waists, lightly touched with embroidery, forming an animated strawberry shortcake on a platter whose dark design is formed of short blue skirts and plain black hose. The columns separate and unite again, till at a signal solid ranks disperse and one great group is formed that fills the stadium.

Twelve thousand women bow their scarlet caps, and the whole field blushes with a rosy light. Arms and heads are raised and the soft, brown tone of healthy flesh is seen (see picture, page 115). Twelve thousand backs are turned and the rich, warm hue of face and throat dissolves to glistening white. They bend to earth and white and red are lost beneath a sea of blue. A whole landscape changes light values in the twinkling of an eye, while the music which directs the movements seems to be expressed in tones of varying waves of light.

A similar effect is produced by the fawn costumes of the men, against which the bright red shirts, adopted out of admiration for the Garibaldian hosts, now burst into flame as the coats are thrown back and red-clad arms are spread, now hide behind as soft a tint as that a young deer wears while hiding from a foe.

So perfect is the precision of these mass drills that one can scarcely believe the statement that these thousands of Sokols, or Falcons, have come from towns widely scattered throughout the republic, and that only one or two mass rehearsals were possible before the grand display. The action of twenty-four thousand feet is so synchronized that their movement on the sand of the stadium is like the lightning hiss of a serpent or the sharp crunch of stones sucked seaward on the shores of the Dead Sea.

So graceful is the general effect that one is surprised to find how awkward some of the individuals are. The flat chest of a mere youth is seen beside the rounded waist of a man of fifty, yet the effect is one of uniform strength and virility. As runners and hurdlers, the Czechs are distinctly inferior in form. But as examples of perfect training and organization, nothing in the world compares with the great mass drills of the men and women Sokols.

A CITY OF ARCADES

Prague is essentially a city for the pedestrian wanderer. A sightseeing bus or a lorgnette would chase away the charm. Formal sights are disappointing except to experts, but to him who likes to loiter among medieval scenes, taking pleasure in watching this old lady whose worn umbrella shelters a slender stock of fruit, or contemplating with leisurely delight the life that surges through the covered passageways lining the cobbled streets of the Mala Strana, few cities so intrigue one's interest.

The Czechs who emigrate to Cleveland ought to feel at home there, for Prague is also a city of arcades. Some of these are low-arched passages that remind one of an Old Chester whose cubist lines are bent to graceful curves, or of the dimly lighted *souks* that usher one into the caravansaries of Bokhara.

Others are great open halls that cut their way through massive modern blocks, their plate-glass walls placarded with posters and pierced by entrances to moving-picture shows and cabarets that love the dark, with hair-dressers' windows full of cheap perfume at high prices, and with a postage-stamp dealer or two. No modern arcade in Prague would be complete without a postage-stamp dealer whose windows are pock-marked with treasures for the philatelist.

In one of the arcades the visitor advances past the frankly informing photograph of the newest dance queen to drop

a coin in the slot and have some tasteless fluid squirted into a glass by hidden forces that earn for the place the name of "Automat." In these days, when paper lucre and uncanceled postage stamps have driven out hard cash, an automat whose vitals need the clink of metal coin to stir them into action, has a hard time.

Popular as canceled postage stamps are in Bohemia, the favorite art production nowadays is an American-made thousand-crown note. This charming piece of art has not, like its hundred-crown brother, been counterfeited; hence, the man who receives one in change does not have to break his neck while holding it between him and the light to see whether it has a waffle-pattern water-mark in it.

Counterfeit money, however, does not seem to bother any one, the holder least of all; for, although a Czech crown will buy three or four Austrian crowns, no one seems to have the slightest respect for the money. Your waiter makes change in the same way that the stage hands in the small-town opera house used to make a snowstorm, dropping one piece of paper after another until his arm was tired, and neither he nor the recipient seems to care whether the season is one of a heavy fall or not.

The computation runs into hundreds of thousands, and when the traveler arrives in a region where money still retains a trace of its former "kick"—say 2.75 per cent—he thinks that somebody is making him a present of his purchases until, too late, he counts the cost.

PRAGUE YIELDS TO FAULTS OF FRIENDS

Having resisted the determined forces of Germanization, Prague is now yielding to the more insidious faults of her friends. America is represented by the American bar and jazz, Great Britain by Scotch whisky, and France by lotteries and cabarets.

With relief efforts still necessary to prevent real suffering throughout the land, one feels that too many reckless souls in the capital are spending their evenings "à la Sans Souci," with the pop of champagne corks and the one-two-three kick of the stage houris as an accompaniment to their ragtime psalm of life.

Photograph from Maynard Owen Williams

THIS STATELY GOTHIC TOWER OF THE CHARLES BRIDGE, ON THE RIGHT BANK OF
THE MOLDAU (VLTAVA), IS CONSIDERED THE MOST BEAUTIFUL TOWER
IN A CITY OF TOWERS AND ARCHES (SEE PAGE 113)

Photograph by Maynard Owen Williams

THE FOURTEENTH CENTURY TYN CHURCH, THE ST. PAUL'S CATHEDRAL
OF PRAGUE

To the left is the famous astronomical clock of Prague, one of the oldest in Europe, with its figures of the Lord and his apostles. It is built into the front of the Town Hall, inside of which are the dungeons where patriots once awaited execution.

"Montmartre" has already established itself and the *"danse intime"* has arrived, if anything, ahead of time. Liberty is a heady wine, as Russia knows, and one gladly leaves, though not without regrets, the charming medieval town turned capital, to visit the real citizens of Czechoslovakia in their small cities and country homes.

A tedious night journey, broken simultaneously with the day and continued on a train that seemed to lack decision as well as speed, brought me to the little junction of Uherské Hradiště (between Bisenzo and Hollein). Rather than wait for the local train that runs to the town between rich Moravian fields, I disturbed the single carriage-driver long enough to ask if he would accept a fare.

The antiquated victoria had long since lost pride in its personal appearance, and an old gray mare, built, or worn down,

SLOVAKIAN GROUP IN PARADE: PRAGUE

Photographs by Rudo Bruner-Dvorak

GIRLS IN NATIONAL COSTUME IN PARADE WELCOMING SOLDIERS RETURNING FROM
SIBERIA TO PRAGUE, CZECHOSLOVAKIA

FOLK-DANCING: PRAGUE, CZECHOSLOVAKIA

Photographs by Rudo Bruner-Dvorak

"NA ZDAR!" MEANING "SUCCESS TO YOU!"

"When a few hundred gaily-colored aprons—bright green, changeable to gold, yellow with a silver overlight, pink, blue, cerise—are displayed in one moving picture, it is a very charming scene."

Photograph by Maynard Owen Williams

A PROUD DAUGHTER OF PRAGUE

It was among the ladies of Prague that Titian is said to have found his ideal of a beautiful feminine head. The German writer who is authority for the statement attributed to the artist quaintly adds, "For the young ladies of 1841, I am ready to give my testimony most unreservedly."

try road was quiet in the early morning sun, and on each side the harvest fields were dotted with tall shocks of heavy grain. Occasionally wagons, drawn by meek-faced cattle, moved slowly by, and peasant women, carrying heavy loads, passed silently along, their barefoot tread muffled in the thick dust of the roadway.

Potential roasts and feather beds reluctantly made way for our chariot, waddling along with the shuffle of a comedian and the gaze of one whose troubles are too great to be disclosed, but stopping now and then to emit a sound evidently learned from an asthmatic motor horn. A goose may lack brains, but it has decision. When it starts for one side of the road, nothing short of an untimely death will deflect it from its purpose.

Out across the broad field, cut into alternate strips of harvest russet and vegetable green, a combined thresher and baler was at work. Asking my grizzled Jehu whether

to gothic lines, was hitched to one side of a two-horse pole. The driver matched his reginal car, being dressed in an old gray suit, baggy at pockets and knees, and a derby with the edge binding worn to a greasy fringe, which was duplicated on his unshaven chin.

But he jumped quickly down and helped so earnestly with my heavy bags that I conceived a liking for him which grew with our acquaintance. The coun-

he had time enough for a detour, we turned aside to bandy words with the women who were doing the major portion of the work on the threshing crew. Then onward to the town, where my driver took me to a modest hotel.

NATIVE COSTUMES FORM A PANSY-BED OF COLOR

The next day, Sunday, dawned "brite and fare," and the street outside my win-

Photograph by Maynard Owen Williams

ON THE ROAD TO KARLSBAD

"Karlsbad," because these are German boys, and the Germans in Czechoslovakia stick to the older nomenclature. Ask a Czech and he would tell you the road led to Karlovy Vary, which illustrates one of the minor but, to the traveler, highly irritating consequences of the war (see page 143).

Photograph by Rudo Bruner-Dvorak

PRAGUE'S HOLIDAY FESTIVALS COMBINE THE FEATURES OF AN AMERICAN STRAW RIDE AND FANCY DRESS BALL

Photograph by Maynard Owen Williams

WHERE THE MARCH ENTERS THE DANUBE

Where the March (Morava) mingles its waters with those of the Danube, a jagged spur of the Little Carpathians juts out, and the rocks can scarcely be distinguished from the ancient castle ruins.

THE VILLAGE VAZSECZ, UNDER THE TATRA MOUNTAINS: CZECHOSLOVAKIA

Photograph by Dr. V. Sixta and Son

The high Tatra Mountains are rivals of the Swiss Alps for scenery. It is here that the Hungarian gentle folk have for generations been accustomed to find their mountain vacation playgrounds.

Photograph by Dr. V. Sixta and Son

PUBLIC SHEPHERDS IN THEIR FIELD COSTUMES

Little wonder is it that the relatively high wages of America put the wanderlust into millions of European hearts. Wages are so low that throughout Bohemia it is cheaper to have stock herded by public shepherds than to build fences (see also pages 144 and 146).

dow, where a crimson blossom bloomed, became a varicolored ribbon of men and women bound for church. Hurriedly dressing, I made my way to the great square. In the farther corner the huge mass of the cathedral rose above its companion buildings, and at the base of the dull stone façade a pansy-bed of women knelt in prayer.

The native costumes of Czechoslovakia are a never-ending delight. From the first day, I tried to discover which was Bohemian, which Moravian, which Silesian, and which Slovakian; but the bewildering array refuses to be so easily classified.

Although each town has its peculiar style, in each there are such differences as are due to individual tastes. In Kyjof or Uherské Hradiště, as in Paris, Mich., or Rome, N. Y., women are nonconformists in matters of dress.

The men of Uherské Hradiště run more or less to type, with high boots, brightly polished but dusty to the ankles, wide white trousers, and a shirt eloquent of wifely toil beside some sylvan stream and nicely embroidered at the wrist and throat with delicate designs which do not suggest the horny-handedness of the women who produced them (see page 130).

A panel of dark material hangs down in front and a gay sash of red and black, much like the Filipino gee string, hangs to the ankles. The vest is thickly braided and has innumerable frogs. These may be hidden behind flaming balls of wool that make one think that the Reds have learned to color shaggy chrysanthemums in the way the sons of Erin color carnations for St. Patrick's Day in the mornin'. The round-topped hat is circled by a very attractive figured hat-band, all black.

AMAZING FEMININE APPAREL

But the women run the whole gamut of color, and when one sees them massed in the mellow light of a great church interior he looks to see what stained-glass window or prism-decorated chandelier has thrown its varicolored beams across the multitude (see page 128).

The women's shoes are stout, high cut, and topped with patent leather trimmed to a scalloped edge, so that they give a

strangely graceful appearance to the stocky legs of the peasant women. Their stockings are for protection as well as for display, some with small square designs knitted into the dull black.

The skirts are plain black, with no trimming except a line of fine embroidery, worn, like the attractive smocking of the Chinese coolie apron, just below the waist, but they are very heavily plaited and are hung above a surprising number of lace-trimmed petticoats.

The waist-length jacket may be quite plain except for an appliqué design of hand-made lace around the bust and on the sleeve from elbow to wrist, but the head-dress and apron are as gay in tone as the *obi* of a Japanese doll of twelve.

WONDERFUL SLAVIC HEAD SHAWLS

Some of the Slavic head shawls, which give a Madonna oval to the broadest of peasant features, are neat white cotton with red polka dots or a dark gray design. Others are shimmery white silk, embroidered with light tints or heavy designs in cream or white. Still others are cut plush, with a heavy knotted fringe, such as makes one think of castanets and a blood-red rose.

There is something about manifold plaits that is as redolent of romance as the speech of a Parisienne, and, when a fringed scarf is added, the shapelessness of the peasant and the high cheek-bones of the Slavic face cannot rob the wearer of a charm which in the half light of evening or from a moderate distance makes mere man want to burst into some sentimental ballad.

And when a few hundred gaily-colored aprons—bright green, changeable to gold, yellow with a silver overlight, pink, blue, cerise—are displayed in one moving picture, it matters not that the wearers lack the classic beauty of a Venus or the form of a Juno. It is a very charming scene.

In Merv I saw the havoc modern commerce has wrought with lovely Oriental rugs.* The same thing is taking place in the peasant costumes of Czechoslovakia, with the same aniline dyes being substituted for vegetable colors, which were

* See "Russia's Orphan Races," by Maynard Owen Williams, in THE GEOGRAPHIC for October, 1918.

CHILDREN IN VAJNORY, NEAR PRESSBURG (BRATISLAVA), CZECHOSLOVAKIA

Photograph by Dr. V. Sixta and Son

The children of the Czech and the Slovak alike are happy of disposition and quick to learn.

not only much softer when new, but which fade into mellow tones no chemical dye can duplicate.

Factories are calling the women from the farms, where they utilized the winter months in working out the designs traced by the village designer or in evolving their own. Thus, gradually the arts of the past are being lost.

THE DECADENCE OF NATIVE COSTUMES

City girls and foreigners, whose sense of art is inferior, have conceived a great liking for these peasant costumes, with the result that there is a market, not only for the product of months or years of loving labor, but also for hurried work, devoid of imagination and machine-like in its mediocrity. The value of a fine costume runs into thousands of crowns, and cheaper ones have to be supplied. The result is deplorable.

Not only are hideous color combinations displayed and machine-made ribbons used in place of better ornament, but the costumes, donned by those to whom they are only a type of fancy dress, lack the dignity which is never lacking when they are used by the real peasant.

Where soft leather boots should be worn, with just a glimpse of knee clad in an honest, heavy stocking, the town girl puts on high-heeled slippers and the sheerest of silk hose; so that a thoroughly modest dress becomes a thing of scorn, and on the street one sees grotesque shapes with the heads of city women, the bodies of peasants, and the limbs of a midnight frolicker.

Even from the tiny Slovak villages young girls are going to the cities. They have no time nor energy to work on delicate needlework of their own, and the small savings from their wages are not sufficient to supply such splendid costumes as their mothers wore; so they are coming more and more to wear white hats with wide, diaphanous brims, spotless white dresses, and white stockings and slippers. Although charm is maintained, all individuality is lost.

Peasant art, which gained an enviable reputation in Austria-Hungary, is being sacrificed for money which buys almost nothing, and the lovely costumes of Czechoslovakia will soon be seen only in museums, along with roc's eggs and the molar apparatus of a whale.

One crosses harvest fields to picture scenes of peasant life only to be addressed in the accents of Cleveland and Youngstown. A sturdy worker bends to lift his fork of grain, and I ask him whether he will take a slightly different position.

"You from America? Sure Mike, you take any kind of picture you want."

However much we may rejoice in matchless landscapes, it *is* convenient to be whisked about the country in a Czech-made motor car whose chauffeur speaks three languages.

One enters the huge Skoda works with resentment at the prosaic monotony of the buildings, but discovers there the long-prophesied miracle of a munition plant turned into a locomotive works and factory of printing presses, and realizes that not only are pruning-hooks and plowshares and car wheels more useful than siege guns, but that here, in a place where one would least expect it, the pen and the press are displacing the sword and the death-dealing monsters of militarism.

The song of romance will not be stilled, and even in the whirring shuttles of Liberec the melody of a modern hymn of joy and freedom is slowly taking shape. The day is bound to come when the white coal of countless streams will wash clean the atmosphere which the fumes of coal have soiled, and the Caliban miners thus released can once more rise to fullest stature beneath the dome of out-of-doors.

The River Vag (Waag) alone could drive dynamos enough to make Slovakia the dwelling-place of light, and many another stream rushing toward the sea will sing a more industrious song when harnessed to its task.

SLOVAKIA, THE "WILD AND WOOLLY EAST END"

In Prague they speak of Slovakia as a New Yorker speaks of Idaho or Arizona—a sort of distant relative with unquestionable charms, but not quite versed in the latest freaks with which great cities try to console themselves for the lack of wide open spaces under heaven's blue vault.

Photograph from Maynard Owen Williams

MORAVIA SLOVAKS

From Moravia to America, and from the Hussites to the Methodists, may seem two long jumps; but the contact was made in both cases in Georgia, U. S. A.; for the followers of Huss founded the Moravian Brethren, whose survivors fled to Saxony after the Thirty Years' War, and thence emigrated to Georgia. There John and Charles Wesley, then engaged with their mission, were deeply influenced by the kindly, persuasive ways of the Moravians.

A RADIANT HUMAN RAINBOW FROM MORAVIA

Her kerchief of red, with yellow flowers; her shawl white, with violet embroidery; her apron blue; her skirt red, and her sash multihued, this charming girl of Czecholand is a competitor of the rainbow.

A YOUNG SLOVAK GIRL FROM RUZOMBEROK IN NATIONAL DRESS

There are enough Slovaks in the United States to populate a city the size of Baltimore. The Slovak home in Czechoslovakia is a model of neatness.

Photographs by Dr. V. Sixta and Son

SUCH PICTURESQUE COSTUMES ARE FAST DISAPPEARING

Each little Slovak village has its peculiarities of dress. The initiated can often tell from what village a girl comes by one glance at the way she folds her kerchief.

Photograph by Maynard Owen Williams

CHILDREN OF PILSEN (PLZEN) IN THE YARD OF THE FAMOUS BREWERY: CZECHOSLOVAKIA

"Plzen" was always Czech for "Pilsen," but now that independence has come to the Czechs, they see no reason why the German language should retain a monopoly of their geographic nomenclature (see page 143).

To the Czechs, Slovakia is the wild and woolly East End of the republic, its intricately engraved egg-shells and bright pottery as exotic as Navajo art to a man from Fourth Avenue. Yet even this former crossroads of commerce between Krakow and Budapest and between Warsaw and Vienna is not immune from the cosmopolitanism that robs beauty spots of distinctive character the while it broadens their horizons.

AT A MOUNTAIN RESORT IN THE HIGH TATRA

Far down the long dining-hall a Hungarian orchestra, which has been playing a barbaric melody, suggestive of a tiny campfire beside some lonely road, with the gaunt, swarthy faces of the men cast into high relief by the ruddy glare and with brightly clad figures of Tsigane women moving about in the deep shadows of towering trees, now plays Handel's "Largo" with delicacy and feeling.

Across the room a woman, whose beauty is as unstudied in effect as it is painstaking in method, is smoking a perfumed Russian cigarette, whose glowing tip only occasionally challenges the sparkle of jewels on fingers and breast. Her companion, an officer in the neat uniform of the Czech army, bows cordially to a serious-faced man, who has devoted much time to tanning his bald head with all the care and enthusiasm with which another would color a meerschaum pipe to the same warm tone.

Nearer at hand there sounds the girlish laughter of lovely twins in evening dress. I had seen them earlier in the day returning from a climb, their fair young faces flushed with exercise.

Their heavy blond hair, now elaborately arranged, was then confined in yellow silk toques and they were dressed in long blue sweaters, trimmed with white angora, knee-length skirts and tan stockings, with another darker pair of heavy wool rolled down to the tops of their business-like Alpine boots.

A BEVY OF MORAVIAN BRIDESMAIDS

The dowry of a Moravian peasant's daughter is often from five to twelve thousand dollars and her wedding festivities cost hundreds of dollars; yet she does not scorn to go barefooted around home or to work in the fields when help is scarce.

Photographs from Maynard Owen Williams

MORAVIAN SLOVAKS, LOVERS OF MUSIC AND DANCING

The wealthier farmers usually keep three horses, and the horse determines a family's social status. A "horse" peasant's family stands higher socially than that of an "ox" peasant.

Photograph by Dr. V. Sixta and Son

THE OX-TEAMS OF CZECHOSLOVAKIA

The oxen of Czechoslovakia have horns that might make any roan Texas longhorn turn green with envy.

Spotless linen, sparkling glassware, silent waiters hurrying back and forth with food and drink, and outside the wide windows the pine-clad slopes and craggy peaks of intimate mountains newly clad in an early coat of snow. Not exactly a pastoral scene, yet we are in the midst of the High Tatra, in the heart of Slovakia.

PROVERBIAL SLOVAK ILLITERACY WAS DELIBERATE

Slovak stupidity was as deliberate as is the sturdy determination of the Czech. Forced by the Magyars to learn a hated tongue or go untaught, the Slovak chose the latter course; hence he is largely illiterate today, his ignorance a tribute to his sense of freedom. One of their number explained it to me in the café of a mountain village. He saw me sitting alone and his first sentence explained why he came over to speak to me.

"You are lonely, I think," he said.

And as we sat there, in that small café, he told me how, rather than submit to Magyarization under the Hapsburg ré-gime, he refused to study till the day when his fond dream came true and he set out for America.

After his arrival in the United States he went to night school, and, judging from the quality of his English, he profited well from the privilege.

"Nobody forced me to learn English," he explained; "I did it because I wanted to. English is a very 'practische' language, and I wanted to be an American through and through, so I worked hard at night to learn. I got my first papers all right, and then I came back for a visit. Then came the war, and I had to stay."

His number is legion and he is remaking the mental atmosphere of Slovakia.

SLOVAKIA IS A MUSEUM OF FOLK ART

One respects the Czechs for what they have done; one loves the Slovaks for what they are. Kindly, hospitable, simple-souled, religious, true Slavs in faults and virtues, the Slovaks represent the conservative element in the new republic. Few of them are yet trained for leadership, but their presence in the State

Photograph by Dr. V. Sixta and Son

TWO SLOVAK COUNTRYMEN FROM KRUPINA, CZECHOSLOVAKIA

The honesty of the Slovak peasant is proverbial, his home bank usually being willing to lend him money, even to go to America, knowing that he or his family will somehow manage to liquidate the debt. Usually the immigrant to America sends back home as much as a whole family makes in Slovakland. No wonder America seems a wage-earner's paradise!

argues well for its permanence, if their instincts are not outraged by too rapid change. They are the common people.

Slovakia is a vast museum of folk art. Songs that have sprung from the hearts of the people and have passed from lip to lip for centuries have a haunting quality which is the soul of art, because it mirrors the soul of the people. Captivating measures that sing of stamping boots and voluminous skirts whirled in the picturesque dances of the Slovakian peasant have long since reached the outside world, though often disguised as "Hungarian rhapsodies."

Pottery with native designs of distinction and purity decorate the walls of many "best rooms" in Slovak villages, and the wonderful products of Slovak needles rank with the most beautiful embroideries and laces in the world.

Sabbath-day Slovakia is a picture which even an Uprka cannot paint—the colorful picture of a people whose culture was not learned in the school-house, but was born in the hearts of hard-working folk, often bowed before the shrines and altars of a very real and intimate religion.

One of the distinctive Slovak villages is Čataj (pronounced Chatai). The tinted walls of the houses have a darker color for some distance from the ground, and there is a narrow line of clean-scrubbed bricks bordering the foundations to keep the water from the thatched roofs from spotting the base.

Not only are the outer walls of several colors, with the windows clearly lined in contrasting tints, but some of the old women of the village have painted original designs on the walls of their kitchens. Some haughty users of ready-made furniture might sneer at these sometimes crude patterns, but they show imagination and meaning as well as care and housewifely pride.

MORE WOMEN THAN MEN VOTE IN CZECHOSLOVAKIA

The painstaking embroideries eloquent of patient skill, the spotless white of the men's costumes, and the stiff white ruffs with which the infants are lovingly provided constitute a great tribute to the energy of the women of Czechoslovakia in the home. That they are not lacking

in political privileges, however, is shown not only by the fact that they have equal franchise but also that they use it. In the June, 1919, elections 2,746,641 women voted in comparison with 2,302,916 men voters. Thirteen of the 302 members of the House of Deputies are women and three of the 150 Senators.

AS A GUEST OF THE MAYOR OF A SLOVAK VILLAGE

We were guests of the hard-working mayor of Čataj and his barefooted wife and mother, and had the privilege of examining his house. One wall of the guest-room was a mass of pottery and the opposite wall was obscured behind the great pile of bedding which is an earnest of the hospitality one finds in the humblest Slovak home.

The long red bench was made in 1856 and bears the picture of Adam and Eve and the serpent. This well-known trio, sometimes painted in all the grotesqueness of early Italian drawing, is found in every home, for Adam and Eve are held as the patrons of marriage.

The mayor showed us his harvest crown, which the people brought to him as a seasonal tribute. It was a huge affair, constructed of the choicest stalks of grain, and that and a beautifully carved piece of common wood, which dated from 1714 and which served as a sort of scepter or badge of office, were the only emoluments the worthy man received.

Returning to a spotless room in Prague after weeks of toilsome travel in which comfort was as scarce as interest was general, I heard that a festival was to take place in the Slovak town of Turčiansky Sv. Martin, nestling at the foot of the Tatra Mountains, only a short distance from the romantic valley of the Vag (see page 155).

The comforts of Prague had a peculiar appeal to one who had almost forgotten what a bed was, but I thought of the sleepy little town with grass growing between the cobblestones of its great square, vacant as a yawn, and with its churches dominating the skyline as they dominated the thoughts of the people. I thought of the lovely valley of the Vag, with ruined castles jutting upward from

THE OLD CASTLE AT ZVOLEN: CZECHOSLOVAKIA

Photograph by Dr. V. Sixta and Sun

These ancient castles in Slovakia were usually built on eminences, to be used as fortresses in case of attack. They are the centers about which many legends have been written.

THE WOODEN RUSSIAN CHURCH: HUKLIJEVO, CZECHOSLOVAKIA

The Podkarpatska Rus (known in America as Ruthenian) element in Czechoslovakia is very devout. Having the forms and ceremonies peculiar to the Greek Orthodox Church, they yet hold allegiance to the Pope at Rome. Note the cross on the tower, with its three transverse pieces, which tells of its Russian rather than Latin origin.

Photograph by Dr. V. Sixta and Son

139

Photograph by Dr. V. Sixta and Son

WHERE ANCIENT NATIONAL COSTUMES AND FOLK-ART HAVE RESISTED THE RAVAGES
OF MODERNISM

The kitchen, the parlor, the pottery, the furniture, and even the exterior of the houses in many Slovak villages are decorated in gay but harmonious colors and ornaments that have characterized them for centuries.

scores of grain-clad hills and with sturdy peasants rafting logs down the turbulent stream to the little factories, which would shape them into furniture destined for France and England and even for America.

I thought of the pleasant riverside villages overflowing with playing children; long, regular rows of lowly huts with steep shingled roofs and fantail gables and flowers in every window; and I thought of the riot of color of the native costumes constantly reshaping themselves into new combinations and harmonized through the necromancy of a kindly sun.

So, to re-acquaint myself with the homely charm of Slovakia, I slipped away from Golden Prague to chase away for a while the memories of Rumania, with its gayety and awful trains; Vienna, wearing a painted smile to cover a starving body and soul; Budapest, where food and room are cheap and everything else is dear; Serbia, with its new station build-ings and bridges taking the place of the wreckage of war; Croatia, the cultural and lovely, which already feels the pressure of Serb supremacy; and German Austria, whose charm exceeds that of Slovakia itself, but whose people do not measure up to the majesty of the landscape.

PREPARING FOR A FESTIVAL

Turčiansky Sv. Martin was a different place. Up and down the streets one could see the citizens digging the sod out from between the cobbles and sweeping the whole town until it was commonplace in its cleanliness.

Every train was bringing in its quota of visitors, many of whom were forced to sleep in barns throughout their stay. If they were peasants, they had disguised the fact under frock coats, ill-fitting derbies, white dresses, and white hats.

The gathering was not the festival which I had expected to find, but a con-

Photograph by Dr. V. Sixta and Son

MEN FROM VOLOVEC, RUTHENIA (PODKARPATSKA RUS), CZECHOSLOVAKIA

By dint of hard work the peasant coaxes the soil to production's limit; but when this does not suffice to satisfy the incredibly small needs of his large family, he sets out as an itinerant, now selling linens in the cities of the plain, now seeking odd jobs as tinker and glazier in all the countries of Europe.

ference of the *intelligentsia* of Slovakia; not the care-free holiday of roistering farmers and their red-faced wives, but the endless speeches of cultured leaders.

GREETINGS FROM MANY WHO FORMERLY DWELT IN AMERICA

But there were enough reasons for feeling at home. The restaurant-keeper at the junction station wore an American suit that spoke new-world clothing dealer from lapel and pocket flap. The first sight that met my eyes as I entered the Slovak Museum was the Stars and Stripes on the main stairway, and the second was the picture of the college professor with whom I was traveling.

I sought to buy some Slovak sweets, only to be offered a box of American chocolates, just arrived. A Slovak pianist from Chicago introduced himself, and several other men came up to tell of their life in "the States."

I went to a furniture factory and was shown around by the foreman, who had

spent several years in America, and to a paper factory, where a most attractive Slovak boy, only one month back from the baseball diamonds, hockey fields, and gridirons of America, showed me around with the aid of another lad, perhaps two years younger, who looked with open envy at the twin-starred service pin on the sweater of his chum.

Half a dozen American Y. W. C. A. girls added a decided charm to the event, and I shared my room with two workers for the Y. M. C. A., which is doing a magnificent work for the soldiers and civilians of Czechoslovakia.

SLOVAK MUSIC, DRAMA, AND POETRY

Disappointing as the gathering was for what it was not, it was highly satisfying for what it was—a meeting of those who are trying to raise Slovakia to a higher plane without robbing it of the peculiar culture which entitles it to an honored place among peoples possessing a love and understanding of art.

Photograph by Rudo Bruner-Dvorak

THE CASTLE OF CHARLES IV, BETWEEN PRAGUE AND PILSEN, CZECHOSLOVAKIA

The Emperor Charles IV built this castle to house the imperial regalia. The picturesque.
little town of Carlstein is situated at its feet.

An excellent orchestra played several spirited numbers newly composed by its leader. The pianist from Chicago was brilliant in his rendition of some Slovak music. Several poems by Slovak poets were read and received with much applause. Each number was an expression, not only of Slovak culture, but of the individuality of the man or woman who offered it for the enthusiastic approval of the audience.

The next evening there was a Slovak drama, somewhat melodramatic for sophisticated tastes, but nevertheless well written and well acted. It was a revelation to find a community that could entertain itself so well with its own productions. There were no "canned goods" among the attractions. Turčiansky Sv. Martin has not yet sold its spare time to a vaudeville agency and a film exchange.

The town was alive with color and movement. The stolid peasant, to be sure, was absent; but every street was touched with color furnished by the national costumes which were worn by many of the women and young girls.

Most of these young women were town-bred and somewhat unaccustomed to the costumes they wore, but what was lacking in fitness was made up in fit. Each woman had chosen from a bewildering array the particular dress that would accentuate her charm. The stairway in the humble Narodny Dom café was a cascade of color which one would not find duplicated in the grand foyer of the opera or in the more bizarre opera house at Tiflis (see page 155).

FESTIVITIES UNABATED AT THE DAWN OF
A NEW DAY

I had come to this little Slovak town to rest. When I retired, long after midnight, the orchestra in the big auditorium was still fresh in its task of driving the whirling dancers round and round the hall. At 5 o'clock in the morning the porter knocked to announce that my train was soon due, but the orchestra was still at work as I passed out, and a hundred bright costumes were being spun around the room by gallant swains who looked as fresh as their bright-eyed partners.

One unaccustomed to Slovak ways should never go there to rest until he finds out whether a festival is in progress. A sedentary life in cities does not fit one for lasting out the frivolities of the country folk of Slovakia.

CZECHOSLOVAKIA HAS RESUMED HER GEO-
GRAPHICAL MAIDEN NAMES

When the Czech divorce from Austria was recognized by the great powers, the first thing the little country did was to go back to its maiden names. This, of course, has given great joy to the people; but why a trade name as valuable as *Karlsbad* should be sacrificed for *Karlovy Vary* is a little hard even for Czechoslovakia's sincerest admirers to understand. It is bad enough having to drink the water without having to learn a name like that, and one is quite surprised to know that *Marienbad* is just as attractive under the impossible cognomen of *Mariansky Lazne* as it was under its German name.

The foreign traveler who is ignorant of Czechish has his choice between using a German time-table, which stops trains at places perfectly understandable on paper, but leaves him "up in the air" when he gets off, or using a Czech time-table, which mentions towns he never heard of, but which are sure enough there when he reaches the station.

But in Czechoslovakia the old high priest of the tourist is sadly discredited and even the person who likes to think he has a touch of intelligence may make the most absurd errors if he places any credence in the statements of the once infallible Karl Baedeker of Leipzig. This is not only true with regard to the statement "R. from 4." which used to mean that a man could get a room in Prague for four crowns, but is even true when he makes a too absent-minded use of the map.

On my arrival in Prague I annexed a German-speaking cabbie and started out on the round of official calls which the post-war prefectures of police and other officials expect from every visitor; and after a time, which seemed an eternity when figured in crowns, I started back to the station for my luggage.

I should have known better; but I made the mistake of thoughtlessly quoting from the map which had been my

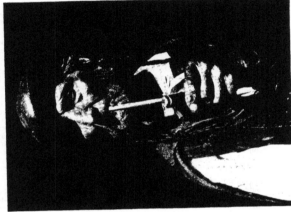

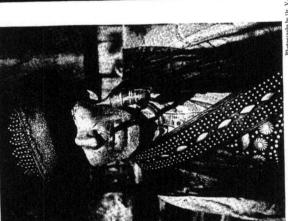

TWO PEASANT TYPES OF SLOVAKIA (SEE ALSO PAGE 126)

To the Czechs, Slovakia is the wild and woolly east end of the republic. It is a veritable museum of folk-art. Many of the shepherds
wear tremendous leather corsets and belts studded with brass,

The shepherds of the Carpathian foothills dress as picturesquely as
the cowboys of our cattle country. Pastoral Slovakia is to industrial
Bohemia what the Scotch highlands are to the midlands of England,

mentor during my wanderings and asked the man to drive me to the "Franz-Josephs Bahnhof."

He made it perfectly plain that Francis Joseph is as dead in spirit as he is in body, and that a new day has dawned in central Europe.

"Franz-Josephs Bahnhof?" he laughed. "There is no such thing. What you want is Wilson Station."

I had known it all the time, for my ticket from Paris had read "Prague, Gare Wilson," but I did not know then, as I know now, that it is a very small Czechoslovak village that does not have its bridge, park, square, or street named after the American President.

THE CZECH IS STINGY WITH HIS VOWELS

The Czech is both generous and hospitable, and to be his guest is to endanger your digestion and sobriety for the rest of your life, but he is very stingy with his vowels. As we passed through a large town which makes one man moisten his lips and another whistle "The Message of the Violet," I saw the sign "Plzen."

"Why waste the 'e'?" I asked.

The Czech assumes that an "e," an "i," and a "u" convey the same sound impression, so he gives *Brunn* the name of *Brno*.

Why he should spell it Berno, Birno, or Burno he can't understand; so he leaves out the vowel and lets the reader take his pick, since it's all the same sound, anyway. But when the unwary traveler carelessly mixes the German name with the Bohemian and calls the Moravian capital "Bruno," he thinks that's a poor joke.

The Czech prides himself that his language is phonetic. Leaving out such little things as pronouncing "rip" "zheep," it is. "Zmrzlina" starts in with a hard freeze, and then flows as fluidly from the tongue as the hoky-poky ice-cream for which it stands.

If the Czech is stingy with his vowels, that does not signify that he does not amply reward a letter for overtime. When you see a Czech letter wearing a service stripe, you may be sure that it is doing double duty. "Cop" would be understandable even to a wharf rat, only the Czech would call it "sop"; but put a service stripe (v) on the first letter and it becomes "chop."

L-u-c-k-y looks lucky, but the Czech spears the "u" with an accent and decorates the "c" with a service stripe, and lúčky becomes "lootchky," like Bolshevik gold, and there you are!

Czechish is a peculiar language, but not half as peculiar as the result attained by those who take a perfectly fine word like "Čech," which is pronounced "check," and spell it "Czech," which is neither phonetic nor intelligible. To start a brave little people, who have troubles of their own, out on the rough road of Central European life with the awful name of "Czechoslovakia" is to put a spirited young colt under an unfair handicap.

But "What's in a name?" asks the Czech.

And the best way you can reply is to counter, "Yes, what is?"

You ultimately learn that the train which runs from Oderberg to Kassa, according to the ticket, and from Bohumin to Košice, according to the time-table, finally gets there, and if that is where you want to go, why bother about names? Only you must be careful to learn the nationality of the man with whom you are talking before you say "Poszony," "Pressburg," or "Bratislava" for Slovakia's capital and Czechoslovakia's port on the Danube. Unless you find out in advance of using the word, the chances are two to one against you.

THE GOVERNOR OF RUTHENIA IS A
PITTSBURGH LAWYER

Both the Professor and I wanted to see primitive conditions; so we went up to what is variously known as Ruthenia, Rusinia, and Podkarpatska Rus and started to chase the ultimate frontier of civilization at forty miles an hour in the governor's motor car. That is not the way to come upon an ultimate frontier. Such game is only stalked on foot or with the aid of a sure-footed little burro.

But we did get to conditions so primitive that six human beings and four cattle, not to count pigs and poultry, lived in a single-room house, and we did experience the feeling which only carbolic soap can quite relieve.

As is now well known, the present gov-

Photograph by Dr. V. Sixta and Son

A YOUTH OF DETTVA, CZECHOSLOVAKIA,
DRESSED IN HIS SUNDAY BEST

A festival scene in Slovakia presents a
veritable cascade of color in the variety of
native costumes.

ernor of Podkarpatska Rus (Ruthenia),
which is an autonomous part of Czechoslovakia, is a Pittsburgh lawyer and an
American citizen. From the little capital
of Uzhorod (Ungvar), he has one of the
most difficult problems of government
in all of Europe.

His people are not only densely ignorant for the most part, many of them
wolfish, low-browed men, who remind
me of the present-generation Ainu of
Japan, but are also miserably poor.

This fact did not prevent the collection
at the church I attended from being remarkably generous, and at least two
women wore as their proud jewelry buttons of the Third Liberty Loan.

The men used to go down into the
Hungarian plains to work and there get
bread for their families; but a political
barrier has been erected where before
there was a thoroughfare.

DIFFICULT DAYS AHEAD

Pastoral Slovakia is to industrial Bohemia what the Scotch highlands are to
the midlands of England. Many of the
men are uncouth shepherds, who wear
tremendous leather corsets studded with
brass, that make the regulation Sam
Browne belt look like the toy accoutre-
ments mounted on cardboard which department stores used to sell for five-year-
olds. But the Slovakian shepherd is a
polished gentleman compared with the
mountaineer of Podkarpatska Rus.

This ill-favored but beautiful land was
long an economic dependent of the rich
Hungarian plain, but Governor Zatko-
vitch does not think that such conditions
need last forever. There is almost twice
as much land under cultivation here this
year as last, and he believes that the little
state can feed itself in time.

Much food used to be obtained in exchange for timber, which was floated
down to the factories on the Hungarian
rivers; but much of the completed furniture and paper used to be shipped back
past the forests where the wood was
grown, and the Ruthenians are now being
encouraged to start furniture and paper
factories of their own, which will elimi-
nate this waste.

Within Podkarpatska Rus itself conditions have also changed. Formerly all

Photograph by Dr. V. Sixta and Son

SLOVAK GIRLS OF THE VALLEY OF THE VAG (WAAG), CZECHOSLOVAKIA (SEE PAGE 149)

The town of Pistyan, of which these young women are natives, possesses mud and sulphur baths which attract thousands of people annually. The whole country around is distinguished for its ruins of feudal castles.

the foremen in the region were Hungarians and they hired their fellow-countrymen, thus forcing the Ruthenians to become farm laborers in Hungary. Now both foremen and laborers are Ruthenians.

In spite of all the optimism one can muster, however, one cannot ignore the fact that here, as elsewhere in post-war Europe, boundary lines have been erected which violate the principles of geography, and Podkarpatska Rus probably has difficult days ahead.

In Uzhorod we saw an interesting side-light on what one writer has called "the United Hates of Central Europe." About all the attractions that end-of-the-earth town could offer on our first day there were a theater, a cabaret, a moving-picture show, and a football game. We voted for football and saw an excellent game between the Uzhorod team and a Hungarian eleven from Budapest.

The Hungarians won, and in the dining-room that evening these huskies celebrated their victory by singing some of

their national songs. Can you imagine the Wisconsin pigskin artists sitting down in the main restaurant of Ann Arbor the night of their victory over Michigan and singing songs of triumph? Yet the audience in Uzhorod not only allowed their late enemies in war and sport to live, but roundly applauded their singing as they had their play. At the same time an eleven composed of Hungarian pickets along the Danube came across the bridge, with its barbed wire entanglements marking the Czech-Hungarian boundary, and played football with the Czechish soccerites.

POLITICIANS ARE ATTEMPTING TO PERPETUATE RACIAL HATREDS

If an American thinks that many of the European boundary squabbles are petty, he has only to look back to the time when Ohio and Michigan mobilized their militias over the question of their boundary line and discover why it is that the members of the Michigan legislature from the northern peninsula go through

Photograph from Maynard Owen Williams

A BOHEMIAN TEAMSTER OF PILSEN (PLZEN), CZECHOSLOVAKIA

The Bohemian driver is very fond of trappings for the harness of his horse.

Milwaukee on their way to Lansing. Propaganda bureaus are doing their best to perpetuate hate and a paper shortage in all the countries of central Europe; but if the politicians aren't careful the people are going to fraternize until it will take a lot of persuading to get another big army into the field.

DANGERS ARE ECONOMIC RATHER THAN POLITICAL

Austria-Hungary was a patchwork of peoples, but the same is true to a lesser degree of every present state from the Baltic to the Ægean.

With the best of intentions, the newly formed governments cannot function without hurting the feelings of such of the minorities as were formerly of the ruling class. There is bound to be some friction. If those who were subject to autocratic power only a few months ago are now humming "I've got my captain working for me now," it is not to be expected that the erstwhile captain is very enthusiastically applauding the effort.

Photograph by Dr. V. Sixta and Son

WOMEN OF TRENTSCHIN (TRENCIN), CZECHOSLOVAKIA

As one ascends the Valley of the Vag from its confluence with the Danube above Budapest, the Magyars disappear and Slovaks become more numerous. These are types of the
women one sees in the High Tatra mountain region.

But, if my observations are correct, the present dangers in central Europe are more economic than political, and the war-weary people would gladly do the lion-and-lamb act together, were it not for the fact that the lambs are so hungry that they are not quite sure but that they would relish a little tender lion meat, and *vice versa*.

The boundary that sticks up like a sore thumb to the casual traveler is even more irksome to the peoples concerned. I count myself a friend of the Czechs, but if I had had to spend many more hours getting permission to take my American films out of their country, I would have felt like declaring war on them myself. Yet my films are no more necessary to me than Czech coal is to Austria. After one has been accustomed to use an open road for many years, he hates to find that somebody, be he friend or foe, has closed it.

When the Czech soldiers, early in the

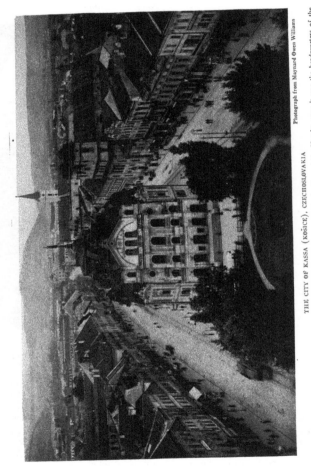

THE CITY OF KASSA (KOŠICE), CZECHOSLOVAKIA

Kassa, the main city in eastern Slovakia, was famous in the Middle Ages. At the time of the Hussite wars it was the headquarters of the Hussite general, Jan Jiskra, who defied the Hungarian king and held most of Slovakia for fifty years. As an exile, the famous Czech educational reformer, Komensky (Comenius), lived here for some time.

THE CHILDREN OF CHOD (CHODSKO), CZECHOSLOVAKIA

Descended from the ancient "walkers," which is the literal meaning of the word "chodove," these Bohemians live in the mountains on the German border, where their ancestors used to guard the frontier. They are a brave, hardy people.

A SLOVAK PEASANT FARMER'S HOME IN THE CARPATHIAN COUNTRY
Their peasant homes may be humble, but they are hospitable.

Photographs by Dr. V. Sixta and Son

SLOVAK EMBROIDERY WORKERS IN DETTVA
As needle-workers, the women folk of Czechoslovakia show a keen appreciation of the
beautiful. They draw their own designs freehand, and then execute them.

AN UNCROWNED PRINCESS OF PILSEN

Pilsen has a claim to fame other than that of being the city that made beer famous. It had the first printing-press in Bohemia. Beer, to the Pilsener, does not suggest "Bohemianism," but a staple industry, occupying an entire quarter of the city, with its cellars burrowing underground for miles.

war, refused to fight the battles of their oppressors and deserted to the enemy by companies, they established a precedent for military sabotage which cannot but deeply affect the whole question of im-p e r i a l i s m and subject nationalities throughout the world.

This voluntary surrender of the Czechs, sanctioned and applauded as it has been by the great powers, rang the death knell for militarism of the Prussian type,

wherever it is found, just as the voluntary support afforded by England's colonies revealed a loyalty which is now bringing its reward in new measures of self - government to hitherto subject classes.

CZECHOSLOVAKIA IS A RELIEF MAP OF ROMANCE

Czechoslovakia is a vast relief map of romance, a treasure-house of art, and a

Photograph by Dr. V. Sixta and Son

A SLOVAK TRADESWOMAN DISPLAYING NATIVE EMBROIDERY

She is a native of Pistyan, in the neighborhood of the ruins of the famous castle of Cachtice, where the mad Elizabeth Bathory lived at the beginning of the seventeenth century. Elizabeth is said to have had executed 300 of her country-women in an insane effort to restore her own beauty and retain her youth.

Photograph by Maynard Owen Williams

TALK ABOUT BIG BROWN EYES!

Sometimes the photographer is able to catch a little one when she is not too scared to smile. This charming miss is a native of Turčiansky Sv. Martin, where many distinguished Slovaks, who had recently returned to their homeland from America, gathered in conference recently on the welfare of their state.

154

Photograph by Maynard Owen Williams

FOUR OF A KIND: TURČIANSKY SV. MARTIN

The short skirt, now a fad in Paris and the United States, is habit in Czechoslovakia. Aprons are not to be removed, but put on, when a maid starts for a promenade. The bodices and the kerchief head-dresses are characteristic. One might suspect the newer republic had felt the influence of prevalent American styles in shirtwaist modes and the "mutton-sleeve" fashion of an earlier date. Chronologically, the reverse is true. The Czech grandmother need not complain that débutantes do not dress as they did when she was a girl. They do. The young woman with the embroidered white apron is a Slovak matron. She is wearing a Cicmamy costume, in many respects the most beautiful to be found in Czechoslovakia.

pansy-bed of interesting peoples in rainbow costumes. It might well be made an international exhibit of much that is interesting in racial development and life.

THE STRUGGLE FOR A SOUND ECONOMIC BASIS

But romance is lighter on the stomach than the most widely advertised breakfast food, and what the little country that stands between Germany and Poland on the north and the two pitiful remnants of proud Austria-Hungary on the south is now trying to do is to put itself upon a sound economic basis, prosaic as the process must be in comparison with the spectacular adventures of the Czechoslovaks during the last few years.

Drawn by A. H. Bumstead

A MAP OF CZECHOSLOVAKIA WHOSE PLACE NAMES PRESENT SERIOUS PROBLEMS FOR
THE STUDENT

Many of this new republic's cities and rivers are known to American readers solely by
their German names. But the Czechs are anxious to have the world know their towns by
their geographical "maiden names," so all official Czechoslovak maps published in the home
country designate Prague as Praha, Pressburg as Bratislava, Brunn as Brno, Marienbad as
Marianske Lazne, Karlsbad as Karlovy Vary, Ungvar as Uzhorod, and Pilsen as Plzen.
Even the familiar province name of Ruthenia becomes the difficult Podkarpatska Rus. From
northwest to southeast, Czechoslovakia has a length of six hundred miles. When its canal
system is completed it will have access to the North, Baltic, and Black Seas through the
Elbe, Oder, and Danube rivers. The Oder, which is not shown on this map, is to be connected with the Danube by a canal running the width of Moravia and entering the latter
river at Pressburg. (See also the New Map of Europe, issued as a supplement with this
number of THE GEOGRAPHIC.)

Czechoslovakia has enough glory to go
around. Her present task is to divide
this glory a little more equally, so that
the soldier of fortune will have something beside his medaled uniform, and
the Magyarized Slovak farmer, recently
released from Hungarian influences, will
take a justifiable pride in being a citizen
of a fine, free race.

CZECHS AND SLOVAKS UNITED BY
OPPRESSION

For three centuries Bohemian national-
ism, fearfully oppressed, cherished the
memory of a national defeat at White
Mountain.

While the Czechs felt oppression from
the north, the Slovaks were being exploited by a hated autocracy and cultoc-

racy to the south; but as they faced their
foes they drew near to each other.

One who has known the whole-hearted
hospitality, the intelligence, the artistic
sense, and the downright democracy of
the Czechs cannot but echo the greeting
and farewell which he hears on every
side—"Na zdar!"

Slowly steaming up the charming
Moldau, with a military band playing on
the upper deck, we looked out to see a
picture of rare grace. Two little peasant
girls, hearing the music, had dropped
their rakes and were dancing lightly
along the bank, keeping perfect time to
the music. As the music stopped, they
waved their hands to us and shouted,
"Success to you!" and our reply to them
was our wish for Czechoslovakia—"Na
zdar!" "To you success!"

INDEX FOR JULY-DECEMBER, 1920, VOLUME READY

Index for Volume XXXVIII (July-December, 1920) will be mailed to members upon request.

THE NEW MAP OF EUROPE*

Showing the Boundaries Established by the Peace Conference at Paris and by Subsequent Decisions
of the Supreme Council of the Allied
and Associated Powers

By Ralph A. Graves

*In order to make the New Map of Europe of service to the largest number
of Geographic readers, it has seemed preferable to retain the familiar forms of
place names rather than adopt those conformable to the tongues of the several
new nations within whose boundaries they now appear. It is the indisputable right
of the inhabitants of a country to say how its geographic names shall be spelled and
pronounced, but it would be a source of confusion to the average student to have
his entire geographical and historical background swept away by the elimination,
for example, of Prague in favor of the Czechoslovak form of Praha; of Warsaw
in favor of its Polish equivalent, Warszawa; of Vilna in favor of the Lithuanian
Vilnius; of Danzig in favor of the Polish Gdansk; and of Fiume in favor of the
Slavic Rieka. (See also pages 143 and 145.)*

THE NEW MAP of Europe is
before us, born of the treaties of
peace with the vanquished Central
Powers. These treaties—those of Versailles with Germany and Hungary, that
of St. Germain with Austria, that of
Neuilly with Bulgaria, and that of Sèvres
with Turkey—purport to erect new
boundary lines between countries of conflicting economic interests, antagonistic
racial distinctions, and rival historic traditions.

How long these boundary lines are re-
spected and how materially, as necessity
arises, they can be modified without resort to force will depend upon the wisdom exercised by those statesmen upon
whom devolved the responsibilities growing out of the greatest conflict of human
history; how long these boundaries can
be made to endure against the assaults of
predatory interests, each nation against
its neighbors, depends upon the firmness

with which the concert of nations exerts
its influence for peace.

Writing in the National Geographic
Magazine for December, 1918, Dr. Edwin A. Grosvenor, in "The Races of Europe," said of the then forthcoming peace
conference:

"For the first time in human experience, the effort is being made by victors
after a great war to trace the new frontiers in accordance with the racial aspirations and affinities of the peoples involved. Because of the impossibility of
defining exactly the limits of a race, many
heart-burnings are inevitable in the new
adjustment of European boundaries."

The results of the Peace Conference at
Paris and of the subsequent conferences
of the Supreme Council are represented
in the Map of Europe which is issued
as a supplement to this number of The
Geographic.†

---

* Additional copies of the New Map of
Europe (30 x 33 inches), with index, may be
obtained from the headquarters of the National
Geographic Society, Washington, D. C., at
$1.00 each, paper edition, and $1.50, mounted
on linen. This is the most legible map of
convenient size issued in America since the
Peace Conference in Paris.

† The student will find the National Geo-
grapic Society's Map of the Races of Europe,
issued as a supplement to the December, 1918,
number, of special interest for comparative
purposes, showing to what extent ethnographic
frontiers have been followed in the revision of
the political map of Europe. Extra copies of the
"Races" map may be obtained from the headquarters of the Society in Washington; paper,
50 cents; linen, $1.00.

Photograph by A. Frankl

KING CHRISTIAN X OF DENMARK ENTERING HIS RESTORED PROVINCE OF
NORTHERN SCHLESWIG

This was one of the parts of the former German Empire which was set aside as a
plebiscite area. The people voted recently to give allegiance once more to the Danish King,
from whom they were taken by Prussia in the War of the Danish Duchies in 1864. Note the
Danish flag.

Photograph by A. Frankl

KING CHRISTIAN X OF DENMARK GREETING HIS PEOPLE IN NORTHERN SCHLESWIG

After forcible incorporation into the German State for more than half a century, the people of
northern Schleswig made their return to Denmark an occasion of great festivity.

But the world entertains no delusions as to the inflexible permanence of this map. As Dr. Grosvenor observed two years ago:

"Neither a year nor a generation will suffice to make it. None of the now living will behold it when made. The Peace Conference will render its august decisions, and its members will depart, but the races remain on the spot, and on them the making of the new Europe devolves.

"Europe, though so old, is for the greater part young and inexperienced in self-government and political duty and opportunity. The gait of more than one newly enfranchised people will resemble the uncertain walk of a just-awakened child."

THE MAP'S STORY OF GERMANY'S LOSSES

Some of these new nations whose names are now blazoned on the map of Europe are not infants, but are sovereign states which are experiencing a national rebirth. Among these the most conspicuous example is Poland. Others, like Jugo-Slavia, Czechoslovakia, Latvia, Es-

thonia, and Lithuania, are received for the first time into the society of civilized states.

The map's story of Germany's territorial losses needs but little textual elaboration. In addition to the loss of 5,600 square miles and nearly 2,000,000 inhabitants by the recession to France of the provinces of Alsace and Lorraine, the former empire of the Kaiser surrendered control of the great Saar Valley coal field, to compensate in part for the coal mines of northern France destroyed or crippled by the invaders.

Fifteen years from the coming into force of the Versailles Treaty, the inhabitants of the Saar Basin shall determine by plebiscite whether they wish to remain under the control, as at present, of a Commission of the League of Nations, become part of the French State, or revert to Germany.

Belgium has acquired full sovereignty over the sections of the tiny area known as Moresnet—Neutral Moresnet and Prussian Moresnet—and of the *Kreise*, or district, in which are the small towns

Photograph by A. Frankl

VIEW OF WILHELMPLATZ, MYSLOWITZ, IN THE PLEBISCITE DISTRICT OF
UPPER SILESIA

An English correspondent in Germany describes Upper Silesia today as a vast battlefield, where the fighting is being done by two big armies of spies, agents, and propagandists. No date has yet been set for the vote to determine whether this, one of the greatest industrial and mining districts in the world, shall become a part of Poland or go to the German Republic.

of Eupen and Malmédy. These have small significance, except in so far as they rectify the political frontier so as to make it accord with the natural boundary line between Germany and Belgium.

Germany has been compelled to surrender to Poland territory equal in area to the State of South Carolina, with a population of 6,000,000, and in order to provide this re-created state with an outlet to the sea, the Germans give up the great Baltic seaport of Danzig, which becomes the "Free City of Danzig," under the protection of the League of Nations. The Memel district, to the northeast of East Prussia, is entrusted to the Allied and Associated Powers pending a final settlement of its sovereignty.

In addition to the Saar Basin, the German treaty designated six areas for plebiscites to determine their eventual ownership—two in East Prussia (Marienwerder and Allenstein), Northern Schleswig, Southern Schleswig, Holstein, and Upper Silesia. All of these plebiscites have been

held save that in Upper Silesia. The people of Holstein and Southern Schleswig elected to become reincorporated into the German State; Northern Schleswig voted to return to Denmark, and East Prussia expressed a preference for German as against Polish absorption.

As a guarantee for the faithful execution of her contracts under the treaty, Germany consents to the military occupation of territory to the west of the Rhine with bridgeheads at Cologne, Coblenz, and Mainz, designated on the map as the "Zone of Allied Occupation." This occupation is to continue for fifteen years.

FORTIFICATIONS OF THE KIEL CANAL
HAVE BEEN DISMANTLED

The Kiel Canal, whose construction in the interest of the development of the imperial navy Bismarck had in mind when he seized the Danish duchies more than fifty years ago, is thrown open to the merchant shipping of all nations at peace with Germany, while its frowning

Photograph by A. Frankl

CHILDREN ON THE BANKS OF THE NOGAT RIVER: EAST PRUSSIA

The Nogat separates the territory of the Free City of Danzig from East Prussia, where a plebiscite was held recently. The East Prussians voted to cast their lot with the Republic of Germany rather than give allegiance to Poland, and these children are among those celebrating the announcement of the result.

battlements, as well as those of its sentinel islet, Helgoland, have been dismantled at German expense.

AUSTRIA-HUNGARY DISAPPEARS AS A POLITICAL STATE

By the Treaty of St. Germain, signed September 10, 1919, in the Stone Age Room of the chateau at St. Germain-en-Laye, a suburb of Paris, the decadent Dual Monarchy of Austria-Hungary, composed of a heterogeneous group of discordant nationalities, passed into history, and from the wreckage are emerging the republics of Austria, Hungary, and Czechoslovakia, while large areas of the territory once controlled by the Hapsburgs have passed under the sovereignty of the kings of Italy, Rumania, and Jugo-Slavia.

Before the World War the Dual Monarchy boasted an area of 260,000 square miles and a population of 50,000,000. The area of Austria proper was 134,000 square miles, with a population of 29,-000,000. By the Treaty of St. Germain, the Austrian Republic becomes an impoverished state of 32,000 square miles (smaller than the State of Maine), with a population not exceeding 6,500,000, nearly a third of whom are crowded into the destitute capital, Vienna.

The only plebiscite district provided for in the St. Germain Treaty was that of the Klagenfurt district, which lies on the border of Croatia, now a part of the new Serb-Croat-Slovene State. By a substantial majority, the inhabitants have voted to remain with Austria.

By its loss of Transylvania, with 22,-000 square miles and nearly 3,000,000 inhabitants, to Rumania; of Croatia, Slavonia, Dalmatia, and portions of Banat to Jugo-Slavia; and of some 25,000 square miles and 3,500,000 inhabitants in Slovakia to Czechoslovakia, Hungary has been reduced from 125,000 square miles and more than 20,000,000 subjects to 36,-000 square miles, with 8,000,000 people.

The treaty by which the Kingdom of

Photograph by Nevin O. Winter

POLAND'S OUTLET TO THE SEA, THE FREE CITY OF DANZIG

By the Treaty of Versailles, Germany surrendered this great Baltic seaport to the custody of the League of Nations. The chief administrative officer is a high commissioner appointed by the League.

HELSINGFORS, FINLAND, CAPITAL OF THE NEW REPUBLIC OF THE NORTH

Like the peoples in the sister republics of Czechoslovakia and Poland, the Finns are anxious to cast aside the Swedish name of their capital city and have it known henceforth as Helsinki. With its fortress, Svenborg, Helsingfors has a population of nearly 200,000 and is the seat of a famous university.

BOLSHEVIST SOLDIERS TAKEN PRISONERS IN A RECENT CAMPAIGN IN LATVIA

Photographs by A. Frankl

THE SAME BOLSHEVIST PRISONERS SHOWN IN THE UPPER ILLUSTRATION, AFTER
RECEIVING AN ALLOTMENT OF CLOTHING DISTRIBUTED
BY AN AMERICAN RED CROSS UNIT

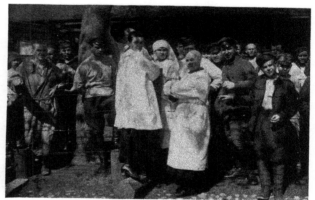

Photograph by A. Frankl

ESTHONIAN NURSES WITH THEIR CONVALESCENT PATIENTS AT A TYPHUS HOSPITAL
IN NARVA

Esthonia is one of the three Baltic States whose independence has not been recognized
by the United States. It set up a republican form of government in February, 1918. Narva
is on the Narova River, the outlet of Lake Peipus into the Gulf of Finland.

the Magyars was so shorn was handed to
the Hungarian delegates at Neuilly, near
Paris, on January 15, 1920, but was not
signed until the following June 4, in the
long gallery of the Grand Trianon at
Versailles.

By the Peace Treaty of Neuilly, signed
November 27, 1919, Bulgaria sustained a
smaller proportion of territorial losses
than any of the other Teutonic allies. Its
principal cessions of territory accrue to
the benefit of the Greeks, who gain Bulgarian Thrace and thereby wrest the important Ægean littoral from their hated
enemy of the Second Balkan War.

To Jugo-Slavia the Bulgars surrender
a strip of territory which includes the
town of Strumitsa; also two fragments
along the West Bulgarian front, one of
which contains the town of Tsaribrod.

The estimated area of Bulgaria before
the war was slightly in excess of 43,000
square miles (about the size of Virginia),
with a population of 4,750,000. The new
boundaries give the kingdom an area of
approximately 41,000 square miles (the
size of Ohio).

SUBJECT PEOPLES RELEASED FROM TURKISH MISRULE

By the provisions of the document
which will be known in history as the
Treaty of Sèvres (not yet ratified by the
Turkish Government), the "Sick Man of
Europe," concerning whose health Tsar
Nicholas I of Russia first expressed grave
alarm nearly three-quarters of a century
ago, is dead. "Turkey in Europe" is now
scarcely more than a name—a small tract
of land (the Chatalja District) west of
Constantinople, embracing the area from
which the Sublime Porte gets its water.

The Dardanelles, the Bosporus, and
the shores of the Sea of Marmora become "The Zone of the Straits," controlled and governed by an Interallied
Commission, and a small area, known as
the Suvla Reservation (Gallipoli Peninsula), is set aside as a cemetery for the
Allies who fell in the attempt to take

RETURNED PRISONERS FROM SIBERIA AT NARVA, ON THE BORDER OF SOVIET RUSSIA

Photographs by A. Frankl

RECENT PRISONERS OF WAR, FORMER SUBJECTS OF THE TSAR, NOW CITIZENS OF ONE
OF THE NEW BALTIC REPUBLICS

Constantinople. Greece receives Turkish Thrace, which lies to the southwest of Constantinople.

Scarcely less drastic has been the dismemberment of the Ottoman Empire in Asia, where provision is made for óne autonomous and four independent states, in addition to complete renunciation of all Turkish interests in Egypt and consent to a British mandatory over Palestine.

The independent states are Syria (temporarily under French mandate), Mesopotamia (temporarily under British mandate), whose boundaries President Wilson has been asked to define, and the Arab Kingdom of the Hedjaz, over which presides the Grand Sherif of Mecca.

The autonomous State of Kurdistan is to comprise the Kurdish area east of the Euphrates and south of the to-be-determined southern frontier of Armenia. The territorial adjustments in Asia Minor will be more comprehensively shown in THE GEOGRAPHIC'S Map of Asia, 28 x 36 inches, to be issued with the May number.

The Allied Supreme Council has provided access to the sea for Armenia, Georgia, Azerbaijan, and Bulgaria by a group of internationalized ports, chief among which are Dedeagatch, Batum, and Trebizond.

In addition to the hitherto Turkish islands of the Ægean, which pass under the sovereignty of Greece, the latter country also assumes the administration of a large area in Asia Minor surrounding the important seaport of Smyrna. This region is to have the privilege of a plebiscite at the end of five years.

One of the unusual settlements growing out of the Treaty of Sèvres is the disposition of the "Dodecanese" (Sporades), twelve islands which Italy occupied during the Turco-Italian War of 1912. Both racially and by historic traditions, the inhabitants of this archipelago lying off the southwest coast of Asia Minor are preponderantly Greek. However, owing to the earlier conflict over their possession, it was arranged that Turkey should cede them to Italy, who in turn, on the same day and at the same place, ceded all of the group except Rhodes to Greece. This double cession recalls a feature of the Treaty of Prague in 1866, when Austria, at the close of the Austro-Prussian War, ceded Lombardo-Venetia to France, who in turn immediately ceded the territory to Italy.

Rhodes, according to a Greek-Italian agreement, is to remain under Italian administration for fifteen years, at the end of which time a plebiscite is to be held; but if, in the meantime, Great Britain decides to relinquish Cyprus in favor of Greece, Rhodes likewise is to be surrendered.

Turkey has renounced all rights to Egypt as from November 5, 1914, and recognizes the British protectorate over that country. While the treaty formally recognizes the annexation of Cyprus by Great Britain, this, the third largest island of the Mediterranean, has been administered by the British for more than forty years; so that, for practical purposes, its status has been little affected as a result of the World War.

While Great Britain at present maintains a protectorate over Egypt, the British Government is considering the Milner Mission's proposal for recognition of Egyptian independence and a treaty of alliance under which Great Britain will undertake to guarantee the existence of Egypt against outside aggression. A British garrison will be maintained in the Suez Canal Zone.

FINLAND PROFITS BY RUSSIA'S BREAK-UP

The Republic of Finland has been born of the war with less travail than any other of the new nations of Europe. After being united to the Russian Empire as an autonomous grand-duchy for more than a century, its House of Representatives proclaimed the state's independence in December, 1917, and it became a republic according to the constitutional law of June 14, 1919. It has been recognized by most of the world powers.

In square miles of territory, Finland equals Czechoslovakia, Austria, Hungary, and Bulgaria combined. Measured by the "yardstick" of American State areas, it approximates the New England and North Atlantic group, including New Hampshire, Vermont, Massachusetts, Connecticut, Rhode Island, New York,

A STREET SCENE IN BELGRADE, CAPITAL OF THE NEWLY ORGANIZED KINGDOM OF
THE SERBS, CROATS, AND SLOVENES

By the amalgamation of the South Slav populations and territories of Serbia, Montenegro,
Bosnia, Herzegovina, Croatia, Slavonia, Dalmatia, and other provinces, Jugo-Slavia becomes
a state three times the size of pre-war Serbia.

Pennsylvania, New Jersey, Delaware, and Maryland. If the new republic should succeed in acquiring title to the region to the northeast designated as the "Pechenga District," 3,500 square miles will be added to its domains.

One of the most serious controversial issues growing out of the establishment of the Finnish Republic is the disposition of the Åland Islands. Formerly the property of Sweden, the 3,000 rocky islands, of which perhaps 80 are inhabited, were ceded with Finland to Russia more than a hundred years ago. The population, which was estimated at 20,000 before the World War, is mainly of Swedish descent.

Sweden has long fretted over Russian ownership of the Åland Archipelago, owing to its strategic command of the entrance to the capital city of Stockholm, and in 1856, by the Treaty of Paris, which brought the Crimean War to a close, the Tsar was prohibited from fortifying the islands—a restriction which was galling to Russia, but one enforced by Great Britain and France at the behest of the Swedes.

It is maintained by Sweden that a plebiscite has shown an overwhelming majority in favor of Swedish sovereignty. Finland, on the other hand, argues that the islands have been administered as part of the Finnish province of Abo-Björneborg for more than a century, and that a majority of the islands lie nearer to the Finnish coast than to the Swedish.

The sovereignty of the islands is still one of the moot points in the territorial settlements of Europe. It has been

Photograph from Frederick Simpich

SELLING CHICKENS IN THE STREETS OF BUCHAREST, CAPITAL OF RUMANIA

Rumania has become a major power of southeastern Europe, as a result of its allotment of Austro-Hungarian and Russian territory.

brought to the attention of the Council of the League of Nations, which has appointed a commission to make an inquiry and submit recommendations as a basis of amicable settlement.

THREE INFANT REPUBLICS ON THE BALTIC

The three Baltic States of Esthonia, Latvia, and Lithuania, fragments of disintegrating Russia, have declared their independence under republican forms of government—Esthonia and Lithuania in February, 1918, and Latvia in November of the same year.

All three have been recognized by most of the European powers, but not by the United States. Their boundaries are as yet extremely indefinite and their respective territorial aspirations overlap in many places.

The frontiers, as defined tentatively on the accompanying map, accord to Esthonia an area about equal to that of New Hampshire and Massachusetts combined, with Latvia half again as large, and Lithuania the largest of the trio, with an area in excess of the combined areas of New Hampshire. Vermont, Massachusetts, and Rhode Island.

THE RE-CREATION OF POLAND

For many decades "The Three Partitions of Poland" * has been a text upon which historians and new-school diplomatists have preached against the pernicious practices of the "Old Order" in Europe—an order which conceived peoples and their home-lands to be mere chattels, to be exchanged, bartered, or purloined by kings and princes. It was inevitable, therefore, that under the promised "New Order," inaugurated upon the conclusion of the World War. Poland should be one of the first sovereign States to be re-created.

As reconstituted, the Republic of Poland, with its seat of government in the ancient capital of Warsaw, derives its territory from the three powers which

* See "Partitioned Poland," by William Joseph Showalter, in the NATIONAL GEOGRAPHIC MAGAZINE for January, 1915.

Photograph by D. W. Iddings

THE ACROPOLIS FROM THE TEMPLE OF ZEUS: ATHENS

Tremendous impetus has been given to the Greater Greece movement by the territory allotted to the Kingdom of the Hellenes through the paring of Turkish domains in Europe. Before the overthrow of Premier Venizelos, the Allied Powers were even reported to view with favor the suggestion that the administration of Constantinople itself be entrusted to the Athens Government.

profited by the three partitions, the last of which took place a century and a quarter ago. From Germany the Allied and Associated Powers took parts of Posen, West Prussia, East Prussia, and Silesia; from Austria-Hungary most of Galicia and a part of the crownland of Bukowina, and from Russia all of Russian Poland.

The eastern boundary of the new Po-land was not definitely fixed by the Allies because of chaotic conditions in Russia— conditions which precluded any possibility of a definitive boundary treaty being negotiated. The friends of Poland in 1919 did, however, draw a line to the east up to which Polish civil administration was approved (see map). By this boundary the new Poland holds sway over an area of 100,000 square miles.

Subsequent to the delimitation of the civil administration area, the Poles and the Bolsheviks, after a bitter warfare, signed an armistice and established vaguely what has become known as the "Polish-Bolshevik Armistice Line of October 12, 1920."

If this latter line approximates the eastern boundary of Poland as it is eventually established, the rehabilitated republic will add an additional 40,-000 square miles to the territory accorded it by the Allies. It will be a nation exceeding by one-third the area of pre-war Italy, and will have a population about equal to that of the Peninsula Kingdom. Poland's chief outlet to the sea is the Free City of Danzig, previously mentioned (see page 160).

In these estimates the plebiscite area of Eastern Galicia is included in Poland, inasmuch as it is to be under Polish administration for a period of 25 years.

Extending like a gigantic wedge nearly 600 miles long and only 150 miles broad at its widest part, the new Republic of Czechoslovakia stretches from eastern Germany to northwestern Rumania—a corridor nation of central Europe (see separate article in this number of THE GEOGRAPHIC, pages 111 to 156). It is composed of three closely related racial elements — the Czechs, the Moravians, and the Slovaks (see map, page 156).

The Czechs (Bohemians) to the west, girded about by a natural frontier of

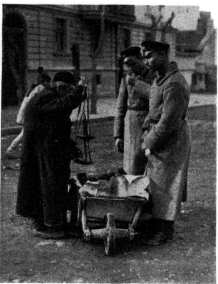

Photograph from Frederick Simpich

SELLING BREAD IN THE STREETS OF SOFIA, THE CAPITAL OF SHORN BULGARIA

With a population of 100,000 before the World War, Sofia is the largest city of Bulgaria and is the seat of a university. It occupies the site of the ancient Roman colony of Serdica, founded by the Emperor Trajan, and was a favorite residence of Constantine the Great.

mountains which separates them from the Germans, are joined to the Slovakians, whose home is to the south of the Carpathian Mountains, by the land of the Moravians. Moravia forms a gap through which passes the great central European route between the Adriatic and the Baltic.

By development of the waterways of the country and the construction of canals, Czechoslovakia, while without a seacoast, will have access to three seas—to the Black Sea by way of the Danube, to the Baltic by way of the Oder, and to

Photograph by International Film Service

AN AIRPLANE VIEW OF CONSTANTINOPLE, CITY OF THE SULTANS

This remarkable photograph, taken from an airplane flying at considerable height, shows the eastern portion of Stamboul, in which section are the most celebrated and important buildings of the city. The points of interest shown in the photograph are: Left foreground, the Mosque of Sultan Achmet I; in the center, the far-famed Church of Sancta Sophia, now a mosque; in front of Sancta Sophia the buildings of the Ministry of Justice; the open space at the point of the peninsula, the Imperial Enclosure and Seraglio. In the left background is the Galata Bridge, and on the other side of the water the quarters of Pera and Galata, where most of the Europeans reside.

the North Sea by way of the Elbe. It is proposed to connect the Danube with the Oder by a canal running from Pressburg to Prerau, and this waterway will be linked by a canal to Pardubitz, on the Elbe.

High hopes are entertained for the eventual prosperity and stability of Czechoslovakia. Bohemia and Moravia were the most important industrial regions of the old Austro-Hungarian Empire, and Slovakia is a rich agricultural land. In population and area the new republic roughly approximates our States of New York, New Jersey, and Delaware combined. Of its people, some 5,000,000 are Bohemians, 3,000,000 are Slovaks, 2,000,000 are Moravians, and more than 2.500,000 are Germans.

THE SOUTH SLAVS ARE AMALGAMATED WITH SERBIA AS THE CENTRAL UNIT

Leaving the land of the Czechoslovaks, the eye travels southward across the twin-born republics of Austria and Hungary to the Jugo (South)-Slav Kingdom of the Serbs, Croats, and Slovenes, built up of groups of South Slavic peoples, among which the Serbians are preponderant numerically.

The apportionment of territory to the Jugo-Slav State has been a problem of world concern since the first sessions of the Peace Conference in Paris. So numerous and diverse were the interests affected and so bitter has been the rivalry between the new state and its "vis-a-vis" nation lying across the Adriatic that there was constant danger of armed conflict between these two over the adjustment of boundaries. Every decision was fraught with Alsace-Lorraine potentialities of danger.

On November 12, 1920, however, the Italians and Jugo-Slavs at the Conference of Rapallo adjusted all differences over such intricate questions as the Pact of London Boundary, the famous "Wilson Line" through Istria, and the extravagant claims of the two contending parties. It is to be hoped that danger of what a noted European statesman once characterized as a "sore boundary" has been obviated.

By the agreement reached at Rapallo (a small winter resort southeast of Ge-

noa), disposition was made of that apple of discord, the seaport of Fiume. The population of Fiume proper is mainly Italian, but the hinterland is almost exclusively Slavic, and the city is the only fully developed and adequately equipped seaport by which the commerce of the Jugo-Slav State has outlet to the Adriatic. Some of the other towns of the Dalmatian coast (Spalato, Cattaro, and Metkovic) have equally good or even better harbors, but they are undeveloped and are not connected with the interior by standard-gauge railways.

The situation in Fiume is analogous to that at Danzig, a city German in population but the sole gateway on the Baltic for Poland's maritime commerce. The compromise effected is likewise similar to that agreed upon for Danzig. The port becomes the "Free State of Fiume," and provision is made for a commission, composed of Italian and Jugo-Slav members, which will settle all technical questions regarding traffic in the port with due regard to the commercial needs of Jugo-Slavia. Sushak, the Croat suburb of Fiume, is to remain to Jugo-Slavia, but with the privilege of joining its port to Fiume if it desires to do so.

In the settlement of the Fiume Question no official cognizance is taken of the disconcerting fact that for more than a year the seaport has been in possession of Italy's soldier-poet and picturesque adventurer-patriot, Gabriele d'Annunzio, who with a small army of followers took possession September 17, 1919. D'Annunzio maintained his dictatorship in the city despite the protests of the Allies and the disavowal of his own government, his slogan being "Annexation of Fiume to Italy or death!"

By the terms of the Rapallo agreement, the Dalmatian coast and islands become a part of the new state of Jugo-Slavia, with the exception of the town of Zara and two or three islands, the most important of which are Cherso and Lagosta, allotted to Italy.

Jugo-Slavia's domain has been affected by a larger number of treaties following the World War than any other state of Europe. By the Austrian peace treaty, Carniola and Dalmatia have been acquired; by the Hungarian treaty, the

OVERLOOKING THE HARBOR OF SMYRNA, CHIEF SEAPORT OF ASIA MINOR

With the consent of the Allied and Associated Powers, the Greeks have occupied Smyrna and its contiguous territory. By the Treaty of Sèvres, which the Turkish Government has not yet ratified and which may be modified as a result of recent political developments in Greece, the Greeks are given the right to administer this territory for a period of five years, at the end of which time a plebiscite is to be held to determine its ownership.

Photograph from Frederick Simpich

provinces of Croatia and Slavonia and part of Banat, and by the treaty with Bulgaria three small areas, including Tsaribrod and Strumitsa, have been added (see page 165). Austria and Hungary jointly surrendered the provinces of Bosnia and Herzegovina, over which the former Emperor-King had extended his sovereignty in 1908.

MONTENEGRO IS ABSORBED; ALBANIA REMAINS INDEPENDENT

In addition, the former Kingdom of Montenegro has been absorbed. As to the former Kingdom of Serbia, it would perhaps be more nearly proper to speak of its expansion to include Jugo-Slavia than of its "absorption" by the new state. It is the Serbian King Peter I who occupies the throne of Jugo-Slavia in the capital city of Belgrade.

The new boundaries make Jugo-Slavia a state three times the size of pre-war Serbia, with a population roughly estimated at 14,500,000, which is more than three times that of Serbia in 1914.

No part of Albania's boundaries is as yet definitely fixed, but as tentatively shown on the accompanying map the state has an area about twice the size of Connecticut, with a population estimated at something less than a million—not greatly at variance from pre-war Albania.* The Austrians overran most of Albania in 1916, but in June, 1917, the general in command of Italian forces in the country proclaimed it an independent state.

Up to a few months ago, it was Italy's expressed desire to establish a protectorate over Albania, but such a policy has been abandoned, and the principal seaport Avlona (Valona) has been evacuated by Italian forces. Italy, however, has retained possession of the island of Saseno, which commands the entrance to the Gulf of Avlona.

ITALY REDEEMS ITS PEOPLE FROM THE AUSTRIAN YOKE

Italy's acquisition of the islands of the Adriatic previously mentioned consti-

tutes only a small part of its territorial gains as a result of the war. "Italia Irredenta" is once more under Italian sovereignty.† The redeemed area includes the Trentino region, Gorizia and the Istrian peninsula, together with the great seaport of Trieste, thus insuring Italian control of the Gulf of Venice and all the north Adriatic littoral. The survey of the boundaries and enumeration of the population are not yet completed, but it is estimated that the area gained is between 15,000 and 18,000 square miles, with nearly 2,000,000 inhabitants.

In addition to this region, Italy is administering the important island of Rhodes for a period of fifteen years, at the end of which time a plebiscite is to determine whether or not it shall be ceded to Greece (see also page 167). The small island of Kastelorizo (Castellorizzo), near Kekova Bay (Asia Minor), is acquired by Italy through a provision of the Turkish Treaty.

GREECE PROFITS GREATLY, WITH POSSIBLE RESERVATIONS

By the time of the distribution of this number of THE GEOGRAPHIC the status of Greece as shown on the accompanying map may have changed materially, owing to the repudiation by the Greek people of their distinguished statesman, Premier Venizelos—a course which the Allied Powers seem inclined to construe as a Germanophile reaction.

But if the territorial gains as originally provided by the Allies are allowed to stand, no nation will have profited more in proportion to its pre-war importance than Greece. In addition to the acquisition of Thrace and numerous islands of the Ægean, the Kingdom of the Hellenes also assumes administration of the important Smyrna district in Asia Minor, with the proviso that a plebiscite be held at the end of five years to determine whether or not it shall remain permanently in Greek hands.

Restoration of Constantinople to Greece after nearly five centuries of Turkish occupation is not a chimerical dream of the

---

* See "Recent Observations in Albania," by Brigadier General George P. Scriven, U. S. A., in the NATIONAL GEOGRAPHIC MAGAZINE for August, 1918.

† See "Austro-Italian Mountain Frontiers" and "Frontier Cities of Italy," by Florence Craig Albrecht, in the NATIONAL GEOGRAPHIC MAGAZINE for April and June, 1915.

Photograph from Frederick Simpich

SMYRNA EXTENDS FROM THE SLOPES OF MOUNT PAGUS TO THE SHORES OF THE GULF OF SMYRNA

With an estimated population of 350,000, the chief seaport of Turkey in Asia is more than twice as large as Athens, the Greek Capital. Many peoples have held sway over Smyrna since its founding as an Æolian colony, more than seven centuries before the Christian era. The city is famous for its rugs and its figs.

176

Greater Greece, provided the recent political events have not permanently alienated Allied sympathies.

Greece also is pressing its claim to the Epirus district, embracing some 2,000 square miles. Albania is the rival claimant here.

RUMANIA DOUBLES ITS AREA AND POPULATION

By its recovery of the fertile province of Bessarabia, which Russia absorbed at the conclusion of the Russo-Turkish War of 1878, and the acquisition of the former Austrian crownland, Bukowina. together with Transylvania, a part of Banat and other provinces from Hungary, Rumania becomes the largest of the Balkan States, with an area equal to the combined areas of Czechoslovakia, Hungary, and Austria and with 17,000,000 inhabitants.[*]

To the northeast, across the Dniester River, lies the nascent republic of the Ukraine, whose territorial limits can as yet be indicated only vaguely. It is a land rich in agricultural resources, especially that portion known as the "Black Soil District." Some statisticians have computed its area to be in the neighborhood of 200,000 square miles (twice that of Jugo-Slavia). with a population of 30,000,000.[†] Ukrainian propagandists lay claim to 330,000 square miles and 45,000,000 population.

PROBLEMS WHICH THE NEW NATIONS FACE

Even after the course and extent of its boundaries have been determined, a new nation has not yet launched its ship of state upon the turbulent seas of international politics and commercial rivalry. Indeed, boundaries are little more than the preliminary plans or drawings, indicating the length, breadth, and tonnage of the proposed "ship."

[*] For accounts of Rumania's history and aspirations, see in the NATIONAL GEOGRAPHIC MAGAZINE "Rumania and Its Rubicon," by John Oliver La Gorce, September, 1916; "Rumania, the Pivotal State," by James Howard Gore, October, 1915, and "Rumania and Her Ambitions." by Frederick Moore, October, 1913.

[†] See "The Ukraine, Past and Present," by Nevin O. Winter, in the NATIONAL GEOGRAPHIC MAGAZINE for August, 1918.

Now begins the great task of constructive organization, the training of officers, the equipping and provisioning for the voyage.

Each of the new states of Europe is beginning its national life with even less capital in experience than had the Thirteen Colonies after the American Revolution. In some instances they lack such machinery of government as customs posts and the trained officers to administer them; their postal systems have been disorganized by violent severance from old governments, and the innumerable new postage stamps in themselves tell a fascinating story; mints have had to be established to provide a complete system of coinage.

NEW STATES ARE WRESTLING WITH FISCAL SYSTEMS

The development of a sound fiscal system is one of the most difficult problems of modern statecraft, especially in a world where normal exchange rates no longer exist, for in the financial world chaos has followed the overthrow of credit. That credit must be reëstablished both at home and abroad before any of these nascent nations can make substantial progress. Parliamentary debate must crystallize into wise legislation. Even the election machinery which enables a people to register their will requires development in some regions, where universal suffrage has never been enjoyed heretofore.

In the restored countries the problem is as difficult as in the new. Poland, for example, has not been called upon to exercise the functions of self-government in more than a century, while in the case of Bohemia (the land of the Czechs) the gap of time between the suppression of the ancient free constitution of the kingdom and the advent of President Masaryk under the new constitution, adopted by the Constituent Assembly in Prague on February 29, 1920; is nearly three hundred years.

How many of these craft of state can sail on

"In spite of rock and tempest's roar,
In spite of false lights on the shore,"

none can prophesy.

Photograph by Herbert Corey

THEY INSISTED UPON BEING PHOTOGRAPHED; THEN THEY INSISTED ON THE TEN
CENTIMES EACH, WHICH IS FAIR WAGE FOR MODELS IN SALONIKI

Even before the World War brought the armies of the Entente Allies to its gates, a walk
through the streets of Saloniki made the visitor think of a congress of nations, so numerous
were the nationalities encountered and so varied and picturesque the costumes to be observed.

THE WHIRLPOOL OF THE BALKANS

BY GEORGE HIGGINS MOSES

United States Senator from New Hampshire, formerly Minister to Greece and Montenegro
AUTHOR OF "GREECE AND MONTENEGRO," "GREECE OF TODAY," ETC., IN THE NATIONAL GEOGRAPHIC MAGAZINE

THE EAST embodies mystery. Whether one regard it in its larger aspects as a whole, as the cradle of the race and the source of human history, or as limited to any of those portions—the Far, the Middle, or the Near East—into which writers, cartographers, and diplomats, for convenience' sake, have separated it, one will find the same shadowy and elusive elements of racial quality and preponderance of mental development and application, of blended religion and politics, which from time out of mind have allured and baffled the explorer, the missionary, and the statesman alike.

Across its whole expanse, from the Euphrates to the Adriatic; over all its seas, from the Hellespont to the Lido; through all its defiles, from the Ural to Tarabosch, have swept the successive tides of racial supremacy—Aryan, Hellenic, Slavic, Latin, and Teutonic.

THE BATTLE EVER TO THE STRONG IN THE EAST

The chief chapters of the history of the East have been written in blood. They are stories of rapine, pillage, and murder. There the battle has ever been to the strong. There, ever,

The good old rule
Sufficeth them, the simple plan,
That they should take who have the power,
And they should keep who can.

The share of the Near East in all this welter of conquest and revolution, savagery and progress, fanaticism and faith, chivalry and cowardice, bravery and butchery, honor and craft, has been typical of the entire region. And yet, perhaps because of, rather than in spite of, the fact that it stands nearest to us, or because it lies beside rather than in the beaten path of travel, it is probable that the Near East is to us of the Western world more elusive and legendary than either the Orient or the Antipodes.

"East is East and West is West, and never the twain shall meet," sings Kipling; but in the Near East meet they do, though they never mingle. Life there is a curious and enticing kaleidoscope of Europe and the Orient, of antiquity and modernity, through which obtrude from time to time the vexing survivals of earlier days, when morals and methods, customs and circumstances, were shaped upon an order existing for the moment and subject at any time to assault and overthrow.

THE LOCALE OF THE NEAR EAST

It is difficult to assign exact territorial limits to the Near East; and as for the Balkans, it may be said, as did Dooley of the Philippines, that before the World War few Americans knew whether they were mountains or canned goods. However, for the present purpose we may assume that the Near East is that stretch of territory which runs away from the head of the Adriatic down the Dalmatian coast and skirts the Ægean to the Bosporus. The mighty Danube well-nigh bisects it from north to south, while in the opposite direction it is cut in twain by the Balkan range. Its shores, from Miramar to the Golden Horn, are washed by smiling seas. To the north the grim Carpathians tower.

Here dwell a dozen peoples, each with its own costumes and customs, speaking a score of tongues and owning half as many religious faiths. At times the common strain of blood or religion, the common hatred of a dread oppressor, a common lust for land or for power, or a common sense of self-preservation has led to a coöperated effort in diplomacy or war; but never, I think, have they all acted in unison; never have even a few of them coöperated for any extended period.

The most obviously predominant characteristic of the Near East, and the one which has probably contributed most to separate peoples who otherwise would find alignment, is the tangle of tongues, which thrusts itself forward even at Trieste; and with it, in many quarters,

Photograph by Katrice Nicolson

EMBARKING FOR A VOYAGE ON LAKE SCUTARI: ALBANIA

Before the World War, Lake Scutari was bisected by the boundary line between Montenegro and Albania. Fringed by mountains, many of which are snow-capped during most of the year, its clear waters dotted with myriad islets, Scutari is one of the most beautiful lakes in Europe, and if railroads and modern hotels should ever make it accessible to tourists, it would be a formidable rival of Lake Como or Geneva. It has an area of 135 square miles and is extremely dangerous for native craft during rough weather.

goes a divided allegiance based upon race, religion, or national policy.

Trieste, for example, remains as much Italian in spirit as in that far-distant day when the Doges of Venice first set there the strong and enduring mark by which one can still trace the progress of Frankish power from the Lido to the Marmora. It has been my fortune to traverse the Adriatic—up or down or across—a baker's dozen of times; and always I have found it a sight of impressive sadness, those sturdy and still useful remains of the Venetian occupation, which stretched its sway so far and did its work so well. At its fountain head there now exist no more than the mediocre activities of a second-class city which until the Peace Treaty of St. Germain belonged to a second-class nation.

These marks of Venetian glory arise on every hand as one journeys eastward.

Along the Dalmatian coast, at Cattaro, throughout the Ionian Isles, crowning the sheer heights at Corinth, guarding the storm-bound gateway to Crete, and all through the blue Ægean are the frowning battlements with which Venice defended her possessions and which successive and successful assailants have not despised to make use of to this day.

For instance, at Cattaro, the Turk, the Montenegrin, and the Austrian have from time to time taken and occupied the massive works of that Gibraltar of the Adriatic; at Corfu the Greek recruits for recent wars have been housed in barracks bearing the lion of St. Mark's, while behind the stern barrier of the Lovčen, a hundred miles inland in Montenegro, stands Spuz, a conical mountain, capped with a Venetian fortress whose taking from the Turks a generation ago, during the Russo-Turkish War, was described

Photograph by Herbert Corey

TREKKING THROUGH MACEDONIA IN WAR TIME

Macedonia has a glorious tradition as the homeland of Philip of Macedon and his son, Alexander the Great.

to me by the royal captor, Nicholas, as I was one day journeying in auto with him to pass a week-end at his villa at Nikshich.

Skirting the Dalmatian coast the little steamer, with its motley ship's company, makes many a port and there may be seen many a remnant of an earlier and an even greater Latin supremacy in this land of changing allegiance; for here the Roman Empire long centuries ago set some of its strongest outposts, and its monuments as they appear today are among the best preserved examples of Rome's art and architecture at one of its finest periods.

The coliseum at Pola is a majestic pile, well worthy to stand with its great proto-type at Rome, while at Spalato the temple and baths built by Diocletian after his abdication are still extant and in use by the thrifty natives as public buildings. Diocletian himself was born in this part of the world, and among the few works of men which antedate the coming of the Montenegrins to the Land of the Black Mountain are the scanty remains of a Roman structure which once adorned the great emperor's birthplace.

This extreme north of the Near East

became Austrian in name by the Treaty of Berlin in 1878; and the yoke of the Hapsburgs sat all too heavily upon the mingled races who people these shores.

THE LAST PATRIARCHAL RULER

Alien rulers this region has had in numbers, but probably none who were held in more cordial hatred than the late Franz Joseph, the visible symbol of whose power was never absent from view, beginning with the huge naval arsenal and docks at Pola, a short distance below Trieste, and ending with the frowning casemates which pierce the hills at Cattaro, where the mountain sides literally bristled with cannon and where the clank of the saber constantly echoed across the pavements of the town which, though Italian in aspect and Austrian in allegiance, remained, nevertheless, Serb in feeling, its heart ever in the highlands beyond the beetling crags which hem in the mountain eyrie of the Montenegrin eagle from the sea.

Here Austrian extension to the southward was halted because thrust in between the Austrian littoral and the shores of autonomous Albania was the narrow

Photograph by Herbert Corey

A WOMAN OF MACEDONIA

pick up the superfluous stones from the earth's surface. He placed them in a bag which burst as he was flying over Montenegro—and certainly the landscape bears out the tale.

THE HOME OF AN UNCONQUERABLE PEOPLE

Yet amid these barren surroundings have dwelt for 500 years an unconquerable people, who have preferred liberty in this desolation to slavery in fat lands, and whose bravery has enriched the legends of mankind with tales of suffering indescribable, and of courage illimitable in defense of that freedom which "of old has sat upon the heights." To preserve their liberty a handful of valiant souls fled into these well-nigh impenetrable hills after the final overthrow of ancient Serb glory on the fatal field of Kossovo, in memory of which disaster the hat of the Montenegrin bears to this day its band of black.

Here at the outbreak of the World War ruled the last of the patriarchs—Nicholas I—the one monarch of Europe who, to my mind, really fitted his tradition. For more than half a century he maintained the ascendancy of the Petrovitch dynasty, had twice doubled the area and population of his realm as the result of his personal leadership in war, granted to his people a constitution, a ministry, and a parliament—and yet remained himself the final source of authority as in those days, not so far in the past, when he sat in his chair beneath the plane tree in front of the palace at Cetinje and personally heard and decided the grievances of his peasants one against another.

SERBIA, THE MOST FERTILE OF THE BALKAN STATES

That part of Jugo-Slavia which constituted the Kingdom of Serbia before the World War is the most fertile of all the Balkan States, and, in addition, possesses water-power and mineral deposits of large value. It was a land of no large fortunes and of no abject poverty, owing to a system of land ownership which assured to every peasant his homestead. It is a country, too, of no large cities (if one excepts those populous Macedonian towns which fell to her as the spoil of the Balkan Wars); but it is filled with numerous small and industrious farming

coast-line of King Nicholas of Montenegro, the patriarchal ruler of that wild and turbulent expanse of naked hills which one writer once graphically characterized as Savage Europe.

And savage indeed it is at first sight. Crossing the steep heights which lead to the interior, the smiling Adriatic is soon shut out from view and the grim landscape of brave Crnagora shocks the eye with its sense of utter desolation—a wild, turbulent ocean of rock, rising and sinking in angry gray waves flecked with white, which seem to leap and rage and battle together like a sea lashed by a storm.

At the Creation, so runs the Montenegrin legend, an angel was sent forth to

A CAMP OF REFUGEES AT SALONIKI, GREECE
In the distance may be seen one of the ancient donkey tracks leading to the Monastir Road.

Photographs by Herbert Corey
LAUNDRY DAY FOR A MACEDONIAN HOUSEHOLD
The younger woman has just lighted a fire of straw and put the pot on to boil.

THE FAMOUS PORTLAND VASE, NOW IN THE BRITISH MUSEUM

One of the most beautiful examples of ancient art is this gem of dark-blue glass, adorned with matchless reliefs in opaque white glass, telling the story of the meeting of Peleus and Thetis. This priceless relic from a Roman tomb was smashed into a hundred pieces by a madman in 1845, but it has been repaired so skillfully that its beauty has been preserved to a remarkable degree.

villages. Through Serbia sweeps the main artery of travel to the East, the line of the Orient Express, which divides at Nish, the ancient Serb capital in the southern hills, and runs, on the one side, to Bulgaria and Constantinople, and on the other through Macedonia to Saloniki.

Alone among the Balkan States, Serbia achieved her independence unaided. The freedom from Turkish thraldom which Greece owed to the Allied fleets at Navarino and which Bulgaria and Rumania took as their share of the fruits of the Russo-Turkish War, Serbia won for herself under the leadership of one of her own chieftains—Black George, the swineherd. And, East and West, the two striking features of the World War which most impressed me were the regeneration of two peoples, the French and the Serbs, who gave renewed evidence that blood will tell, and whose valorous defense of their own soil included an abiding faith that, whatever changes this war might work upon the map of the world, its future and enduring boundaries should be cast upon the firm basis of recognized and recognizable spirit of nationality expressed in blood, tongue, and religious faith.

MACEDONIA'S CRY FOR HELP THROUGH THE CENTURIES

For centuries Macedonia cried for help, and on occasion did not hesitate to help herself. Across her soil have surged the waves of successive racial domination, and the first great name to emerge from the shadowy tradition with which Macedonian beginnings are invested is that of Philip, under whom Macedonia and Bulgaria were united in a short-lived supremacy which fell with the death of Alexander the Great, son of Philip. It left, however, a legacy of racial rivalry which persisted to this day between Greek and Slav and which was further complicated by the victories of the great hero of the Serbs, Stephen Dushan, whose empire was scarcely exceeded by Alexander's in that part of the world.

Is Macedonia by blood preponderantly Greek, Bulgarian, or Serbian? This question, submerged, but not settled, by the Turkish Conquest, for 500 years has formed a central problem of the Near East. It has produced periodic outbursts of arson, pillage, and murder. Again and again it has caused the plains of the fair province to run red with the blood of slaughtered peasants who have refused to foreswear their nationality.

It has prevented the union of those who are bound by the tie of a common faith and a common suffering; it brought about a central schism in the Orthodox Greek Church, and in it the Turkish conqueror always found the surest basis for his continued misrule.

It was laid aside only once—in 1912—and then only long enough to dispossess

the Ottoman tyrant; but it broke out once more in the blundering and criminal war among the Balkan Allies.

SALONIKI COMMANDS THE ÆGEAN

Macedonia's plains are fertile, its hills filled with ore or underlaid with oil, its cities are populous and its ports important. Saloniki, its chief city, has been a center of activity since time out of mind. It stands in a commanding position at the head of the Ægean Sea; behind it lies a contributing territory of great and practically untouched richness, which now, freed from the paralyzing touch of Turkish tyranny, is destined to become a busy and prosperous province.

Saloniki is now Greek in rule as it long has been in race; and to the imaginative Hellenes the city has been sealed to them with the blood of their martyred King George I, who was assassinated here in 1913. But, like all the large towns of the region, its population is a varied one, and one of its picturesque features is that large body of Spanish Jews who fled thither from the Inquisition and who have still preserved their individuality, which they proclaimed by their garb.*

MOUNT ATHOS ONE OF THE WORLD'S RICHEST MONASTERIES

East of Saloniki the Chalcidicean Peninsula spraddles its ungainly shape down into the waters of the Ægean and here are found some of the most interesting of the world's monastic institutions—the famous monasteries of Mount Athos,† exercising both a civil and an ecclesiastical domination, with many quaint regulations—one being to the effect that no woman shall ever set foot upon the monastic possessions. This rule is strictly enforced, and one may well imagine the consternation of the fathers when the Queen of the Greeks proposed to pay a state visit to Mount Athos after the fall of Saloniki.

Mount Athos is among the richest of the world's monkish establishments, not alone in the amount of its revenues and

---

* See "Saloniki," by H. G. Dwight, in the NATIONAL GEOGRAPHIC MAGAZINE for September, 1916.

† See "The Monasteries of Mount Athos," by H. G. Dwight, in the NATIONAL GEOGRAPHIC MAGAZINE for September, 1916.

Photograph by Mortimer J. Fox

THE THESEION, BEST-PRESERVED EDIFICE OF THE ANCIENT GREEK WORLD: ATHENS

Under the shadow of the mighty Acropolis stands the classic Temple of Theseus, which has served various purposes throughout the centuries—pagan temple, Christian church, and Turkish palace.

186

the fertility of its estates, but in its rare library of manuscripts of the early fathers of the church, few of which have yet been opened to the scrutiny of scholars and in which there doubtless, is to be found a mine of information touching both faith and doctrine as they were delivered to the saints.

MONASTERIES HAVE PLAYED AN IMPORTANT RÔLE IN NEAR EAST HISTORY

The orthodox monasteries of the Near East have played a most important part in the history of that region. Their origins, naturally, are for the most part shrouded in mystery, but their place in the affections of the people is clear—and justly so.

It was within these cloistered walls that the Christian faith was kept alive through the long night of Turkish rule in the Balkans. Here, too, was maintained the tongue of whatever race the pious brothers claimed and today Greek, Bulgar, and Serb, whatever else they may differ upon—and their quarrels have been many and desperate—unite in a veneration and love for the church as an institution which I have never seen equaled elsewhere.

About the Balkan monasteries cluster many of the finest traditions of the Near East. For example, the beautiful cloisters of the Metropolitan Monastery at Cetinje are a venerated sanctuary of faith and freedom in the Black Mountain dating back for a full half millennium. It stands upon the spot where Ivan the Black established his seat of government when he fled from the Turk and where he set up the first Slavonic printing press in the world. Often besieged, it once capitulated to the Turk, but it was soon retaken by the sturdy warriors who descended in force from the heights of the Lovčen, whither the Moslem had been unable to follow them.

Once the monastery was blown up by the monks themselves, who perished with their precious books and documents rather than see their sacred walls degraded by the Mohammedan foe.

In its present form the structure dates only from the eighteenth century, but its quaint clock tower and its shaded cloisters give it an aspect of much greater age.

Here rest many of the valiant Vladi-

kas, or prince-bishops, who so long ruled the land with a combination of church and state, and here are to be found the cannon captured from the Turk on many an historic field. Here, too, is preserved a page from the first Gospel issued from the famous press (whose type were afterward melted down to make bullets) ; and it is little wonder that the Montenegrin peasant making his way to market at Cetinje pauses as he glimpses the shrine from afar and crosses himself devoutly as he whispers a prayer for his country.

Above it rises the Tower of the Skulls, the old-time citadel of the monkish defenders, which takes its name from the fact that up to a short time ago it bore a gristly fringe of Turkish heads impaled upon its ramparts. These grim reminders of a gory past were dear to Montenegrin veterans; and many were the murmurs of disapproval when the Gospodar concluded to remove them.

The monasteries of Greece have had a varied fortune. Some of them, waxing fat in lands and income, have been taken over by the government and their acres distributed, the enclosures of the American and British schools of archæology at Athens standing on ground which had been sequestered from the brothers of a monastery nearby. But others, like the famous shrine at Kalavrita, set high in the hills and overlooking the smiling waters of the Gulf of Corinth, are held in continued favor.

KALAVRITA, BIRTHPLACE OF GREEK INDEPENDENCE

It was at Kalavrita that the beginnings of the Greek War for Independence were made, and the tattered banner which the Archbishop Germanus took with him from his cell when he sallied forth to begin the contest is still kept as the sacred war banner of Hellenism and was brought out in much state at the beginning of the late war with Turkey.

Another well-known shrine in Greece, and one which is most frequently visited by tourists, is that at Meteora, where the giant needles of rock are capped with extensive buildings to which the venturesome, may ascend either by rickety ladders set in the interior crevices or by

Photograph by D. W. Iddings

A MAT MARKET IN BULGARIA, NEAR THE RUMANIAN FRONTIER

Most of the floor coverings used in the houses of the Balkan States are hand-woven from grass fiber. The men who weave the fiber rugs also sell them. The middleman is the exception and not the rule here.

A VERY OLD CHRISTIAN CHURCH IN ATHENS

means of a net drawn up by a creaking and primitive windlass.

BULGARIA'S BIG SHARE IN BALKAN HISTORY

Bulgaria bulks large in Balkan history. In one generation of freedom she made incredible progress and crowned her achievements with exceeding prowess in the First Balkan War. From this glowing pinnacle a hideous mistake brought her to a sad repute in no wise due to the qualities of her people, but rather to one headstrong and chauvinistic statesman whose fateful counsel undid in a single month all that the founders of the Balkan Alliance had worked for two years to accomplish and whose mad folly destroyed, for the majority of the world, an impression of Bulgarian wisdom and capacity which had been toilsomely built up from such meager beginnings. A second error, entailing even greater disaster, was made when the Bulgar King cast the lot of his people with the Teutonic Allies in the World War.

The history of Bulgaria differs little from that of her sister Balkan States; the successive chapters are written in blood. Herodotus, the father of history, was the first to notice the wild Thracian and Illyrian tribes who inhabited that portion of the peninsula and what he said of them centuries ago has a poignant emphasis in these last sad days of Bulgarian experience: "If they were only ruled by one man and could only agree among themselves, they would be the greatest of all nations."

THE ADVENT OF THE SLAV IN THE BALKANS

These ancient Bulgars, however, were doubtless of another strain than those who now claim the name and who are purely Slav—more characteristically so than the Russians even.

Just when the Slav first set his mark

Photograph by Frederick Moore

THE OLD COOK OF RILA MONASTERY: BULGARIA

This patriarchal-looking individual was an important personage for the poor families that used to visit the shrine at Rila. With this enormous spoon he would ladle out soup to the hungry.

their golden age or to the present day—when less than forty years sufficed to mark the passage of the country from a state of awful servitude to a place of power and prosperity. And now, eight years after the triumphant conclusion of the First Balkan War, it is again a shorn and shattered nation.

By reason of their closer proximity to the on-marching forces of the Prophet, the Bulgarians fell earlier captive to the Turk than the other Christian peoples of the Balkans; and the Turkish supremacy in Bulgaria, which began with the fifteenth century and lasted well into the nineteenth is the gloomiest epoch in the national annals.

There, as ever where the Turkish foot had trod in triumph, freedom vanished, learning languished and the memories of past glories all but disappeared. Even the character of the people seemed to change, and had it not been for the priests and the brigands it is probable that the thread of Bulgarian national life would have been definitely sundered. But in their mountain fastnesses this strange combination of the monk and the marauder kept alive the national feeling.

in this region is difficult to say, but there he has been for more than a thousand years, spreading out from the parent center in a brood which at length has covered much of the territory from the Euxine to the Adriatic. He early embraced Christianity and from the first Boris down to the last, whose coming of age I helped to celebrate several years ago, religion has highly colored the politics of Bulgaria.

Gibbon, in a famous passage, has remarked that "the glory of the Bulgarians was confined to a narrow scope both of time and place"; and true it is, whether one speaks of that remoter era when the Emperor Simeon gave to the Bulgars

BRIGANDS ARE POPULAR HEROES IN THE BALKANS

The brigands of the Balkans have ever been the popular heroes. In Serbia they appeared under the name of Haiduks, in Bulgaria as Haidutin, and in Greece as Klephts, the most famous of the latter being perhaps Marco Bozzaris, though another of them, Kolokotronis, later won renown as a regular officer and an equestrian statue has been erected to him in the square near the Parliament House in Athens.

Like Robin Hood, the Balkan brigands are always represented as the protectors of the poor and the weak, the friend of all Christians and the ruthless scourge of the Ottoman oppressor. Thousands of legends and songs have grown out of their exploits; and had they made war against the common foe only instead, as they too often did, among themselves, their fame would rest upon a far firmer foundation.

Nevertheless, among all the agencies which contributed in the end to the winning of Bulgarian independence, the brigands were by far the most continuously active; and the long centuries of Turkish misrule were constantly broken by a series of abortive revolts, which were suppressed with increasing cruelty, until the brutal massacres of 1875 inspired Mr. Gladstone to the famous Midlothian campaign, gave to the Tsar a convenient handle against the Sultan, and brought on the Russo-Turkish War.

Photograph by Hester Donaldson Jenkins

A VILLAGE LASS OF BULGARIA CARRYING WATER FOR HER MOTHER

Hers is the land of the attar of roses. The world's supply of this perfume comes from southern Bulgaria.

THE TREATY OF SAN STEFANO CURBED BULGARIA

That war was ended by the Treaty of San Stefano, which essayed to establish a big Bulgaria; but, thanks to Disraeli, British influence brought about the Congress of Berlin, and it was a little Bulgaria which finally secured a place at the world's council table.

A lowly place it was, but with splendid courage the Bulgarian set out to make it better, and the story of Bulgarian development in a single generation finds few parallels among modern nations. Except for the humiliating war with her one-time allies in the Balkans and her subsequent suicidal espousal of pan-Germanism, Bulgaria's advance has been constant and remarkable in the face of grave ob-

A SERBIAN CHIEF FULLY ACCOUTRED

Alone among the Balkan States, Serbia attained her independence unaided. She won it from the Turks under the leadership of one of her own chieftains, Black George, the swineherd.

novo, the ancient seat of government, with many smaller centers, are towns of which any nation might be proud.

The country possesses great wheat fields, extensive forests, rich mines—all of which have been made to respond to that patient industry for which the Bulgarian peasant is the model for all his Balkan neighbors. A unique product—and the most profitable—is the attar of roses, the world's supply of which comes from southern Bulgaria and which has enriched the landed peasants of that quarter beyond the wildest of their dreams.

BULGARIA NOW RID OF BURDENSOME ARMY EXPENSE

Training for the Bulgarian army, before the signing of the Treaty of Neuilly in 1919, began in boyhood, and the former Bulgarian military force remains, in my mind, as among the most effective of the world's fighting machines. Created in the first instance by Russian genius and designed to coöperate with Russia's forces in the Russian advance to the southern seas, it was maintained by the Bulgarian people at a cost of taxation and toil incalculable.

But the memories of Slivnitza, of Lule-Burgas, of Kirk-Killisseh, and of Adrianople should not be dimmed by recent events brought about by a misguided and fatal division of opinion among a few leaders wherein the selfish and the stubborn had their way—a way which has led Bulgaria to so sorry a place in the world's esteem, but from which the patient perseverance of her people will one

stacles. Three monarchs have sat upon her troublous throne. Under one of them a Bulgarian army, deserted by its Russian tutors on the eve of battle, reached the gates of Belgrade; under another the Bulgarian banner was brought to the outer defenses of Constantinople. Under the third Bulgaria has again begun the slow march upward.

Sofia, the capital; Philippopolis, the largest city; Varna, the chief port; Tir-

day again bring her forth triumphant in those victories which peace possesses no less than war.

RUMANIA, THE ARGENTINE OF THE NEAR EAST

Rumania is in, rather than of, the Balkan States. Claiming a Roman origin, speaking a Latin tongue, ruled in her formative days by a German king more Hohenzollern than the Kaiser even, the kingdom has nothing in common with her neighbors except a formal adherence to the Orthodox faith.

Rumania is the Argentine of the Near East, a land where fortunes have been amassed with incredible rapidity; where it is unfashionable to live within one's income; where society has a gayer tone than one can easily depict.

Bucharest, the capital, vaunts itself as the Paris of the Near East, and other Rumanian cities have grown rapidly. Sinaia, in the Carpathian foothills, is the summer capital, gay with hotels and villas surrounding the palace of the King, and thither betake the court, the diplomatic circle, and the rich upon the approach of the summer heat, which makes Bucharest intolerable.

On the shores of the Black Sea, at the mouth of the Danube, and elsewhere are other charming resorts; but Rumania as a whole is a vast wheat field, the granary indeed of the Near East, from which feeds many a mouth from Constantza to Cattaro.

Photograph by H. G. Dwight

A TWO-MAN SAWMILL IN CONSTANTINOPLE

There are four cities in the world that belong to the whole world rather than to any one nation, according to Viscount Bryce, the distinguished British statesman. They are Jerusalem, Athens, Rome, and Constantinople. For fifteen hundred years Constantinople has been a seat of empire, and for an even longer period the emporium of a commerce to which the events of our own time give a growing magnitude.

Come we now to Greece, the land of song and story, the birthplace of modern history, the cradle of philosophy, the home of art and architecture, the scene of varied human fortunes, where cluster the finest traditions of our race—the land which I know best of all the foreign world and which has won and retains my constant and increasing admiration.

To separate the life of modern Greece from the splendors of its classic or Byzantine days is not easy, and the Greeks themselves would be the first to resent it.

A VIEW OF CONSTANTINOPLE FROM A CITY OF THE DEAD

"It is in Constantinople that the currents which cause the whirlpool of the Balkans have both
their origin and their end."

They, of a truth, deem themselves the direct descendants of the worthies of classic days, and certain it is that their life has shown a persistent continuity which warrants the claim.

Whether their land has been ruled by a Roman emperor, a Frankish duke, a Venetian bailli, or a Turkish pasha, the thread of Hellenic existence has remained unbroken. In the monasteries have been preserved their religion, their tongue, their traditions; mothers have taught their children the glories of the Greek heritage in defiance of the infidel interdiction, and today the Greek people stand forth, in character at least, exactly as they did in days of yore, as Aristophanes pictured them, as St. Paul described them, on Mars' Hill, and as every scholar has learned to regard them.

MODERN ATHENS IS A BRILLIANT CAPITAL

In many of its aspects Greek life remains unchanged from its classic features. Modern Athens, to be sure, is a brilliant capital, advancing its claim to be known as the Paris of the Levant. Less than a century ago it finally passed from Turkish possession, and it was then a handful of hovels huddled together beneath the Acropolis. Today it is a city of wide and gay streets, dotted with small parks and adorned with handsome public buildings, many of them the gifts of rich

Greeks, who have delighted to spend upon the mother country the fortunes which they have gained abroad.

To such generosity Athens owes the noble group of buildings which comprises the University, the National Library, and the fine classic reproduction which houses the Academy of Science, and, above all, the noble Stadium, built upon the old foundations and along the old lines and ingeniously carrying in its fabric every fragment of the old structure which could be found.

In the midst of all this modernity stand the remnants of the golden days of Athens, sedulously preserved and open to inspection and study with a freedom nowhere equaled. The focus, of course, is the Acropolis, with its Parthenon, incomparable even in its ruins, its cliffs and grottoes still the home of legend and of fable (see pages 170 and 186).

Within a narrow circle all the phases of ancient Athenian life are represented. Under the shadow of the mighty rock stands the classic Temple of Theseus, best preserved of all the ancient monuments and serving a varied purpose throughout the centuries as pagan temple, Christian church, and Turkish palace.

Only a few steps away rise the well-kept walls of the Stoa of Hadrian, which, with other works of Latin origin, speak of that distant day when a Roman emperor ruled the violet-crowned city.

DILAPIDATED FRESCOES RELICS OF EARLY CHRISTIAN ERA

Of the early Christian era there are many dilapidated frescoes upon the walls of nearly all the classic structures, and the first expressions of the architectural aspirations of our faith are to be found in the beautiful little Byzantine churches, the most striking of which is that of St. Theodore, set down in the midst of one of the great business streets of the city, its foundations already sunken beneath the detritus of the centuries, yet scrupulously guarded against commercial encroachment.

Of Turkish days there are but few distinct traces of structural importance; but the bazars, as typified by the Lane of the Little Red Shoes, the home of the cobblers, or Hephaistos street, the quarter of the coppersmiths, are far more Oriental than either Hellenic or European.

In this land of changing allegiance the marks of the Venetian occupation, as elsewhere, were set deep and strong. Corfu is today, in its externals at least, as much Italian as either Venice or Naples; while Nauplia, Patras, and many of the island seaports still make use of the battlemented fortresses erected by the Latin rulers.

THE PIRÆUS, ONE OF THE BUSIEST PORTS OF THE MEDITERRANEAN

As of old, the Greeks swarm the seas. The Piræus is one of the busiest and most crowded of Mediterranean ports; it is indeed the center of transshipment for all the East, and the Greek merchant marine has multiplied its fleet from year to year. The Corinthian Canal, after many financial vicissitudes, now seems in the way of becoming each year a more and more useful route between the Ionian and the Ægean Seas, and its sheer walls are eloquent of the persistence with which an ancient dream has been fostered and brought to realization.

The Greeks are essentially a town people—made so doubtless by the necessity, in Turkish days, of coming together in masses for self-defense. But, whatever the reason, one-tenth of the entire population is to be found in Athens and the Piræus.

The drain of emigration from the rural districts has been enormous. In the striking words of one of the cabinet who discussed the question with me, it constitutes "a grave national hemorrhage." Indeed, in some villages of the Peloponnesus there remain scarcely enough men to fill the offices.

In one sense, however, the emigration has been of benefit to the country; for large sums of money are sent back each year, especially from America, to the families which have been left at home.

But, while Athens and a few of the larger towns have taken on the aspect of today, country life in Greece remains in most of its fundamentals as it has been for ages. Within two hours' drive of Athens I have seen peasants plowing their fields with crooked sticks, exactly as they did, I imagine, in the days of Homer.

The shepherd boys of today manage their flocks—and I may remark in passing that there are said to be more goats than Greeks in Greece—with a crook fashioned upon the same lines as that which Corydon carried. And in Thessaly one still finds in daily use the solid wheeled cart that has come down without substantial change from the days of Jason.

The distaff remains the chief instrument in preparing wool for the hand looms on which are woven the coarse and shaggy stuffs worn by the peasants, and one rarely finds it absent from the busy fingers of the older dames who sit and work in the sun. Nausicaa and her maids gathered at the fountain on that day when Ulysses came to port have their modern counterpart in almost every public square of Hellas.

A LAND OF SUNSHINE

Greece is a land of much sunshine and life is followed much in the open. The family oven is invariably to be found in the courtyard, and it is heated with dried twigs brought from the country districts in huge loads upon the patient little donkeys, who vie with the goats in being the most useful members of the household.

Market day, of course, brings all the community together and is generally an occasion of much gaiety, while the feast days, which are numerous, are literally observed. Fasting, too, is frequent and severe.

On feast days there is always dancing, the most famous to be seen at Megara during Easter week—a survival, and the only one, of the olden pan-Athenaic pageants of classic times.

Megara prides itself on being a pure Hellenic community in the midst of the Albanian flood which once overran the Attic Plain. It was once famous as a marriage mart during the Easter dancing season. This is no longer true, because, as the maidens sigh, so many of the men have gone off to America.

At Megara the native costume is seen at its best. It is rarely worn anywhere nowadays and has almost wholly disappeared from the cities. But for the Ev-

zones, or household troops, at the ugly barracks which the Greeks call the "big" palace, the fustanella would be almost as rare a sight in Athens as the classic garb, which is worn there only by American dancers.

RAILROAD COMMUNICATION IS INADEQUATE

It is not yet easy to go about in Greece. The railroad lines are meager, the roads are not good, and the hotels leave much to be desired. But the famous battlefield where "mountains look on Marathon and Marathon looks on the sea" is easily accessible to Athens. So, too, Olympia, where archæologists have unearthed remnants of the great temple, with its incomparable Hermes, the masterpiece of Praxiteles and clearly the finest sculpture which has yet come from human hands, is a favorite shrine for lovers of the beautiful.

But the most accessible of all the great centers of classic life is Delphi, a fitting shrine for an oracle, with its massive and somber cliffs and its majestic hills looking out to the gulf. Here the French savants have done a wonderful piece of excavation and have brought to light the ancient city with its treasures, its famed Castalian Spring, its theater, its treasuries, and its Sacred Way.

The seat of the Sybil has been identified, and that it still retains its oracular powers I can testify; for when I was last there my Dutch colleague stood upon the spot where the tripod had stood and I asked, "Who will be the next President of the United States?" And the answer came, Delphically enough, "The best man will win."

The most commanding figure in the Hellenic world today, clearly the first statesman of the Balkans, and a man worthy to rank with the best of the world's ruling geniuses, is the recently deposed Greek Prime Minister. Eleutherios Venizelos—Athenian in blood, Cretan by training, but thoroughly cosmopolitan in his breadth of view and grasp of affairs. To him is ascribed—and rightly—the credit for the Balkan Alliance, which astounded Europe by its successes, and most of the gains in territory and prestige for Greece in the Peace Treaty which the Allies forced the representatives of Turkey to sign at Sèvres (see page 167).

It is at Constantinople that the problems of the Near East have always centered in their acutest form. There, where teeming thousands throng the Bridge of Galata; where twenty races meet and clash with differences of blood and faith never yet cloaked beneath even a pretense of friendliness; where fanaticism and intrigue play constantly beneath the surface of oriental phlegmatism and sporadically break forth in eddies of barbaric reaction; where all the Great Powers of Europe have for generations practiced the arts of a devious diplomacy—there, I say, has always been found the real

storm-center of the danger zone of Europe.

There it is that the currents which cause the whirlpool of the Balkans have both their origin and their end. This imperial city, for nearly two thousand years a seat of power, still clutches the key to commerce for both the East and the West.

Strategically and commercially a coign of vantage, Constantinople in capable hands means most of all that for which armies and navies nowadays contend. He who can foretell its fate can read a wider future than any of us can now imagine; for, as the fall of Constantinople five centuries ago produced, so the wise disposition of it may calm the whirlpool of the Balkans.

THE ORKNEYS AND SHETLANDS—A MYSTERIOUS GROUP OF ISLANDS

By Charles S. Olcott

WHEN the great fleet of Admiral Jellicoe rushed to the support of Admiral Beatty, in the most stupendous naval conflict of history, it left a mysterious base, presumably at Kirkwall, in the Orkneys. When Lord Kitchener went down to his too-early death, it was in a ship off the coast of the Orkneys. When the American minesweeping squadron, under Rear Admiral Joseph Strauss, U. S. N., undertook the unprecedented task of clearing the North Sea of its mine barrage, the base of operations was in the Orkneys.[*]

If you were to ask the passengers on any transatlantic steamship, you would find but few who could tell, very definitely, even the position of the islands on the map—whether they were nearer Iceland than Scotland; whether Shetland was the northernmost or southernmost of the group; whether there were two islands or one hundred and fifty. And

[*] See, in the NATIONAL GEOGRAPHIC MAGAZINE, "The North Sea Mine Barrage," by Captain Reginald R. Belknap, U. S. N., in February, 1919, and "The Removal of the North Sea Mine Barrage," by Lieutenant-Commander Noel Davis, U. S. N., in February, 1920.

if you were to inquire as to the chief business of the northernmost group, the entire company would agree, most likely, that it is the raising of Shetland ponies rather than the curing and packing of herring (see supplement, Map of Europe).

The mystery surrounding the islands and the general lack of knowledge concerning them are both quite consistent with their history. When the fierce Norsemen descended upon the shores of the Orkneys, they found no foe to contest their coming. Their exploits were sung triumphantly by those who took part in them, and repeated through succeeding generations, until the narratives were gathered together in the Orkneyinga Saga, forming the history of three centuries of violence and bloodshed.

WHAT BECAME OF THE ORKNEY PICTS?

Yet nowhere is there mention of a native population to be overcome, though numerous mounds, brochs, or towers, graves, stone implements, and other relics told of the existence there of a large population. It is known, too, that missionaries or monks visited the islands to

Photograph by Thomas Kent

A STREET IN KIRKWALL, ORKNEY ISLANDS, FAMILIAR TO MEN OF THE AMERICAN
NAVY WHO SWEPT THE NORTH SEA OF ITS MINE BARRAGE (SEE PAGE 197)

"A very narrow lane, called Bridge Street, leads back from the steamship landing. It is
paved with flagstones, and when a team passes the pedestrians have to stand close to the
walls or enter the doorways."

SHORE STREET, KIRKWALL, AT MIDNIGHT

Photographs by Thomas Kent

ALPERT STREET, KIRKWALL, WHICH HAS THE UNIQUE DISTINCTION OF POSSESSING
A SOLITARY TREE

THE STONES OF STENNESS: ORKNEY ISLANDS

Part of the Circle of the Sun. Only fifteen stones now remain standing; there were originally thirty-five or forty (see page 208).

Photographs by Charles S. Olcott

MAESHOWE, A CIRCULAR, GRASS-COVERED MOUND, 90 FEET WIDE AND 30 FEET HIGH

This mound was probably several hundred years old when the Norsemen first landed in the Orkneys. The most remarkable feature of the mound is the size of some of the stones of which it is built. One of these is estimated to weigh eight tons (see page 201).

minister to the Picts, as the earliest known inhabitants were called.

The first mystery to be solved is, therefore, what became of these Picts, who had so completely disappeared by the end of the eighth century that the Norsemen found only their dwellings and graves and no human beings except, perhaps, an occasional Culdee hermit?

Although the question cannot be answered, the visitor of today may step inside of a very complete "house" or dwelling-place which was there when the Norsemen first arrived, more than eleven centuries ago.

This is the mound known as Maeshowe, on the island of Pomona, or the mainland of the Orkney group. It is a circular, grass-covered mound 90 feet in diameter and 30 feet in height, on one side of which is a narrow doorway about four feet high. We found the custodian at a neighboring farm-house and were conducted into a passageway 54 feet long, through which we walked in a stooping posture until at last we were able to stand erect in a room 15 feet square and 13 feet high (see page 200).

The most surprising feature of this mound is the size of some of the stones of which it is built. One of them is estimated to weigh eight tons!

In an age when men possessed no iron tools, no drills, no means of blasting, no derricks, no wagons nor trucks—none of the things now deemed indispensable in the quarrying, transportation, and placing

Photograph by Charles S. Olcott

THE WATCH-STONE, OR SENTINEL, ERECTED BY A PEOPLE OF MYSTERY

This is one of the famous Stones of Stenness, which stand about two miles from the mound of Maeshowe (see page 200).

of much smaller monoliths—how was it possible to cut such rocks out of the earth's surface, shape them into rectangular building stones, carry them long distances, and finally set them up to form a humble dwelling-place?

The huge stones which in Egypt formed the tombs of the ancient kings were quarried and built into pyramids by the labor of myriads of slaves; but no such conditions existed here. It is possible that Maeshowe was a typical Pictish dwelling. It has three lateral chambers, each large enough for two or three persons to lie in, and on the floor in front of each is a stone of the exact

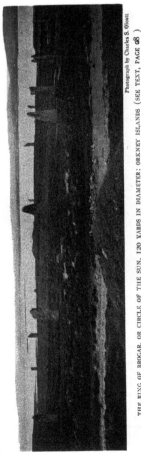

THE RING OF BROGAR, OR CIRCLE OF THE SUN, 120 YARDS IN DIAMETER: ORKNEY ISLANDS (SEE TEXT, PAGE 208)

Photograph by Charles S. Olcott

size required to close the opening, but too huge to be lifted. This fact suggests that the mound may have been intended for a tomb. It is quite possible that it was originally a dwelling and later used for burial purposes.

RUNIC INSCRIPTIONS TELL OF CRUSADERS

On the walls are many Runic inscriptions. Twenty-six of them have been translated, mostly trivial scribblings, as "Hermund Hardaxe carved these Runes," or "Ingigerthr is of women the most beautiful." The only one of importance reads: "The Jorsalafarers broke open the Orkahaug in the lifetime of the blessed earl," and further intimates that they expected to find treasure there and were disappointed.

The "blessed earl," who led an expedition to Jerusalem in 1152, was Rögnvald, and many of the names scratched on the rocks are mentioned in the Saga as living in his time. The sagas also record that the "Orkahaug," or Maeshowe, was the scene of a big Yuleday carouse by Earl Harold and his men, who visited Orkney while Rögnvald was in the Holy Land.

A better form of Pictish dwelling is the "broch," of which more than seventy have been found in the Orkneys and as many in the Shetlands.

The most complete specimen now extant is Mousa, in Shetland, which Sir Walter Scott appropriated as the strange dwelling of Norna of the Fitful Head, transporting it in imagination from a rocky island some ten miles south of Lerwick to the northwestern extremity of the mainland.

These "brochs," or "burghs," are built of loose stones and without cement or mortar of any kind. Though the builders knew nothing of roofs or arches, yet they constructed chambers within the walls, one row above another, encircling the tower for several stories, and connecting the floors with a rude circular stairway or inclined plane.

The windows were all on the inside of the tower, the only outside opening being a single low door.

These rude castles or forts served well for defense against an enemy. The occupant could find sleeping quarters for his family and retainers in the numerous

THE STONES OF STENNESS, THE CIRCLE OF THE MOON: ORKNEY ISLANDS

The rude table of four rough, irregular stones may have served as an altar (see text, page 208).

Photograph by Charles S. Olcott

Photograph by Charles S. Olcott

THE SOUTH AISLE OF ST. MAGNUS CATHEDRAL, KIRKWALL, ORKNEY ISLANDS

The arched opening on the left is said to have led to an underground passage which connected the Cathedral with the Bishop's Palace across the street—a fact which may have given to Sir Walter Scott the suggestion of the mysterious disappearance of Norna, as told in "The Pirate."

chambers within the walls, could accommodate his sheep or cattle in the circular inclosure, and could repel an assault from the parapet some fifty or sixty feet above the ground.

LOVE IN A TOWER A THOUSAND YEARS AGO

It is related in the Saga that Erland Ungi, having obtained from the King of Scotland a grant of one-half the earldom of Caithness, which he was to hold jointly with Earl Harold, the "Orkahaug" carouser, suddenly took a notion to elope with Margaret of Athole, Harold's mother, who was still a beautiful woman though of doubtful character. They fortified themselves in the tower of Mousa, and, though attacked by Harold, successfully resisted first an assault and then a siege.

It is curious that Mousa had been occupied by another runaway couple more than two centuries earlier, or about 900 A. D. One of the sagas tells of the elopement of a certain Norseman with a girl whom his father would not permit

him to marry. They were wrecked on the island of Mousa, but, like Robinson Crusoe, managed to carry their cargo ashore and, finding a ready-made dwelling handy, lived there all the winter.

Some excavations made within the past twenty years seem to indicate that these brochs were built in groups in such a way as to furnish a place of refuge and means of defense for a large population.

SCENE OF THE SHIPWRECK IN SCOTT'S "THE PIRATE"

On the southern extremity of the mainland of Shetland is a high, rocky promontory called Sumburgh. A lighthouse marks the point and serves as a warning to navigators, for the tides from two oceans meet in the "Roost" of Sumburgh and make a dangerous current.

It was here that Scott placed the scene of the shipwreck in "The Pirate." On the rocky coast stands an old ruined castle called Jarlshof, which Scott makes the residence of Basil Mertoun and his son. It was occupied in the sixteenth century

ORCADIAN TOMBSTONES, ST. MAGNUS CATHEDRAL, ORKNEY ISLANDS

These two stones, now set on end in St. Magnus Cathedral, were originally laid in the floor. The full inscription of the one on the left is reproduced on page 217. On most of these stones the sculpture shows the skull greatly enlarged on the left side because of the Orcadian belief that the spirit took its departure through the left ear.

by Earl Patrick Stewart, who, with his father, Earl Robert, cruelly oppressed the islanders for half a century.

In 1897 Mr. John Bruce, the owner at that time, was visiting the place with some friends when they discovered the unmistakable evidences of masonry in the mound beneath the Jarlshof, so ancient as to make the sixteenth-century castle seem quite modern. Some recent storms had washed away the seaward part of the mound, exposing the ends of

walls, the existence of which had been previously unsuspected.

This led to extensive excavations, with the result that the remains of a large broch were unearthed, one-half of which had been washed away by the sea. Close by were three structures, shaped like beehives, the largest of which was oval in shape, 34 feet long and 19 feet wide, and contained five chambers, one of which was 5½ feet wide at the front and 10½ feet at the rear.

Photograph by Thomas Kent

ORCADIAN WOMEN WHEELING DRIED PEAT FOR FUEL

Photograph by Charles S. Olcott

GATHERING PEAT IN THE SHETLAND ISLANDS

There are practically no trees on the Shetland Islands; therefore no wood for fuel. As in Ireland, peat is used as a substitute. It burns with considerable smoke and little flame, but gives much heat.

PICKING LIMPETS OFF THE ROCKS

Photographs by Thomas Kent

PIEBALD SHETLAND PONIES

Not only are the Shetland ponies diminutive, but the island's breed of cows is also small. The Shetlanders pluck the wool from their sheep instead of shearing them, maintaining that this insures a finer second crop.

Photograph by Thomas Kent

GRINDING GRAIN WITH THE QUERN IN THE ORKNEYS

'Commerce and Science may march forward, but the Shetlands and Orkneys, after ages of turmoil, are taking a vacation."

These chambers were built of overlapping stones, gradually closing toward the top and each surmounted with one large stone. At the western edge of the mound a great wall of dry stone was unearthed. It extended back about 70 feet and was from 10 to 20 feet thick.

The position and shape of this wall indicated that it had once been a part of a huge circular wall inclosing a group of buildings of which the broch was the center and largest, while the "beehive" structures were subsidiary.

If such were the case, the whole must have constituted a fortress of great strength, sufficient to accommodate a large population and furnish an adequate

defense for the extreme southern point of the island.

MYSTERIOUS CIRCLES OF MOON AND SUN

Other interesting relics of that unknown people, who vanished so mysteriously, may be seen not two miles from the mound of Maeshowe. These are the so-called Stones of Stenness.

Separating the Locks of Stenness and Harray is a narrow neck of land, known as the Bridge of Brogar, at the southern entrance to which is a huge stone, 18 feet high, popularly designated as the "Watch Stone," or "Sentinel."

In a field at the right are the remains of a circle of similar stones, not quite so large as the Watch Stone, in the midst of which is a rude table, or altar, made of three short stones standing on end, and surmounted by a large flat stone or slab.

This is the Ring of Stenness, or the Circle of the Moon (see page 203).

Across the bridge, a quarter of a mile away, is a larger group, properly designated as the Ring of Brogar, but commonly known as the Circle of the Sun. It is about 120 yards in diameter and was surrounded by a fosse or trench at least 6 feet deep, the outlines of which are distinctly traceable. The stones of this larger circle are from 8 to 16 feet high; one is from 5 to 6 feet wide, and all are crude and irregular in shape (pp. 201-2).

Fifteen remain standing, although the group originally contained thirty-five or forty. They have a strange, shaggy covering of unusually long lichens, like an ancient sheep whose coat has become scraggly with age and exposure to the weather.

North of the Ring of Stenness was the famous stone of Odin, which differed from the others chiefly in having a hole through it.

A STRANGE WEDDING CEREMONY

It was once the custom of the people living near by to gather on the first of each new year at the Kirk of Stenness for a celebration of feasting and dancing which continued several days, or as long as the provisions lasted. This inspired many of the young people to get married, and they would slip away to the Circle of the Moon, where the woman knelt down and prayed to Odin, or Woden, to help her to be faithful to the man, after which they went to the Circle of the Sun, where the man performed a similar ceremony. Then they repaired to the Stone of Odin, clasped hands through the hole, and pledged mutual fidelity.

Such a marriage was considered so binding that even after the death of one of the parties the survivor could obtain release only by touching the dead body—a somewhat inconvenient requirement in case of prior separation; but the only alternative was that if the survivor married again he or she would be obliged to entertain the former spouse's ghost at the wedding. There was, however, another way of escape, provided the couple agreed to disagree. They had only to go into the kirk and walk out, one by the south door and one by the north, and the tie was effectually dissolved.

The Stone of Odin was visited by Sir Walter Scott in 1814, and he made a romantic use of it in "The Pirate." In the same year a neighboring farmer broke it up, with several other stones from the Ring of Stenness, to build a foundation

Photograph by Thomas Kent

A DESCENDANT OF THE VIKINGS

In the summer, when the fishing business is active, the Orcadian is a busy individual, but during the long winter time hangs heavily on his hands.

for his cow-house, for which act of vandalism he was properly boycotted and driven out of the country.

There is nothing about the stones themselves that is wonderful. Anybody could erect similar circles with modern appliances. But these monuments were here when the Norsemen landed and were probably at least three centuries old even then. They have stood for 1,400 years. They were doubtless quarried with stone implements and set in place by the exertion of sheer brute force. Their history is shrouded in obscurity, their very purpose a mere conjecture. They are a part of the mystery of the islands.

Photograph by Thomas Kent

KNITTING SHETLAND SHAWLS

Note the "Crusie," or old lamp, on the wall, which burns fish oil as an illuminant. The
flame is fed by the dried pith of a rush instead of a wick.

Photograph by Thomas Kent

A LOBSTER FISHER OF BIRSAY, ORKNEY ISLANDS

Out of a population of 30,000, some 2,000 Orcadians are engaged in the fisheries, the chief catches being herring, cod, ling, lobsters, and crabs.

Photograph by Thomas Kent

THE LIVING-ROOM, DINING-ROOM, AND KITCHEN IN A COTTAGE IN THE ORKNEYS
The dinner pot is on the boil and the kettle is getting warmed up.

The Orkney sages tell the adventures of a long line of 'warlike, tricky, and murderous earls, who plundered the coasts, burning, killing, or stealing what came within their power, quarreling among themselves, ruling as absolute sovereigns so long as their power lasted, and usually dying a tragic death for much the same reason that Nature decrees a violent end for most of the wild beasts.

It is curious, therefore, to find among these dauntless leaders and ruthless conquerors two whose piety was so marked that their names were afterward included in the catalogue of saints. The history of these two men, St. Magnus and St. Rögnvald, together with that of Swein Asliefson, "the last of the Vikings," must become known to every visitor to the Orkneys, for it is through them that the islands came into possession of their greatest monument, the Cathedral of St. Magnus in Kirkwall.

THE STORY OF ST. MAGNUS

Magnus, the son of Erlend, and Hakon, the son of Paul, became joint earls of Orkney in 1103. Magnus is described as "a man of noble presence and intellectual countenance. He was of blameless life, victorious in battles, wise, eloquent, strong-minded, liberal and magnanimous, sagacious in counsel, and more beloved than any other man."

For a time the cousins were friendly and peace and prosperity came to the islands. But Hakon became jealous of his kinsman's greatness and, after the two had nearly reached the point of open warfare, treacherously suggested a meeting for the purpose of peace and reconciliation. This was to be held on the island of Egilsey, where there is today the ruin of an ancient church which was built before the first Norsemen came to Orkney.

Each man was to have two ships and a stipulated number of followers. Magnus arrived first, with the proper quota, but when Hakon came, it was seen that he had eight ships and a large army, and Magnus knew at once that his hour had come. The saga relates that he met his fate with noble resignation, facing death with the cheerful courage of a Christian martyr.

It was said that the place where he was slain, though previously covered with moss and stones, became at once a beautiful greensward, typifying the entrance of a saint into the "beauty and verdure of Paradise."

Hakon, after this murder, violated all the rules of poetic justice by becoming a pretty good ruler, and eventually died in his bed. His sons, Paul and Harold, succeeded, but, as usual, quarreled, until one day Harold insisted on taking for himself a splendid garment that his mother and her sister had made for Paul. It turned out to be poisoned, and so Harold promptly curled up and died and Paul reigned alone.

RÖGNVALD, THE POET ADVENTURER

But a new claimant now arose in the person of Kali, a very popular young man with red hair, who wrote poetry. He obtained from King Sigurd of Norway the grant of one-half the islands, and was permitted to change his name to Rögnvald, after one of the most accomplished of the Orkney earls. This was intended to bring good luck.

Rögnvald's father was Kol, a very foxy old gentleman, who lived in Norway. He tried various schemes for securing to his son the half of the islands, which Earl Paul flatly refused to surrender, but nothing came of them and Paul seemed stronger than ever. At last he hit upon the right solution. Rögnvald must pray to St. Magnus, who was his mother's brother and to whom the possessions rightfully belonged, and ask his permission to enjoy them. He must promise that if successful he would build a stone minster at Kirkinvåg (Kirkwall) "more magnificent than any other in these lands," dedicating it to Earl Magnus the Holy.

Rögnvald promptly made the vow.

Paul had arranged beacons on the islands as signals of the enemy's approach, that on Fair Island to be lighted first, the others to be lighted when this was seen.

Kol cunningly contrived to deceive the keeper of the first beacon by pretending to approach with a great fleet. The beacons were all fired and the country aroused, but Kol quietly retired. Then

A PLOWING SCENE IN BIRSAY: THE WOMAN IS SPREADING SEAWEED IN THE FURROW IN FRONT OF THE PLOW

Photograph by Thomas Kent

one of his henchmen landed and, becoming friendly with the keeper, managed to find a secret opportunity to pour water on the wood of a freshly built beacon so it could not burn.

By such methods Rögnvald contrived to gain a foothold in Orkney, singing his rhymes as he went.

Then Swein Asliefson, the great Viking, came upon the scene. In a barge, accompanied by thirty men, he sailed across the Pentland Firth from Scotland, where he had been in hiding because of some of his crimes. Seeing some men on a headland, where they were hunting otters, he caused twenty of his men to lie down and conceal themselves. The hunters, mistaking the ship for a merchantman, called to him to bring his wares ashore to Earl Paul, who was one of their party.

Swein's men came to land, killed nineteen of the party, and seized Paul, carrying him away to Scotland, whence he never returned. Swein became a powerful man in the earldom and lived and died "the holy Earl Rögnvald's henchman."

Rögnvald, having thus obtained possession of the lands in 1136, proceeded to perform his vow in the very next year, his father Kol superintending the building of the Cathedral.

Rögnvald became a "Jorsalafarer" in 1151 and, according to the sagas, recited poetry all the way to Jerusalem. We have already had a glimpse of how his men raided the mound of Maeshowe in search of treasure. He was murdered in 1158, after ruling twenty-two years; his remains were interred in the Cathedral and his name went on the calendar of saints before the end of the century.

THE "SHOW-PLACE" OF THE ORKNEYS

The Cathedral of St. Magnus is distinctly the "show-place" of the Orkneys. It is not remarkable so much for its length and breadth and height as for its fine state of preservation, despite its age. Melrose Abbey, in Scotland, was founded about the same time (1136), but in 1544 was a ruin, and suffered still further in the Reformation. Dryburgh, built almost simultaneously, shared the same fate.

But while the reformers were pound-

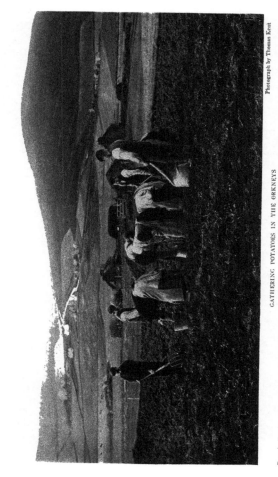

Photograph by Thomas Kent

GATHERING POTATOES IN THE ORKNEYS

Practically the only grain crops in the Orkneys are oats and barley. Turnips and potatoes are the principal root crops. The horses of the Orkneys are hardy, active, and small,-but larger than the famous Shetland ponies.

THE INGLE-NOOK IN AN ORKNEY HOME

Note the dried fish suspended like hams from the rafters.

Photograph by Thomas Kent

ing to pieces the fine old abbeys of Scotland. St. Magnus was not only protected by the people of Kirkwall, but was improved by the addition of some of its finest details. It did receive some harsh treatment about this time at the hands of the Roundheads, and a little later. in 1671, what was once a handsome spire was destroyed by lightning. The attractiveness of the building is now greatly marred by a stumpy little tower.

The interior is distinguished by some massive round pillars, seven on each side, in the Norman style, built of red and yellow sandstone, and a multiplicity of curious old tombstones. Although Earl Rögnvald lies buried there, his saintly uncle, in whose memory the Cathedral was built, was not so honored.

The floors of the nave and aisles were formerly paved with tombstones, the oldest of which seems to be dated 1582. Among them is one in memory of a certain Villiam-Vrving, who must have died a violent death, "Being Schot out' of ye Castel." This good man. who passed out of this life in September, 1614, the month when the Earl of Caithness was besieging Robert Stewart in the castle, was doubtless an ancestor of our own Washington Irving, whose father was born in the island of Shapinsay, across the sound from Kirkwall, and emigrated to New York in 1763.

The tombstones no longer pave the Cathedral, but many of them have been set up along the walls. They are frequently decorated with the skull and cross-bones, the skulls being invariably greatly enlarged on the left side, because of the Orcadian belief that the spirit took its departure through the left ear. A typical stone has the following inscription:

P. P. M. G.

Hier rests the corps
of Patrick Prince mer
Chand in Kirkwall
Sometime espoused to
Margaret Groat who
Left with her Eward
Harie Magnus Helen
& Catherine Princess

This monument
Doth heir present
A' subject to your eye
For Patrick Prince
Is now gone hence

And so above did flye
He left behind
5 Children Kynde
withall a mother deare
To him and them ·
It well become
A mother and a pheare
Obiit 9 March 1673
Aetatis 31

It is difficult to tell what is meant by a "pheare," unless it was put in to rhyme with "deare." The picture represents Death breaking the urn of Life with an arrow. A flame bursts forth from the punctured vessel, from the tip of which the soul flies away in a northwesterly direction. An hour-glass with the sand run out, a sun-dial, two spades, and a coffin complete the doleful ideogram (see illustration, page 205).

Across the street from the Cathedral are the remains of the Bishop's Palace. a building with a large hall and a great round tower. The latter was built in 1540, but the hall is much older, for here Hakon, the last of the great sea-kings of Norway, died in 1263.

This event occurred just after the battle of Largs, on the coast of Scotland. which, though a mere skirmish, was fateful because it gave to Scotland her first claim to the islands, resulting two centuries later in their annexation. Hakon, sick and weary, came to Kirkwall hoping to be restored to health by St. Magnus; but the saint did not intervene and Hakon's body was temporarily interred in the choir of the Cathedral.

BLACK PATE, THE GREATEST TYRANT OF THE ISLANDS

Near by is the Earl's Palace. built by Patrick Stewart, known as "Black Pate." the greatest tyrant the islands ever knew. worse even than his father. Robert, who invented new ways of plundering the people, such as the old Norse earls never practiced.

In 1564 Lord Robert Stewart. an illegitimate son of King James V. obtained through Mary Queen of Scots the grant of all the crown lands of Orkney and Shetland; and in addition (what Mary did not own and therefore had no right to bestow), the lands and services of the free land-owners.

As this could not be immediately acted

A BROUGH IN THE ORKNEYS

A brough is an island, sometimes accessible by foot from the mainland at low water.

A brough is an island, sometimes accessible by foot from the mainland at low water.

A BROUGH IN THE ORKNEYS

Photograph by Charles S. Olcott

SUMBURGH HEAD, AT THE EXTREME SOUTHERN END OF THE "MAINLAND": SHETLAND ISLANDS

Because of the meeting of the tides from two oceans, a swift current is developed here, causing many shipwrecks.

FISHING-BOATS BECALMED IN KIRKWALL BAY

Photograph by Thomas Kent

Kirkwall, which is situated at the head of this bay, received a charter from James III of Scotland six years before Columbus set sail for America.

220

THE HARBOR OF SCALLOWAY: SHETLAND ISLANDS

Photograph by Charles S. Olcott

Scalloway Castle is the ruin of the building erected by forced labor for the infamous Patrick Stewart, the tyrant of the islands. It has a ring near the top, to which offenders were hanged, and it is said that in Patrick's time "the ring seldom lacked a tassel."

THE OLDEST COTTAGE IN THE ORKNEYS

This is a type of "crofter's cottage," now fortunately almost obsolete. It is several hundreds of years old. The woman in the doorway was born here, as were her father and grandfather. A crofter in the Scottish Highlands and adjacent islands is one who rents and tills a croft, or small agricultural holding.

Photographs by Charles S. Olcott

THE HARBOR OF LERWICK, SHETLAND ISLANDS

More than eleven hundred fishing-boats are moored in the harbor every Saturday night during the season. On Monday scarcely one can be seen. Lerwick is the center of a great herring fishery.

upon because of the conflicting appointment of one Gilbert Balfour, Lord Robert was compensated by being created Abbot of Holyrood. He then traded his holy office for the Bishopric of Orkney, about as two boys would swap jack-knives, and when Balfour came to grief because of his loyalty to Queen Mary, Lord Robert took possession of the two groups of islands, church revenues, crown lands, and all.

He proceeded to live upon the inhabitants, levying taxes, exacting service, appropriating lands under all kinds of pretexts, imprisoning, banishing, or executing those who were inconveniently in his way and otherwise twisting to suit his own whim the laws of property which had been recognized for centuries. This continued for twenty-three years, until his death, when his son Patrick not only adopted his father's methods, but invented other and more ingenious schemes of fraud and crime, which he was permitted to employ for seventeen years.

The splendid Earl's Palace in Kirkwall was built by forced labor. The people quarried the stone, transported it, and constructed the building, not by contract, but by compulsion. It was very much the same way with the Palace of Scalloway, which still has a ring at the top to which Black Pate hanged those who objected to following his commands, and the palace at Birsay, which his father built by the same method.

KIRKWALL, BASE OF THE AMERICAN NAVY'S MINE-SWEEPING SQUADRON

Kirkwall, the largest town of the Orkneys, the base of the American Navy's mine-sweeping squadron, which operated in the North Sea in 1919, is a quaint place, and in ordinary times is quiet enough. A very narrow lane, called Bridge Street, leads back from the steamship landing. It is paved with flagstones, and when a team passes the pedestrians have to stand close to the walls or enter the doorways.

At the head of this curious thoroughfare is Albert Street, which has the unique distinction of possessing a single tree. Further on, the street widens into a broad plaza opposite the Cathedral.

Stromness, the second town in importance, lies on the opposite side of the island and is distinctly more picturesque. It stands on the slope of a hill, overlooking a beautiful harbor, and its single street twists and turns through it for about a mile. Our motor car occupied its whole width, but, as there was no other car on the island, this did not greatly concern us.

STROMNESS, HOME OF JOHN GOW, THE FAMOUS PIRATE

Stromness was the home of John Gow, the famous pirate, whose career suggested to Sir Walter Scott the character of Cleveland, in "The Pirate." Here also lived Bessie Millie, an old hag who sold "favoring winds" to the mariners and from whom Scott developed the idea of Norna of the Fitful Head.

From the hill back of Stromness we had a fine view of the island of Hoy, the highest land in the Orkneys. On one side of the hill is the celebrated Dwarfie Stone, another of those mysterious relics, though by no means so old as the stones of Stenness. It is a wedgelike stone, about 30 feet long and 15 feet wide, in which is an opening 3 feet square and 7 feet deep.

At the inner end the opening widens so as to make two short beds, cut out of the solid rock. It is commonly believed to have been the abode of a goblin of evil repute.

In this rambling of the Shetlands and Orkneys, intended to point out some of the curious objects of interest which have thrown a glamor of mystery over the islands, we have left unmentioned the largest and most commercial city of the archipelago, because its importance is entirely modern and its place in history so small as to be scarcely worth mentioning. This is Lerwick, the capital of Shetland. It is far more picturesque, as well as more imposing, than its southern rival, Kirkwall.

Until the arrival of the British fleet during the World War, the harbor of Kirkwall was almost deserted. But Lerwick is the center of a vast fishing industry, and from Saturday to Monday, in the season, its harbor is crowded with

Photograph by Thomas Kent

A VIEW OF KIRKWALL FROM THE SOUTHWEST

"steam-drifters," the modern style of fishing-boats which now control the herring industry.

Lerwick is on high ground, with a road running along the shore, as at Stromness. It is a narrow street, though wide enough for vehicles to pass, and much busier than the streets of Kirkwall.

Narrow lanes, for pedestrians only, lead off the main thoroughfare up the slope of the hill, and these are curiously provided with ropes along the buildings to prevent slipping in icy weather.

The city has a large fish market, where the boats dispose of their catch by auction, a handsome town hall, and many substantially built churches and dwellings.

TWENTY HOURS OF SUNSHINE FOR THE MONTH OF DECEMBER

In the summer, when the fishing business is active, it is a throbbing, wide-awake, bustling city, its streets crowded with men of many nations; but when the fishing season is over and winter settles down there is little to do. The women continue their household duties and knit shawls out of wool, which they card and spin themselves, for the old rhyme applies here as elsewhere—

"Man may work from dawn to the setting of
    the sun,
But woman's work is never done."

When we remember that in midwinter the dawn arrives only shortly before noon and the sunset comes early in the afternoon, it will be apparent that the men of Lerwick have a mean advantage. In fact, the average number of hours of sunshine in December is only about twenty. The idle fishermen spend much of their time in drinking, card-playing, and other amusements, frittering away the earnings of the summer during the long, dark winter.

In midsummer in these high latitudes the sun must get up so early that he thinks it hardly worth while to go to bed.

Shetland, it must be remembered, lies north of the 60th parallel. Trace the line around the globe and you will see that it touches Greenland, passes above the northernmost extremity of Labrador, goes through the upper half of Hudson Bay, skirts the shore of Alaska, and trav-

Photograph by Thomas Kent

STROMNESS, THE SECOND TOWN OF THE ORKNEYS, OVERLOOKS A BEAUTIFUL HARBOR

This was the home of the famous pirate John Gow, whose career suggested to Sir Walter Scott the character of Cleveland in "The Pirate."

ONE OF THE OLDER TYPES OF ORKNEY FARM BUILDINGS, WITH A PEATSTACK TO THE EXTREME RIGHT

For the interiors of such structures, see the illustrations on pages 212 and 216.

A VIEW OF THE CLIFFS NEAR BIRSAY: ORKNEY ISLANDS

Photograph by Thomas Kent

This headland is 287 feet high, and on it will be erected a monument to the late Earl Kitchener, who perished about two miles off-shore. Note the smallness of the figure on cliff at the extreme right.

Photograph by Charles S. Olcott

THE BANQUETRY HALL OF THE EARL'S CASTLE, KIRKWALL, ORKNEY ISLANDS

On the fireplace at the left may be seen the initials of Earl Patrick Stewart. The room apparently contained two great fireplaces and several large arched doors (see text, page 217).

erses the dreary wastes of Siberia—all of which gives one a kind of icy shiver.

But the climate of the islands is so modified by the sea that they are neither excessively cold in winter nor warm in summer. A strong wind blows across them most of the time, and this has interfered with vegetation to such an extent that few trees are to be found. The inland scenery is, therefore, not attractive, but the rugged outlines of the coast, cut up by the action of the sea into numerous inlets, or voes, and carved into fantastic "stacks" and "castles," like the Old Man of Hoy, have a wild beauty of their own.

THE ORKNEYS ARE TAKING A VACATION

As we sailed down the coast at midnight, it was with the feeling that we were leaving a land that was strangely fascinating, where every rock and cave and sheltered voe, every mound and broch and ruined castle or church, seemed to speak of a race of men who had long since disappeared from the face of the earth.

They were men of tremendous activity, giants in stature, fierce, resistless, relentless, yet capable of love and romance, warriors by profession, yet occasionally statesmen, poets, or saints. They came to supersede a mysterious race of whom history can tell us nothing, and when their allotted time was elapsed, they departed as mysteriously as they came, leaving the islands to the keeping of their less ferocious but not more scrupulous brethren of the south.

Since then the centuries have passed, and while the rest of the world has progressed in learning and industry, they have been content to be let alone, to enjoy the peace of obscurity.

Civilization, Commerce, Science, and Art may march forward with proud and determined mien, but the Shetlands and Orkneys, after ages of turmoil, are now taking their vacation.

Vol. XXXIX, No. 3   WASHINGTON   March, 1921

THE
NATIONAL
GEOGRAPHIC
MAGAZINE
COPYRIGHT, 1921, BY NATIONAL GEOGRAPHIC SOCIETY, WASHINGTON, D. C.

FROM LONDON TO AUSTRALIA BY AËROPLANE

A Personal Narrative of the First Aërial Voyage Half Around the World

By Sir Ross Smith, K. B. E.*

DURING the latter phase of the war, while I was flying with the Number One Australian Flying Squadron in Palestine, a Handley-Page aëroplane was flown out from England by Brigadier - General A. E. Borton, C. M. G., D. S. O., A. F. C., to take part in Allenby's last offensive. It was intended that this monster aëroplane should be chiefly employed in carrying out active night bombing operations against the enemy. I hailed as good fortune the orders that detailed me to fly it. The remarkable success eventually achieved by this terrible engine of destruction, and its unfailing reliability during the ensuing long-distance flights, inspired in me great confidence and opened my eyes to the possibilities of modern aëroplanes and their application to commercial uses.

A CHALLENGE IN JEST

It is in a large measure due to the extensive experiences gained while piloting this Handley-Page machine that I was induced to embark upon and carry to a successful issue the first aërial voyage from London to Australia. In a lesser degree, the undertaking was suggested in a joke. One day General Borton visited our squadron and informed me that he

was planning a flight in order to link up the forces in Palestine with the army in Mesopotamia. He invited me to join him.

There was a further proposal, that after reaching Bagdad we should shape a route to India, "to see," as he jocularly remarked, "the Viceroy's Cup run in Calcutta."

"Then, after that," I replied, "let us fly on to Australia and see the Melbourne Cup," little thinking at the time that I should ever embark upon such a project.

Just after the Armistice was signed, General Borton decided to start out in the Handley-Page for India. Major-General Sir W. G. H. Salmond, K. C. M. G., C. B., D. S. O., commanding the Royal Air Force in the Middle East, would accompany us and carry out a tour of inspection.

On November 29, 1918, we took our departure from Cairo, accompanied by my two air mechanics, Sergeant J. M. Bennett, A. F. M., M. S. M., and Sergeant W. H. Shiers, A. F. M., both of No. 1 Squadron. It took just three weeks to pioneer a route to India, where we arrived, without mishap, on December 10, 1918, scarcely a month after the signing of the Armistice.

*Copyright, 1921, by Sir Ross Smith.

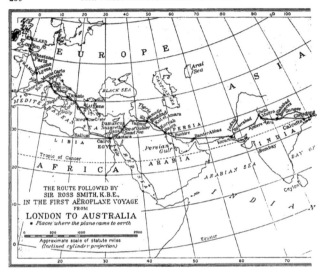

THE ROUTE FOLLOWED BY
SIR ROSS SMITH, K.B.E.,
IN THE FIRST AEROPLANE VOYAGE
FROM
LONDON TO AUSTRALIA
● Places where the plane came to earth

Approximate scale of statute miles
(Inclined cylinder projection)

Major - General Salmond was very proud of this achievement, for it demonstrated that the new arm of the service, the Royal Air Force, had begun to concentrate its efforts on peaceful developments and the establishment of long-distance commercial air routes.

This was the longest flight that had ever been made up to this time, and it convinced me that a machine, properly attended and equipped, was capable of flying anywhere, provided suitable landing grounds existed.

EXPLORING THE ROUTE

After our arrival in India, General Borton communicated with the Air Ministry and asked for permission to charter a steamer to enable him to proceed to Australia to explore the route and arrange suitable landing grounds.

I was to accompany General Borton on this expedition as his staff captain, and it was our intention, after surveying out the route, to return to India, join up with our machine, and continue the flight to Australia over the established course.

The Air Ministry acceded to General Borton's wishes, and the Indian Government accordingly placed at our disposal the R. I. M. S. Sphinx. On February 10, 1919, we sailed from Calcutta, our hold stowed tight with stores and equipment and 7,000 gallons of petrol. We intended to dump 200 gallons of petrol at each landing place for the anticipated flight. But all our well-laid schemes ended in smoke.

THE "SPHINX" BLOWS UP

Two days later, just after leaving Chittagong, in East Bengal, our first port of call, the Sphinx caught fire and blew up.

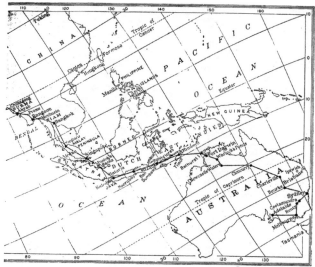

Drawn by A. H. Bumstead

We narrowly escaped going up with it. We lost everything but our lives.

After this mishap we were compelled to return to India to refit. The Indian Government generously lent us another vessel, the R. I. M. S. *Minto*. This time we carried no petrol.

The expedition was rewarded with splendid success during the period of three months we were engaged upon it. We visited Burma, the Federated Malay States, the Netherlands Indies, Borneo, and Siam.

Upon our return to India we were chagrined to find that our machine had been taken up to the northwest frontier to participate in a bombing offensive against the Afghans and had been crashed in a storm.

However, our heart-pangs were mitigated when we learned that the Australian Commonwealth Government . had offered a prize of £10,000 for the first machine (manned by Australians) to fly from London to Australia in 30 days.

Hearing this, I knew there would be many competitors, and the spirit of rivalry grew tense. It stimulated in me a keenness—more than ever—to attempt the flight. My difficulty was how to reach England in time.

SECURING A MACHINE FOR AUSTRALIA

Shortly afterward General Borton was instructed to return to London to report on the route. This opened the avenue of transport for myself and my two mechanics. General Borton himself was very keen to join in the flight to Australia, but, unfortunately, not being an Australian, he was debarred from entering the competition. He very kindly approached Messrs. Vickers Ltd. and asked them if they would supply a machine for

Photograph by Sir Ross Smith

THE PLANE WHICH MADE THE LONDON-TO-AUSTRALIA FLIGHT

This standard Vickers-Vimy bombing plane, equipped with two Rolls-Royce "Eagle VIII" engines of 360 horsepower each, is the same type of machine with which Sir John Alcock made the non-stop transatlantic flight.

the flight. This, at first, they refused to do, but after General Borton pointed out that I had already done a considerable amount of long-distance flying and had been over nearly the whole route, as well as assisted in pioneering it, they finally consented.

My brother Keith was at the time in England awaiting repatriation to Australia. During the latter part of the war he had been flying with the Royal Air Force and had gained extensive and varied air experience. I therefore decided that he would be the best man to take as assistant air pilot and navigator.

Sergeants Bennett and Shiers, in view of their excellent services and the knowledge of machines that they gained in the flight from Cairo to Calcutta, were to accompany us as air mechanics, thus making a total crew of four.

Vickers did not definitely decide to enter the machine for the competition until October, and as we left London on November 12, it will be seen that the time to prepare for such an undertaking was very limited. Our preparations were doubly hurried, first by the knowledge that four other machines had entered the competition and were actually ready to start before the Vickers Company had handed over their machine to us, and,

FILLING THE PETROL TANKS OF THE VIMY WITH "SHELL"

When fully loaded, the great flying-machine, with its wing spread of 67 feet, weighed 13,000 pounds. Its petrol tanks had a capacity sufficient to keep the plane in the air for 13 hours, traveling at a speed of 80 miles an hour (see text, page 237).

second, by the fact that winter was fast approaching and the season might break at any time, thus rendering long-distance flying extremely hazardous.

Once Vickers had decided to enter the machine, however, they threw themselves whole-heartedly into the project and practically gave me a free hand to make whatever arrangements I deemed essential. I had gone minutely into all the intricate details of equipment, the question of supplies, fuel, etc., during my return voyage to England.

The "Shell" Marketing Co. agreed to have our petrol supplies at the required depots to tabulated dates, and Messrs. Wakefield Ltd. in a similar capacity undertook to arrange for lubricating oils.

MAPPING OUT THE ROUTE

The route I decided upon was, roughly, England, France, Italy, Crete, Egypt, Palestine, Mesopotamia, Persia, India, Burma, Siam, Federated Malay States, Dutch East Indies to Port Darwin.

With the route from Port Darwin to our ultimate destination we were unconcerned, for we had received intimation that the Defense Department of Australia had made all necessary arrangements. The great thing was to reach Australia, and, if possible, land our machine there under thirty days.

For my convenience, I divided the route into four stages: First, London to Cairo; second, Cairo to Calcutta; third, Calcutta to Singapore; fourth, Singapore to Australia.

I had been over the entire route with the exception of the first stage, and so was fairly cognizant of the existing conditions—the weather, climate, and the nature of the landing grounds. General Borton had pioneered the first stage in August, 1918; his generous advice, directions, charts, and photographs were invaluable.

SIR ROSS SMITH, K. B. E., M. C., D. F. C., A. F. C., PILOT OF THE VIMY

THE SECOND IN COMMAND, SIR KEITH SMITH, K. B. E.

Sir Ross Smith selected his brother as assistant air pilot and navigator for the historic London-to-Australia flight. The former, during the World War, had been flying with the Australian Flying Squadron in Palestine, the latter with the Royal Air Force.

For the first two stages bad weather was my only apprehension.

As far as Calcutta, passable aérodromes existed, and I could rest assured of Royal Air Force assistance at almost every landing place.

From Calcutta onward we would be entirely dependent on our own arrangements. I considered these last two stages the most hazardous of the flight. Owing to the dense jungles and rough ground, landing places were few and far between, and even those at which we contemplated stopping were very small and unsuited to landing a big machine.

After leaving Calcutta, I proposed landing on the race-course at Rangoon, from which I would fly across the mountain ranges to the Siamese aérodrome at Bangkok. I then proposed to skirt southward down the coast of the Malay Peninsula to Singapore, where once more a landing would be made on a race-course.

The next stop would be made at the hangars of the Dutch Flying School, near Batavia. There would then be no further aérodromes until Port Darwin was reached, a distance of 1,750 miles. I knew that the Vickers Vimy was quite capable of carrying out a non-stop flight of that distance, for this had been demonstrated by the late Captain Sir John Alcock, K. B. E., D. S. C., on his famous transatlantic flight; but I was also aware that to attempt such a long flight with engines that by that time would have done over 100 hours running and covered nearly 10,000 miles would be much to expect.

I therefore decided that, in order to make more nearly certain my chances of success, an aérodrome must be constructed midway. General Borton had selected an admirable site at Bima, on the island of Sumbawa, in the Dutch East Indies. If a landing could be made there, the long stage of 1,750 miles would be halved and the possibility of success more than doubled.

A VALUABLE ALLY

When on my previous visit to Java, I had had the honor of a lengthy interview with His Excellency the Governor-General, Count Van Limberg Stirum, concerning the aérodromes which General Borton and I were selecting in the Netherlands Indies for the proposed aërial route to Australia. His Excellency was most enthusiastic over the development of commercial aviation, and I found him particularly well informed on all aërial matters. He also stated that any aërial route passing over the Netherlands Indies would receive his whole-hearted support and the assistance of his government.

In the course of the conversation I mentioned that I hoped, personally, to attempt the flight from England to Australia a few months later. He said that he would be gratified to assist in any capacity. Remembering this while in London, I decided to ask His Excellency if he would prepare an aërodrome at the selected site at Bima, and sent off a private cable.

LENGTH OF THE LONGEST NON-STOP STAGE REDUCED TO 1,000 MILES

Ten valuable days elapsed before I received a reply, but when it came I was overjoyed to learn that he was not only having Bima prepared, but also another aërodrome at Atambœa, in the island of Timor. This greatly eased my mind, for it meant that instead of having to accommodate our machine with a petrol capacity for 1,750 miles, we need only install tanks for a non-stop flight of 1,000 miles. This greatly added to the buoyancy of the machine, and, through the saving in space, to our personal comfort.

The machine was an ordinary Standard Vickers Vimy bomber, similar to that used by Sir John Alcock for the transatlantic flight, and, apart from the installing of an extra petrol tank, we made practically no alterations.

The machine was powered by two Rolls-Royce "Eagle VIII" engines, each of 360 horse-power. The wing-spread was a little over 67 feet and the total weight, loaded, was six and a half tons. Vickers' factory, the home of the "Vimy," is at Weybridge, about 20 miles distant from London, and is built by the side of the famous Brooklands Motorracing track. After completing the office work in London, the four of us moved to Weybridge and practically lived on the machine.

Photograph by Sir Ross Smith

THE MEN WHO KEPT THE ENGINES GOING, SERGEANT W. H. SHIERS AT THE LEFT AND
SERGEANT J. M. BENNETT AT THE RIGHT

Photograph by Aerofilms, Limited

TOWER BRIDGE AND THE TOWER OF LONDON FROM AN ELEVATION OF 5,000 FEET

The fitting, testing, and final adjusting were thoroughly interesting, and great enthusiasm was shown by the employees of Vickers. It was gratifying to observe that these same men and women, who had produced the great machine flown by Sir John Alcock, felt that their efforts were something more than mere labor. They were producing an ideal from their factory to uphold national prestige. Every man and woman did his or her best, and wished us God-speed.

Thus we were able to place the deepest confidence in the machine; we feared no frailties in its manufacture, and hundreds of times during the flight we had occasion to pay tribute to and praise the sterling efforts of those British workers.

Our petrol capacity would carry us for 13 hours at a cruising speed of 80 miles an hour—ample for the longest stages between aërodromes.

CUTTING DOWN THE BAGGAGE

The question of "spares" was of vital importance and one into which I had previously gone minutely. As we intended starting almost immediately, I decided that it would be useless to ship "spares" ahead, so that the only course left was to carry them with us. This added considerably to the weight of the machine; but the absence of a certain spare part, should we require it, might delay us for weeks, and so put us out of the competition.

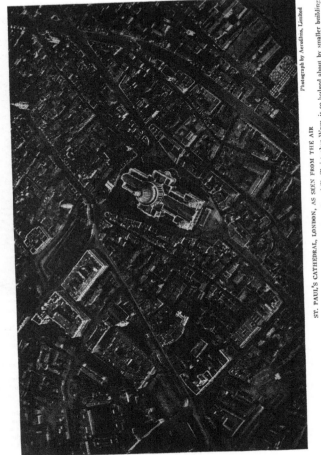

Photograph by Aerofilms, Limited

ST. PAUL'S CATHEDRAL, LONDON, AS SEEN FROM THE AIR

Owing to the fact that this magnificent architectural pile, a monument to the genius of Sir Christopher Wren, is so hedged about by smaller buildings, it remained for aerial photography to give the world a correct conception of the dignity and impressiveness of its exterior.

Eventually the spare parts, personal kit, and miscellaneous gear were assembled and weighed. I decided to limit the total weight of our machine when fully loaded to 13,000 pounds.

I was aware that the deadweight of Sir John Alcock's machine in the transatlantic flight was over 14,000 pounds, but in the vastly greater distance that lay before us, I intended to give my engines as little work as possible.

We discovered that, after the "weighing in," there was an excess of 300 pounds; so something had to go. Our "spares" were indispensable, and so we drastically attacked our personal kit. It was easy enough to cut down our kit—so soon as we were unanimous in deciding to go without any—and so it eventuated that we left England in the garments we wore and with the proverbial toothbrush apiece.

As my brother was navigator, all arrangements concerning maps, etc., were left entirely to him. Wherever possible, we would fly our course by maps and direct observations of features on the ground; but when cloudy or misty weather rendered terrestrial observation obscure, we would rely solely on navigation. For this purpose we carried an Admiralty compass, a ground-speed and drift indicator, and we had our own flying experience to fall back on.

DELAYED BY WEATHER

The machine was at last ready, and, after being flown and tested by Sir John Alcock, was pronounced fit for the undertaking. I considered it advisable to remain another week in England in order to give our supplies of fuel and oil sufficient time to reach some of the more remote aërodromes.

It was galling to have to idle in England while every day we read in the press of the progress of Monsieur Poulet, who had left Paris on October 14 and had by now reached Mesopotamia. The Sopwith machine, piloted by Captain Mathews, had also left England some time previously.

The weather during this week's stay was abominable. Winter was fast closing in with typical English November fogs. Driving sleet and pelting rains fell almost without intermission. One afternoon there was a brief lull, and I managed to get the machine into the air for about an hour and make a final test.

Our machine was still at Weybridge, and the official starting place for the competitive flight was the Hounslow aërodrome.

"WE'RE OFF!"

I had intended flying over to Hounslow on November 13 and starting off on the flight the following morning. On November 11 we were pottering around our machine when the rain suddenly ceased and the fog lifted. It was too good an opportunity to miss! We ran the machine out of its hangar, and I was just about to start up when the clouds closed down again and snow fell heavily.

The weather was very capricious, for in half an hour the clouds rolled away, clearing the air and giving promise of a bright, fine evening. The engines were started up, we climbed into our seats, and took off from Weybridge. As far as we were concerned, the flight to Australia had begun!

During the voyage to Hounslow the machine in every part worked to my entire satisfaction and we landed at the official starting ground without difficulty.

Hounslow is the main "civilian" aërodrome of London, and all commercial machines inward and outward bound from or to the continent start from or land there.

So soon as the machine was in its hangar, I got in touch with Vickers and informed them that I intended starting next morning.

On the morning of November 12 we were called at 4.30, and I was delighted to find a clear, frosty morning. However, at 6.30 a dense ground haze appeared, and weather reports sent by the Air Ministry forecasted bad weather in the southeast of London and the north of France.

The machine was run out from the hangars and Commander Perrin, of the Royal Aëro Club, marked and sealed five parts of it, in accordance with the rules of the competition. It was necessary to produce three of the marked parts upon arrival in Australia, in order to identify the machine.

At 8 o'clock another report stated that

Photograph by Aerofilms, Limited

FOLKESTONE, ON THE STRAIT OF DOVER, AS SEEN FROM THE AIR

Immediately after leaving the flying field at Hounslow, the starting point in the London-to-Australia flight, the world below was lost in fog, but just before reaching the outskirts of Folkestone a rift in the mist enabled the aviators to catch a farewell glimpse of English shores (see text below).

the forecast was Class V, or totally unfit for flying. This was not very reassuring, but our minds were made up and, come fair, come foul, we were determined to start.

A few friends had gathered to bid us God-speed, and, with their kindly expressions and cheers sounding in our ears, we climbed into our seats and took off from the snow-covered aërodrome.

THE RACE BEGINS IN EARNEST

We climbed slowly upward through the cheerless, mist-laden skies, our engines well throttled back and running perfectly. So as to make sure that all was in thorough working order, we circled for ten minutes above Hounslow, then set off.

At 2,000 feet we suddenly emerged from the fog belt into brilliant sunshine, but the world below was lost to sight, screened by the dense pall of mist. Accordingly, we set a compass course for Folkestone, and just before reaching the outskirts a rift in the mists enabled us to pick up the grand old coast-line, every inch of which is measured by history; and so we checked our bearings.

GOOD-BYE, OLD ENGLAND!

There was a certain amount of sentiment, mingled with regrets, in leaving old England; the land of our fathers. Stormy seas were sweeping up channel, lashing white foam against the gaunt, gray cliffs that peered through the mists in the winter light, phantom-like and unreal.

The frigid breath of winter stung our faces and chilled us through; its garb of white had fallen across the land, making the prospect inexpressibly drear. The roadways, etched in dark relief, stood out like pencil-lines on the snow-clad landscape, all converging on Folkestone.

I looked over the side as the town itself, which had played such an important part in the war, came under us. Thither the legions of the Empire, in ceaseless tides, had passed to and from the grim red fields of East and West, all acclaiming thy might, great land of our fathers!

It seemed hard to realize that we had at last started out on the long flight for which we had been planning and working so long, and as I glanced over the machine and the instruments, I wondered what the issue of it all might be—if the fates would be so kind as to smile on us ever so little and allow us to reach the goal of our ambitions, Australia, in thirty days.

The machine was flying stately and steady as a rock. All the bracing wires were tuned to a nicety; the dope on the huge planes glinted and glistened in the sunlight; I was filled with admiration. The engines, which were throttled down to about three-quarters of their possible speed, had settled down to their task and were purring away in perfect unison and harmony.

THE JOY OF FLYING

A small machine is ideal for short flights, joy riding the heavens, or sightseeing among the clouds; but there is something more majestic and stable about the big bombers which a pilot begins to love. An exquisite community grows up between machine and pilot; each, as it were, merges into the other. The machine is rudimentary and the pilot the intellectual force. The levers and controls are the nervous system of the machine, through which the will of the pilot may be expressed—and expressed to an infinitely fine degree. A flying-machine is something entirely apart from and above all other contrivances of man's ingenuity.

The aëroplane is the nearest thing to animate life that man has created. In the air a machine ceases indeed to be a mere piece of mechanism; it becomes animate and is capable not only of primary guidance and control, but actually of expressing a pilot's temperament.

The lungs of the machine, its engines, are again the crux of man's wisdom. Their marvelous reliability and great intricacy are almost as awesome as the human anatomy. When both engines are going well and synchronized to the same

speed, the roar of the exhausts develops into one long - sustained rhythmical boom—boom—boom. It is a song of pleasant harmony to the pilot; a duet of contentment that sings of perfect firing in both engines and says that all is well. This melody of power boomed pleasantly in my ears, and my mind sought to probe the inscrutable future, as we swept over the coast of England at 90 miles per hour.

THE WEATHER PROPHET

And then the sun came out brightly and the channel, all flecked with white tops, spread beneath us. Two torpedo-boats, looking like toys, went northward. And now, midway, how narrow and constricted the straits appeared, with the gray-white cliffs of old England growing misty behind, and ahead—Gris Nez—France, growing in detail each moment!

The weather was glorious, and I was beginning to think that the official prophet, who had predicted bad conditions at our start, was fallible after all. It was not until we reached the coast of France that the oracle justified itself; for, stretching away as far as we could see, there lay a sea of cloud. Thinking it might be only a local belt, we plunged into the compacted margin, only to discover a dense wall of nimbus cloud, heavily surcharged with snow.

The machine speedily became deluged by sleet and snow. It clotted up our goggles and the wind screen and covered our faces with a mushy, semi-frozen mask.

Advance was impossible, and so we turned the machine about and came out into the bright sunshine again.

We were then flying at 4,000 feet, and the clouds were so densely compacted as to appear like mighty snow cliffs, towering miles into the air. There was no gap, or pass anywhere, so I shut off the engines and glided down, hoping to fly under them. Below the clouds snow was falling heavily, blotting out all observation beyond a few yards.

HOW AN AËRIAL COURSE IS SET

Once more we became frozen up, and, as our low elevation made flying extremely hazardous and availed us noth-

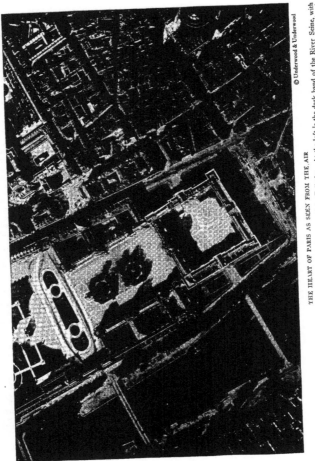

© Underwood & Underwood

THE HEART OF PARIS AS SEEN FROM THE AIR

In the center of the photograph is the Louvre and one end of the court of the Tuileries. At the left is the dark band of the River Seine, with three of its bridges. The building with the dome, to the right, is the Bourse du Commerce. In the upper right corner is the Palais Royal. A blinding snowstorm obscured the cities and landscape of France from the pilot of the Vimy from the time he crossed the English Channel until he came from above the clouds at Roanne, near Lyons, the first landing place (see text, pages 243-245).

ing. I determined to climb above the cloud-mass and, once above it, set a compass course for Lyons.

Aërial navigation is similar to navigation at sea, excepting that the indispensable sextant is of little use in the air, owing to the high speed of travel and the consequent rapid change from place to place and for other technical reasons. Allowances have also to be made for the drift of the machine when side winds are blowing—an extremely difficult factor to determine accurately.

As the medium on which the machine travels is air, any active motion of that medium must necessarily have a direct influence on the machine. If, for instance, the medium on which we are traveling is a wind of 40 miles per hour, blowing directly toward our destination, and the velocity of the machine is 80 miles per hour, then the speed which the machine will travel in relation to the ground would be 120 miles per hour. If we had to forge directly ahead into the same wind, then our speed would obviously be only 40 miles per hour.

To determine the speed of a machine in relation to the ground, an instrument is fitted, called a ground-speed indicator. In side winds the machine makes leeway in addition to its forward movement, and it is the ratio of the one to the other that provides the greatest problem of aërial navigation, especially when flying above clouds or when land features are obscured.

On this particular occasion the Air Ministry had furnished us with charts indicating the trend of the winds and their approximate force at various altitudes, and so we knew, roughly, what allowances to make in our dead reckoning if we lost sight of the ground.

INTO CLOUDLAND

So we climbed steadily in a wide, ascending spiral, until we reached an altitude of 9,000 feet, and were then just above the clouds. Below us the snowstorm raged, but we had entered another world—a strange world, all our own, with bright, dazzling sunshine.

It might have been a vision of the polar regions; it undoubtedly felt like it. The mighty cloud ocean over which we were scudding resembled a polar landscape covered with snow. The rounded cloud contours might have been the domes of snow-merged summits. It was hard to conceive that that amorphous expanse was not actual, solid. Here and there flocculent towers and ramps heaved up, piled like mighty snow dumps, toppling and crushing into one another. Everything was so tremendous, so vast, that one's sense of proportion swayed uncontrolled.

Then there were tiny wisps, more delicate and frail than feathers. Chasms thousands of feet deep, sheer columns, and banks extended almost beyond eyereach. Between us and the sun stretched isolated towers of cumulus, thrown up as if erupted from the chaos below. The sunlight, filtering through their shapeless bulk, was scattered into every conceivable gradation and shade in monotone. Round the margins the sun's rays played, outlining all with edgings of silver.

A BEWILDERING SCENE

The scene was one of utter bewilderment and extravagance. Below, the shadow of our machine pursued us, skipping from crest to crest, jumping gulfs and ridges like a bewitched phantom. Around the shadow circled a gorgeous halo, a complete flat rainbow. I have never seen anything in all my life so unreal as the solitudes of this upper world through which my companions and I were now fleeting.

My brother worked out our course, and I headed the machine on to the compass bearing for Lyons; and so away we went, riding the silver-edged sea and chased by our dancing shadow. For three hours we had no glimpse of the earth, so we navigated solely by our compass, hoping eventually to run into clear weather, or at least a break in the cloud, so that we might check our position from the world below. My brother marked our assumed position off on the chart, by dead reckoning, every fifteen minutes.

The cold grew more intense. Our hands and feet lost all feeling and our bodies became well-nigh frozen. The icy wind penetrated our thick clothing and it was with greatest difficulty that I could work the machine. Our breaths con-

Photograph by Sir Ross Smith

CROSSING THE SOUTHERN ALPS
"Eastward the Alps reared up, serrating the horizon with a maze of glistening snow-peaks"
(see text, page 253).

densed on our faces and face-masks and iced up our goggles and our helmets.

Occasionally immense cloud barriers rose high above the lower cloud strata, and there was no circumventing them; these barriers were invariably charged with snow, and as I plunged the machine into them, the wings and fuselage were quickly armored with ice. Our air-speed indicator became choked, and we ourselves were soon covered white by an accumulating layer of driving snow.

Goggles were useless, owing to the ice, and we suffered much agony through being compelled to keep a lookout with unprotected eyes—straining into the 90-miles-an-hour snow-blast.

A FROZEN LUNCH

About 1 p. m. I suggested to my brother that we should have some sandwiches for lunch. On taking them from the cupboard we discovered they were frozen hard. Fortunately, we carried a thermos flask of hot coffee, and the *pièce de résistance* was a few sticks of chocolate, which was part of our emergency rations. I have never felt so cold or miserable in my life as I did during those few hours. My diary is terse, if not explicit:

"This sort of flying is a rotten game. The cold is hell, and I am a silly ass for having ever embarked on the flight."

To add to our discomfort and anxiety, we were quite uncertain as to our location, and I had visions of what would happen if we encountered a heavy side wind and got blown into the wild Atlantic.

The only really cheerful objects of the whole outfit were our two engines. They roared away and sang a deep-throated song, filled with contentment and gladness; it did not worry them that their radiator blinds, which we kept shut, were thickly coated with frozen snow.

I regarded those engines with envy. They had nice hot water circulating around them, and well, indeed, they might be happy. It seemed anomalous, too, that those engines needed water flowing around their cylinders to keep them cool, while we were sitting just a few feet away semi-frozen. I was envious! I have often thought of that day since and smiled about it—at that diary entry, and at my allusion to the two engines and my envy of their warmth.

The situation was becoming desperate. My limbs were so dead with cold that the machine was almost getting beyond my control. We must check our position and find out where we were at any cost.

A PASSAGE THROUGH THE CLOUDS

Ahead loomed up a beautiful dome-shaped cloud, lined with silver edges. It was symbolical; and when all seemed dark, this rekindled in me the spark of hope. By the side of the "cloud with the silver lining" there extended a gulf about two miles across. As we burst out over it I looked down into its abysmal depths.

At the bottom lay the world. As far as the eye could reach, in every direction stretched the illimitable cloud sea, and the only break now lay beneath us. It resembled a tremendous crater, with sides clean cut as a shaft. Down this wonderful cloud avenue I headed the Vimy, slowly descending in a wide spiral. The escape through this marvelous gateway, seven thousand feet deep, that seemed to link the realms of the infinite with the lower world of mortals, was the most soul-stirring episode of the whole voyage.

Snow was falling heavily from the clouds that encircled us, yet down, down we went in an almost snow-free atmosphere. The omen was good; fair Fortune rode with us. The landscape was covered deep in snow, but we picked out a fairly large town, which my brother at once said was Roanne. This indicated that we were directly on our route; but it seemed too good to be true, for we had been flying at over 80 miles per hour for three hours by "blind navigation," and had been unable to check our course.

THE END OF THE FIRST LAP

At 1,000 feet I circled above the town. Our maps informed us it was Roanne! Lyons, our destination, was only 40 miles away. Exquisitely indeed is the human mind constituted: for, now that we knew where we were, we all experienced that strange mental stimulus—the reaction, after mental anxiety and physical tribulation. We forgot the cold, the snow, the gloom; everything grew bright and warm with the flame of hope and success. And so eventually we reached Lyons and landed.

I have always regarded the journey from Hounslow to Lyons as the worst stage of the flight, on account of the winter weather conditions. We had flown 510 miles on a day officially reported "unfit for all flying." Furthermore, we had convinced ourselves that, by careful navigation, we could fly anywhere in any sort of weather, and, what was still more, we had gained absolute confidence in our machine and engines.

We were so stiff with cold when we climbed out of the machine that we could hardly walk. But what did it matter? Our spirits ran high; we had covered the worst stage; the past would soon be forgotten, and new adventures lay awaiting us in the near, the rosy, future.

The French flying officers were very surprised when they learned we had come from London. They looked up at the weather, at the machine, then at us, and slowly shook their heads. It was an eloquent, silent expression. They were still more surprised when they learned that we intended leaving for Rome the next morning.

Not one of us could speak French very

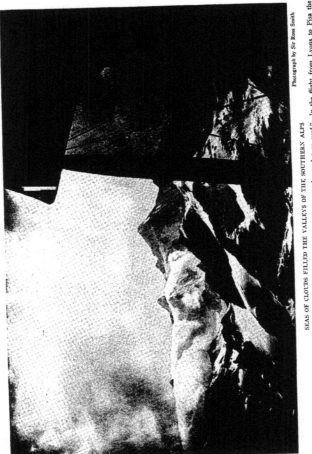

Photograph by Sir Ross Smith

SEAS OF CLOUDS FILLED THE VALLEYS OF THE SOUTHERN ALPS

"Innumerable rocky pinnacles piercing through gave the whole scene the appearance of a rock-torn surf." In the flight from Lyons to Pisa the Vimy was piloted across the River Durance, above the city of Aix, and over the French and Italian Riviera.

A GLIMPSE OF THE RIVIERA: NATURE'S GREAT MAP UNROLLED

"Five thousand feet below us the Mediterranean was laving the cliffs of innumerable little bays and inlets, embroidering a thin white edging of surf round their rugged bases—a narrow white boundary separating green-topped cliffs from deep-blue waters" (see text, page 253).

Photograph by Sir Ross Smith

MONTE CARLO AND THE CASINO

There seemed to be no suitable place for landing here, however, so the aviators had to forego the desire to test Dame Fortune and see if she would be as kind to them at the tables as in the air.

There seemed to be no suitable place for landing here, however, as the aviators had to forego the desire to test Dame Fortune and see if she would be as kind to them at the tables as in the air.

Photograph from Press Illustrating Service

THE ETERNAL CITY AND THE MOST FAMOUS CHRISTIAN CHURCH AS PHOTOGRAPHED BY A MODERN SKY PILOT

The majestic dome of St. Peter's in Rome overlooks the Vatican. From the elevation at which this picture was made, the piazza, with its obelisk, resembles a great sun-dial.

© Kadel and Herbert

BEAUTIFUL ROME PHOTOGRAPHED FROM THE AIR: LOOKING DOWN ON THE FAMOUS
OBELISK IN THE CENTER OF THE PIAZZA OF ST. PETER'S

Tradition tells us that when the obelisk was being erected on the present site, in 1586, the
engineer, Fontana, failed to take into consideration the tension which would result from the
great weight (320 tons) on the elevating ropes. As a consequence, the monolith just failed
of being raised to a vertical position. At this critical moment, although silence had been
imposed upon the bystanders under pain of death, one of the workmen, Bresca, a sailor, cried,
"Wet the ropes!" This was done, causing a shrinkage and thus preventing a catastrophe.

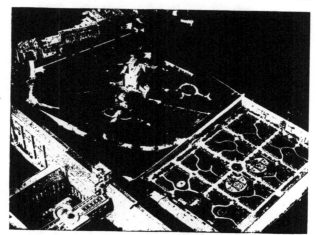

© Kadel and Herbert

THE PAPAL RESIDENCE AND GARDENS IN ROME
At the right can be seen the insignia of Pope Pius X and Pope Benedict XV.

well, and we had considerable difficulty in arranging for petrol supplies to be delivered to the machine by next morning. Sergeants Bennett and Shiers just had time to look over the engines before the winter darkness settled down. We all turned into bed very early, very tired, but very happy.

On opening my personal kit that night I found it, too, had suffered the rigors of the sky journey. It was still frozen stiff—my solitary tooth-brush!

THE SECOND DAY OPENS

Next morning was November 13. I always hold that such a date should be banned from the months of the calendar. Daylight 6.30, cold and frosty. The petrol had not arrived at the machine, so I sent my brother Keith in search of it; his French was even less eloquent than mine. A couple of hours later he returned, looking very grim, followed by 300 gallons of very servile spirit.

I explained in execrable French to a mechanic that I required 24 gallons of hot water for our radiators. It had been necessary to drain the water from the radiators the night before, owing to the low temperature; otherwise the circulating water would have frozen into a solid block and burst the radiators. Ten minutes later the mechanic returned bearing a small jug of hot water. Our faces had been too sore to shave that morning, so I suppose he gathered from our appearance that we wanted the hot water for that purpose.

My brother Keith then had a try in that Australian tongue, famed alike for its potency and rhetoric, and universally understood throughout the breadth of the battlefields. That mechanic bowed most politely and profusely and returned in great haste, bearing triumphantly a second jug of hot water. My brother's growth, like his temper, is much more bristly than mine. While we both were

A DETOUR WAS MADE TO NAPLES (SEE PAGE 260)

But clouds and mists robbed the aviators of the opportunity to take an air view of this superb bay and the city enthroned upon the encircling hills.

Photograph by Earle Harrison

252

literally "losing our hair," my indispensable Bennett and Shiers had filled several petrol tins with water and had borrowed a large blow-lamp. Thus was the water heated and our tempers cooled. .

II. From Lyons—Across the Mediterranean—to Cairo

We had planned overnight to leave Lyons immediately after an early breakfast, and we hoped to land at Rome well before the day closed. The delay in securing warm water for our radiators, however, meant that we were not in the air till 10 o'clock.

It was a frosty daybreak, and for a short time we encountered some clouds; but as we progressed these drifted away, clearing the atmosphere and unfolding a scene of bewildering beauty. Eastward the Alps reared up, serrating the horizon with a maze of glistening snow-peaks. Seas of cloud filled the valleys, with innumerable dark, rocky pinnacles piercing through and giving the whole scene the appearance of a rock-torn surf. Charming villas, set amidst lawns and gardens, lay tucked away over the hillsides. White roadways streaked the landscape, and close by the coast ran the thin lines of steel along which a toylike train was passing with its burden of sightseers to Monte Carlo and the playground of Europe.

ACROSS THE RIVIERA

The air was keen-edged and the cold was still severe, but after the icy blasts and the spear-pointed showers of the previous day, the going was excellent. We were freed, too, from the anxiety of shaping our course by sheer navigation. Nature's great map was no longer obscured. It lay unrolled below, an enlarged edition of our own tiny charts, on which we checked its features. Picking up the River Durance quite easily, we crossed it and passed above the city of Aix; then swung east, heading for the coast and Cannes—across the famous Riviera.

Soon we caught sight of the sea. Five thousand feet below us the Mediterranean was laving the cliffs of innumerable little bays and inlets, embroidering a thin white edging of surf round their rugged bases—a narrow, white boundary-line separating green-topped cliffs from deep-blue waters.

Nice soon lay below us. The city, with its fine buildings and avenues of palms, encircled by high hills, rests on the shores of a sea of wondrous blue. It is a place of ineffable charm and peace.

A large crowd had collected on the Promenade des Anglais to witness our flight and cheer us up. We flew low enough to distinguish the doll-like figures, and though we could not return their greetings we appreciated them none the less. Then onward again with a following breeze, white-cresting the blue sea that stretched away from beneath us to the southern horizon. We circled above Monte Carlo and the famous Casino, admiring the wonderful terraces and gardens, which looked like a skillfully carved and colored model rather than a real palace and its gardens.

We swept round, looking for a landing-place, for I was inclined to test Dame Fortune and see if she would be as kind to us at the tables as she had been to us in the air. There seemed to be no suitable spot on which to land, however, so we headed on to our course again, and soon our regrets faded in admiration of the glorious coast-line over which we were speeding. Suddenly I remembered it was the 13th; Fortune had been kind to us after all.

FAREWELL TO FRANCE AT MENTONE; ITALY'S BORDER CROSSED

Mentone, nestling in its bay, was the last glimpse we had of France; then, still following the railway line that runs along the coast, we crossed the border into Italy without trouble from the customs officials. Less than half an hour later we passed San Remo, and, instead of following the coast-line north, I kept the Vimy headed almost due east, and, crossing the Gulf of Genoa, picked up the coast again at Spezia and turned south once more. Here we met a strong head wind, and this, added to the handicap of our delayed start, made it evident that we could not reach Rome before dark.

I knew that there was an aërodrome at Pisa, since it was one of the stations on

Photograph from H. N. Hampton

A UNIQUE PHOTOGRAPH OF MOUNT VESUVIUS

This picture was made from a plane flying 500 feet over the crater of the famous volcano. Clouds of smoke are seen issuing from the abyss. The difficulties of obtaining a photograph of this kind are great, owing to the very bumpy condition of the atmosphere to which the volcano gives rise.

the air route to Egypt, so decided to spend the night there and go on to Rome early next day. It was well down the afternoon when we picked out the aerodrome, and the ground looked very wet and desolate as we circled above it. But we landed successfully through a whirl of mud and water, whisked up by the propellers.

A CORDIAL WELCOME AT PISA

As we taxied across the slippery 'drome toward the hangars, several Italian flying officers came out to greet us. They were profusely polite, and while our scholarship boasted "little French and less Italian," there was no doubt about their cordial welcome and their curiosity. By means of that universal language of gesture, in which these Latins are so accomplished, they made us at home and indicated that an English officer was stationed in Pisa and that we might reach him by telephone.

After considerable trouble I managed to have him called up and asked him to come down to the aerodrome. I was delighted to find that the officer was Captain Horne, of the Royal Air Force, who had been appointed to the air-route station. Accommodation for our party was promptly arranged, and after attending to the machine we motored into Pisa and stayed the night at an hotel.

Heavy rain set in, and when we were awakened in the morning it was still pouring, with a strong slant from the south. In spite of the unsuitable conditions, we decided to go down to the aerodrome and, if possible, get up and on to Rome that day.

On our arrival at the hangars we

Photograph by S. P. Stambach

THE "SHIP OF ULYSSES," ISLAND OF CORFU, GREECE

Pontikonisi ("Mouse Island"), which lies at the entrance to the harbor of Corfu, according to tradition, was the Phæacian ship which brought Ulysses to Ithaca, and was afterward turned into stone by the infuriated Poseidon. "First we flew east to the heel of Italy, and then headed across the open sea to the Island of Corfu. . . . Almost before we realized it, Corfu loomed up in the mist" (see text, page 265).

found, to our dismay, that the aërodrome looked more like a lake than a landing ground. However, I started up the engines and endeavored to taxi into the wind, but the machine became badly bogged, the wheels refusing to budge an inch.

A force of thirty Italian mechanics came to help us, but it took us an hour and a half to extricate the machine. Our difficulties in getting anything like "teamwork" were increased by our lack of knowledge of Italian, and Sergeant Bennett amused us greatly by breaking into Arabic, with all the French he knew sifted in. A second attempt also resulted in failure, and by the time the machine had been dug out I came to the conclusion that it was hopeless to try to leave that day. It was still raining, so we covered up the engines and reluctantly returned to the town, soaking wet and grimed with mud.

SEEING PISA

Late in the afternoon the rain ceased, so my brother and I went sight-seeing. We visited the usual hackneyed tourist sites, including the famous Leaning Tower. Elections were in progress and the whole town was swayed with excitement. We attracted much attention walking about in uniform; for, besides Captain Horne, we were the only British officers in Pisa.

BENNETT'S FLYING LEAP

We were cheerful, for we had hopes that the water would drain off the aërodrome by the following morning, but once more we awoke to disappointment. Drizzling rain and a cold south wind ushered in the new day. However, we went down to the aërodrome, determined to get the machine into the air somehow.

My brother and I walked over the aërodrome, stamping in the mud to try to find a hard track for the machine. We got very wet, but managed to find a pathway with a fairly hard surface.

All went well until I swung the machine round, just preparatory to opening the engines full out for getting off. In doing this sharp turn, one wheel became a pivot in the mud and stuck fast; so once more we were badly bogged. Our

THE HARBOR OF CANEA, CRETE, WHERE THE VIMY PAUSED IN ITS FLIGHT ACROSS THE MEDITERRANEAN

"We found Canea to be an extremely picturesque and interesting old place, with its massive castle walls, narrow, cobbled streets, and its quaint, old-fashioned buildings, reminiscent of a bygone age." From this point to Salḥm, on the African coast, is 250 miles as the plane flies.

THE SPHINX AND THE PYRAMID OF CHEOPS

© Donald McLeish

"We headed direct for Cairo, across the gray brown sea of sand. . . . We were not sorry to descry those landmarks of the ages, the Pyramids, and soon we could pick out the minarets and mosques of the Egyptian capital itself. We had come through from Suda Bay (Crete), a distance of 680 miles, in a non-stop flight of seven and one-half hours, thus completing the first and worst of the four stages into which I had divided the total journey" (see text, page 266).

A STREET IN CAIRO, WITH THE MOSQUES OF AGHA AND KHEIRBEK IN THE
BACKGROUND

There are more than 260 mosques in Cairo, the air voyageurs' only "port of call" in Africa.

Italian friends came to the rescue again, and by digging and pulling got the machine out of the hole which it had made for itself. The ground was so soft that the wheels began to sink in slowly, and I realized that if we were to get off at all it must be at once.

I opened out the engines, but the machine would not move forward, as the wheels had become embedded in the mud; on the other hand, the tail lifted off the ground and there was the danger of the machine standing up on its nose. To overcome this difficulty, Sergeant Bennett applied the whole of his weight on to the tail-plane, and I once more opened the engines full out. Some of the Italian mechanics pulled forward on the wing-tips, and this time the machine started to move forward slowly. I suddenly realized that Bennett was not on board, but as I had got the machine moving at last, I was afraid to stop her again.

I felt sure that he would clamber on board somehow, as I had previously told him that as soon as the machine started to move he would have to make a flying jump for it or else take the next train to Rome.

We gathered way very rapidly, and, after leaving the ground, I was delighted to see Sergeant Bennett on board when I looked round. The take-off was very exciting and hazardous, as the Vimy had to plow her way through soft mud and water. The water was sucked up and whirled around by the propellers, so that we became soaked through and plastered with liquid mud. I am sure that in a cinema picture our performance would resemble the take-off of a seaplane more than that of a land machine rising from an aërodrome. We were tremendously relieved to find the freedom of our wings again, and though we laughed at our discomfiture, it was certainly a providential take-off and one that I should not care to repeat. We afterwards learned that we had been doubly lucky, for the rain continued to fall in torrents for the next week and the aërodrome was temporarily impossible.

A ROUGH PASSAGE

Our flight toward Rome was one long battle against heavy head winds and through dense clouds. We had been in the air barely an hour when the oil-gauge on one of the engines dropped to zero. .

Thinking that something had gone wrong with the lubricating system, I switched off this engine and flew along close to the ground on the other engine, looking closely for a place to land. Fortunately we were not far from the Italian aërodrome at Venturina, and there I landed.

Sergeant Shiers quickly discovered that the fault was in the gauge itself, and not in the lubricating system, and it was only a matter of minutes before we were in the air again. The wind had increased, and the rest of the voyage to Rome was boisterous and unpleasant. Our average ground speed was a bare fifty miles an hour, so that it was not till late in the afternoon that we were above the city of the Cæsars.

STIRRED BY THE BEAUTY OF ROME

In spite of the fatigue induced by our strenuous experiences of the day and our eagerness to get down to earth, I could not help being stirred by the beauty of the historic city. The sun was peering through the space between the clouds and the distant mountain tops and, slanting across the city, gave it an appearance of majestic splendor. In this soft evening light, Rome reflected something of its old glory. Details were subdued, so that much of the ugliness of its modern constructions was softened. Below, "the Yellow Tiber," spanned by numerous bridges, curved a silvery course out into the twilight and to the sea.

In the brief space of a few minutes we had circled the city within the walls, and it was with feelings of relief that we landed at the Centocelle aërodrome. A hospitable welcome was accorded us by the commandant of the Italian Flying Corps and by the British air attaché. The latter kindly attended to our wants, had a military guard placed over the machine, and acted as interpreter.

My original plan was to make the next stage a non-stop flight from Rome to Athens, thence to Cairo in another flight. This decision was the result of a report received in England that the aërodrome at Suda Bay, on the northern side of

LATEEN-SAIL CRAFT ON THE RIVER NILE

Crete, was flooded and would be unfit for landing till after winter. The air attaché at Rome, however, told me that the Suda Bay 'drome was still in good condition, but that I could make sure by dropping down at Taranto and inquiring at the British aërodrome there.

A glance at the map will show that the Cretan route saves a considerable distance, Suda Bay providing a half-way house. I therefore decided at once to take the Taranto course and try to save the long stretch of Mediterranean from Athens to Cairo.

HOW WE FAILED TO "SEE NAPLES AND DIE"

After daylight, we left Rome in very bad weather. Our route for the first few miles followed the Appian Way, and as we were flying low we had a fine view of this ancient highway. The landscape for the most part was obscured by broken clouds, but through the rifts we had fleeting glimpses of the wild and spectacular nature below us.

Naples was not directly on our course to Taranto, but having visited it previously as a tourist, I made a detour in order to photograph and gaze down upon its wondrous bay from the sky. To my intense disappointment, clouds and mist robbed us of my desire, and even the mighty Vesuvius was buried somewhere beneath the sea of clouds; so, reluctantly, I turned away and resumed our course to Taranto.

Our course now lay almost due east across the Apennines; but here again the clouds had banked against the mountains, and only an occasional peak peered through them. Owing to the clouds and my scant knowledge of the country, I determined to fly low, following, more or less, the course of the valleys, which were nearly cloud-free.

From breaks in the cloud, the sun beamed down on to vales of great loveliness. Numerous small waterfalls dashed down the mountain sides, and streams like silver threads rippled away through

Photograph by Sir Ross Smith

THE SUEZ CANAL AT KANTARA AS SEEN FROM ABOVE

"Kantara now lay below us—that vast series of store-dumps, a mushroom city beneath canvas, which had sprung into being since the British occupation of Palestine" (see text, p. 277).

the valleys. The lower steps of the mountains were terraced, and wherever a flat stretch of soil presented itself small homesteads nestled, surrounded by cultivation. Sometimes we would be only a few hundred feet above the ground when crossing the crest of a ridge; then we would burst out over a valley several thousand feet deep.

Flying became extremely difficult at this stage, owing to the bumpy nature of the atmosphere. At times the machine was literally tossed about like a leaf, and for three-quarters of an hour we experienced some of the roughest flying conditions of the whole journey. On one occasion our altimeter did a drop of 1,000 feet, and bumps of 400 and 500 feet, both upward and downward, were frequent. I can only attribute this aërial disturbance to the rough nature of the country and the proximity of clouds to the mountain tops.

A strong following wind was blowing, and I was very much relieved when we got clear of the mountains and were following the coast down to Taranto.

THE HEEL OF ITALY

The town of Taranto presents a busy scene from the air. A great number of ships and transports were anchored off shore, and as the air had now cleared somewhat, we had a glorious view of this great Mediterranean seaport, which played such an important part in the Eastern campaign. We could still discern long lines of tents in the British camp, and everywhere there was the great activity which characterizes a military center.

The town is small and picturesquely situated at the head of a little inland bay, which forms a magnificent natural harbor. Below us the boom protecting the

JERUSALEM FROM ON HIGH

The student of Biblical history will find many points of interest to study in this photograph, both within and outside the ancient walls. The conspicuous domed structure in the center of the vast open court is the famous Dome of the Rock. Near the upper right corner of the photograph, beyond the walls, is the Garden of Gethsemane.

AN AIRMAN'S VIEW OF ONE OF THE WORLD'S MOST HISTORIC CITIES—DAMASCUS

Photograph by Sir Ross Smith

In their voyage from Cairo to Damascus, Sir Ross Smith and his party, in forty minutes, flew over the region in which the Children of Israel wandered for forty years. "Damascus, a miraged streak on the horizon of a desert wilderness, grew into a hand, assuming height and breadth. Color crept in, detail resolved, developed, enlarged; a city arose from out the waste of sands, an oasis, glorious, magical, enchanting— this was Damascus—a city almost ethereal in its beauty, rearing a forest of splendid minarets and cupolas" (see text, page 280).

Photograph from British Air Ministry

THE VILLAGE OF HIT, AT THE HEAD OF NAVIGATION ON THE EUPHRATES

Practically every native who owns a boat on the Euphrates has followed the injunction given to Noah, to "pitch it within and without with pitch"; and much of the pitch comes from the bitumen springs of Hit.

entrance from submarines was clearly discernible.

When we landed we were greeted by a number of officers of the Royal Air Force who were stationed there, as Taranto at that time was one of the main aërodromes on the route from London to Cairo.

The machine was pegged down and lashed, and after an excellent lunch at the officers' mess we spent the afternoon working on the engines and preparing for the flight across the sea to Crete the following day. The British camp was particularly well kept, and in front of the headquarters there was a fine garden with the chrysanthemums in full bloom.

Here I met many old comrades with whom I had been associated during the war. This meeting was a pleasant relaxation from the mental strain of the past few days, and I gleaned much valuable information about the aërodrome at Suda Bay. I was delighted to learn that it was still in good condition and was in charge of Royal Air Force personnel. This information finally decided me to cancel the idea of flying on to Athens. I now determined to fly on to Suda Bay, thus cutting the long sea flight of the Mediterranean into two shorter sections and saving upward of 200 miles.

After a good night's rest in comfortable beds, we were up at our usual hour and made an early start for Suda Bay.

Once again the weather was cruel to us. First, we flew east to the heel of Italy, and then headed across the open sea to the island of Corfu. Low cloud and rain forced us down to 800 feet above the sea. The flight was miserable. The driving rain cut our faces and obscured all distant vision. Almost before we realized it, Corfu loomed up in the mist, and so I altered the course to southeast and flew down the coast of Greece.

The bad weather made our voyage down this rugged coast very hazardous, and on one occasion, after passing through a particularly low bank of cloud, I was terrified to observe a rocky island loom up in the mist directly ahead. It was only by turning sharply at right angles that I avoided crashing the machine against its precipitous sides. All this time we were flying at a height of no more than 800 feet, and so it was with intense relief that we reached Cape Matea, the most southern point of Greece, and headed across the sea to Crete.

The clouds now lifted, and the mists dissipated, unfolding a scene of rare enchantment. The high ranges of Crete soon loomed up before us. A layer of cloud encircled the island like a great wreath. The mountains rose nobly above it, and the coasts, rocky and surf-beaten, could be seen below. All this, set in a sea of wondrous blue, bathed in bright sunshine, lay before us. It was a gladsome and welcome sight. •

Wheeling above the town of Canea, which is on the opposite side of a narrow neck to Suda Bay, we soon located the aërodrome and circled above it preparatory to landing.

THE LAND OF HISTORY AND LEGEND

The aërodrome is not of the best and is rather a tricky place for negotiating a landing, being surrounded on three sides by high, rocky hills; but we succeeded in making a good landing. Here, too, we were welcomed by an officer of the Royal Air Force and a small crowd of inhabitants, who gathered round the machine, examining it—and us—with curious interest.

With the knowledge that on the mor-row our longest oversea flight, in this half of the voyage, awaited us, we spent most of the afternoon in a particularly thorough overhaul of the machine, and then accepted our R. A. F. friend's invitation to look over the town and take tea at his house. We found Canea to be an extremely picturesque and interesting old place. Its massive castle walls, its narrow, cobbled streets, and its quaint, old-fashioned, but substantial, buildings, reminiscent of a bygone age, are all in keeping with its history, which runs back of the Christian era, and its legends, which run back a league or two further.

Our pilot excited our admiration by the expert way in which he steered us through a maze of rough-surfaced alleyways, our Ford causing a great scattering of children and dogs—both of which appear to thrive here in large numbers.

AN ATTACK OF "PRICKLY HEAT"

Eventually he conducted us to a quaint little café—a sort of tavern, at which the people seem by custom to foregather for a cup of coffee before dinner. The café-au-lait was excellent, and, as our host racily recounted his experiences, I came to the conclusion that life in Canea, small and isolated as it is, holds compensations, and is not nearly as dull as it appears at first glance.

The short run home to our R. A. F. friend's house was certainly not monotonous, but we arrived undamaged and undamaging. Since the house was rather small to accommodate unexpected guests, we cheerfully agreed to sleep in the small British hospital close by. We turned in early, planning to take a good night's rest and get away betimes in the morning.

A few minutes after putting out the lights, I heard my brother Keith tossing about in bed, and called out to know if anything ailed him. "Yes," he said, "I fancy I'm getting prickly heat." A few minutes later I got a touch of it myself, and, bounding out of bed, reached for the candle. The beds were full of prickly heat! "Prickly heat" held the fort in large and hungry battalions.

We retreated and spent the night curled up on the floor of an adjoining room. When we turned out we found that it had

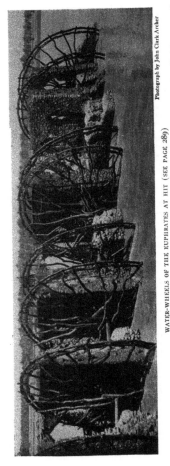

WATER-WHEELS OF THE EUPHRATES AT HIT (SEE PAGE 289)

These irrigation devices, 30 feet in diameter, are operated by the river current. They are equipped with crude cups, which spill their contents of river water into a sluiceway as they make the turnover.

been raining heavily and the air was still thick with drizzle. The prospect was not good for crossing the island, which, though only a few miles wide, is intersected by an irregular range of mountains, of which the famous Mount Ida is one of several peaks. But, with our experience of the muddy aërodrome at Pisa fresh in our minds, we decided to get aloft as soon as possible rather than risk the ground which was already becoming soft, degenerating into a bog.

THROUGH THE MOUNTAIN PASS

We took off quite easily, and soon after leaving the ground encountered a layer of cloud, but pushed through and out—only to find ourselves beneath another stratum. Our charted route lay southeast, then south, with the southernmost point of the island as the objective, and I had been told that it was easy to follow a rough track leading from Canea through a pass in the mountains; but, with clouds above and below, it was not so easy.

I decided to try to locate the pass in the hope of getting through without the necessity of climbing above the mountains, and so wasting valuable time. Fortune favored us. I found the pass and to my joy discovered that there was just sufficient room for us to scrape over the top without entering the cloud. We appeared to be only a few feet above the rocks when we cleared the crest, but it was preferable to having to barge blindly through the clouds, running the consequent risk of hitting a mountain crag.

On the southern side of the ranges the air was much clearer, and we were soon flying over the coast-line. We took observations and set a compass course for Sallum, on the African coast. Two hundred and fifty miles of open sea had to be crossed. Be-

THE RUINS OF THE ANCIENT CITY OF CTESIPHON, NEAR BAGDAD (SEE PAGE 291)

In the center of the picture is the august throne hall, the most splendid example of Sassanian architecture in existence. Through the high arched entrance may be seen the throne on which mighty kings once sat. Nestling at the base of the towering ruin is a native village. This photo-graph was taken from an elevation of 2,000 feet.

BAGDAD, THE MAGIC CITY OF HAROUN-AL-RASCHID, OF ALADDIN, AND OF SINDBAD THE SAILOR

It is today a city of squalor and decay, but its name still conjures in the imagination pictures of oriental opulence, romance, and intrigue (see text, page 291).

fore we started, Bennett and Shiers had given a final look over the engines, which had been running perfectly, and almost the last thing they did before climbing aboard was to inflate the four spare inner tubes of our landing wheels; they would make first-class life-buoys if we had to come down between Crete and Africa.

I would have preferred flying at about 5,000 feet, but our enemies, the clouds, which ever harassed us, forced us to fly at an altitude of from 1,500 to 2,000 feet above the face of the sea. There was a light, favoring wind, and the going was smooth and even; but as the land dropped behind, and mile after mile was flown, one began to realize the meaning of the term, "a waste of waters."

On and on we flew, yet, save for the wind of our own passage through the air, could scarcely tell that we were moving; for, unlike the flight across the land and down the seacoast, there was nothing by which to gauge our movement. The cloud roof was dull and uninteresting; the sea-floor gray, desolate, and empty as far as the eye could reach.

SHIPS OF THE SEA LEFT FAR BEHIND

My brother took out his case and began writing letters. I studied the charts and the compass and kept the machine on the course. Then, suddenly, a little to the right of the course, appeared a minute object that separated into two as we drew nearer, and finally resolved itself into a pair of vessels linked together with a tow-line. Very tiny they looked down there and very lonely.

We were heading for Sallum, on the African coast, 250 miles from Crete, as the 'plane flies. I wondered if these ships were making the same port, and how long it would take them to do the journey that we were counting on accomplishing in about four hours! I felt quite sorry for the poor midgets toiling along with their tow-rope, and speculated on what would happen if a big sea got up. Doubtless they looked up at us—they must have heard our engines booming—and wondered, too. Perhaps they envied us our wings; perhaps they pitied us and congratulated themselves on the sound decks beneath their feet.

Ten minutes and they were far behind

us; another ten and they were out of sight; but they had, without knowing it, cheered us immensely. They proved the only speck of life we saw on all that area of waters. Once more we entered the loneliness of sea and sky, but we had the sense of having passed a definite point, and now we kept a keen lookout for land.

ON AFRICAN SHORES

Our first glimpse of Africa was of a barren, desert coast-line, but it was a welcome sight none the less. On reaching Sallum we turned and flew along the coast as far as Mersa Matruh. The land below was flat and uninteresting desert, with nothing to relieve the monotony. Without landing at Mersa Matruh, we headed direct for Cairo, across the gray-brown sea of sand, passing over Wadi Natrum, which is merely a cluster of straggling palms beside a salt-pan.

We were not sorry to descry those landmarks of the ages, the Pyramids, and soon we could pick out the minarets and mosques of the Egyptian capital itself. Now we were winging our way over Old Father Nile and across landmarks that were as familiar to me as the Heliopolis aerodrome itself, to which destination I was guiding the Vimy.

No wonder I glanced affectionately over the silent engines as we came to rest. I felt extremely happy as we sat there a moment or two, waiting for the fellows to come up and welcome us. We had come through from Suda Bay, a distance of 680 miles, in a non-stop flight of seven and a half hours, thus completing the first and worst of the four stages into which I had divided the total journey.

THE LAST STAGE

That bit of route from London to Cairo—pioneered in 1917 by my old commanding officer, General Borton—had taken its toll, and I had been more than a little afraid of it on account of the possibility of bad weather and my ignorance of the country and the aerodromes. And here we were, safe, with our machine as sound as when she started.

A familiar stage, with all the prospects of fine weather, lay before us. There was some excuse for a flash of thankfulness and exultation. Then the boys were

Photograph from Boston Photo News Co.

THE TIGRIS RIVER, WINDING LIKE A HUGE SNAKE, WITH THE FAMED CITY OF BAGDAD
IN THE LOWER FOREGROUND

The most distant point on the River Tigris shown in the picture is 50 miles away.

greeting us, and a rousing welcome it was from men with whom I had served during the war. Our mechanics, too, found old comrades who hauled them off to celebrate the occasion before attending to the engines.

It was quite like old times to climb into a car, to spin through well-known thoroughfares to Shepheard's, to sink luxuriously into the arms of a great and familiar lounge chair, and to yarn over the events that had happened since last I occupied it.

A NIGHT IN CAIRO

My friends tried to persuade me to attend a dance that was being held there that night, but I needed all the sleep I could get, and so declined reluctantly. But for an hour or more I sat in an easy chair on the well-known veranda, listened to the sweet strains of the music

inside, and that other strange blend of street cries—a veritable kaleidoscope of sound—that may be heard nowhere save in Cairo. I noted, too, the beauty and the chivalry coming in, and watched the curious procession of all sorts passing by.

I had to shake myself to be assured that it was not a splendid and fantastic dream. As we lounged there a messenger boy brought a cable for me—we had sent our own messages off long before. It was from General Borton, congratulating us on our safe arrival in Egypt and wishing us good luck for the next stage.

While I was reading this kind remembrance from my old O. C., an Arab paper-boy came crying his wares, and I bought a news-sheet and read with amused interest the story of our doings during the last few days. I also read, with a shock of keen regret, of the accident that had befallen our gallant competitors, Lieuten-

Photograph by Sir Ross Smith

A SCENE NEAR BASRA

Clusters of date palms and a scant belt of vegetation fringe the banks of the Shatt-el-Arab, formed by the confluence of the Tigris and the Euphrates. "All this was once the Garden of Eden" (see text, page 293).

ants Douglas and Ross, who had both been killed practically at the starting-post, just a few days after we left, through the crashing of their machine. Then we turned to the column that recorded the progress of Monsieur Poulet, who had left Paris thirty days before and who, we saw by the cables, was now in India.

We had certainly gained a good deal on the Frenchman, but he still held a big lead, and we were keen to get on with the next stage. We turned in that night feeling happier and more rested than at any moment since we left England, and we slept like proverbial tops.

III. FROM CAIRO ACROSS PALESTINE AND MESOPOTAMIA

We had intended staying a few days in Cairo to rest, but, owing to the day we lost at Pisa, we were now one day behind our scheduled time; so I decided

that it must be made up. There had been a heavy fog overnight, and on our arrival at the aërodrome the weather conditions were not at all enticing. Telegraphic reports from Palestine indicated "Weather conditions unsuited for flying."

LEAVING EGYPT

My inclinations wavered. We were at a hospitable aërodrome, surrounded by old friends; rain had begun to fall and we were all very tired. The Vimy, however, had been overhauled the night before and everything stood ready. Perhaps at the end of the journey we would be more limb-weary, and a single day might discount the success of the venture; so I made up my mind to proceed.

We took off from Heliopolis aërodrome with the cheers of my old war comrades sounding above our engines. For fifty miles we followed the Ismailia Canal to Tel-el-Kebir. The banks were bordered

A GLIMPSE OF THE TEEMING LIFE ON THE CANAL, EL'ASSAR AT BASRA

Photograph by John Clark Archer

This important city, 70 miles from the Persian Gulf, is the port of Bagdad. The type of boat seen here is the bellem. It suggests the lines of a dugout, a canoe, and a gondola. Basra is the chief date port of the world (see text, page 293).

AN ARAB COFFEE-HOUSE IN BASRA

The settled population of Basra is estimated at not more than 50,000, but it has a heterogeneous mixture of all the peoples of the Near East.

Photograph by John Clark Archer

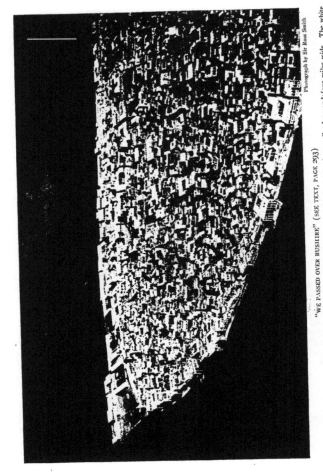

Photograph by Sir Ross Smith

"WE PASSED OVER BUSHIRE" (SEE TEXT, PAGE 293)

This is the most populous port on the Persian Gulf. It occupies the extremity of a peninsula eleven miles long and four miles wide. The white appearance of the city is due to the fact that most of the houses are built of a stone composed of shells and coral.

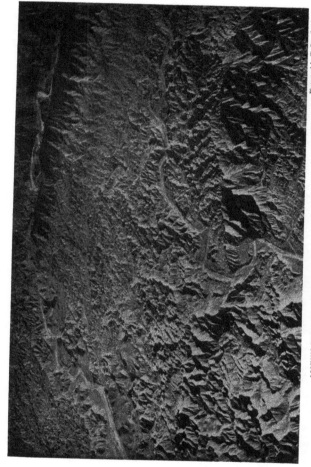

LOOKING DOWN UPON THE RUGGED PERSIAN GULF COUNTRY ON THE PERSIAN SIDE

Photograph by Sir Ross Smith

"Some of the country presents a remarkable sight, and appears as **if** a mighty harrow had torn down the mountain sides into abysmal furrows. Fantastically shaped ridges and razorbacks rise precipitously from deep valleys barren of vegetation and desolate of life" (see text, page 293).

275

THE PERSIAN GOVERNOR, BRITISH CONSUL, AND A CONCOURSE OF NATIVES WELCOMED
THE AVIATORS AT BANDER ABBAS, PERSIA (SEE PAGE 295)

To reach Bander Abbas from Basra necessitated a non-stop flight of 650 miles down
the Persian Gulf along a route that provided no opportunity for a safe landing in the event
one of the motors had failed.

by a patchwork of densely cultivated and
irrigated lands; beyond, arid barrenness,
sand, and nothing.

On the canal the great white lateen
sails of dhows and feluccas in large number resembled a model yacht regatta. It
was all very beautiful and wonderful.
Northward the waterways, canals, and
lakes of the Nile delta stood out like
silver threads woven around the margins
of patches in a patch-quilt, for the sun

had now burst through the clouds, and
all the world sprang into life and light.
From aloft, without the sun, the world
is a gloomy-looking place, doleful and
dead.

Over the famous old battlefield of
1882—Tel-el-Kebir—where Arabi Pasha
suffered ignominiously by the valor of
British arms, even now there is a camp
of British and Indian cavalry.

And soon to Ismailia and the canal that

DELHI, THE FUTURE CAPITAL CITY OF INDIA, AS SEEN FROM THE VIMY (SEE P. 297)

The Ross Smith party reached Delhi thirteen days after leaving London, having flown 5,870 miles. The journey of 2,100 miles from Basra had been accomplished in 25 hours and ten minutes flying time.

links north with south—a straight cut of deep-blue water, running to the horizon transversely to our course—and ahead the gray desert sands, only limited by the blue sky.

Below, a P. and O. steamer, heading south, passes down the Suez Canal. Perhaps she is bound for Australia; she will call in at Adelaide, my home and destination! With a smile, I contrasted the old and the new methods of transportation, and a throb of exultation thrilled us all. Still, we wondered—unspoken the thoughts—who would reach Australia first.

Kantara now lay below us, that vast series of store-dumps—a mushroom city beneath canvas—which had sprung into being since the British occupation of Palestine, and from which practically all commissariat and munition supplies were drawn. As we passed over Kantara, feelings of confidence, mingled with no small satisfaction, filled me. We were now entering upon country I knew as well as my own homeland, for I had spent six months traversing it with the Australian Light Horse before I started flying; furthermore, I had been over the entire air route which now lay before us, as far as Java.

A BRIGHT PROSPECT

The section from Hounslow to Cairo I had always regarded with some trepidation, on account of the winter storms and bad weather. Now we could look forward to improving atmospheric conditions and good aerodromes as far as Calcutta at least. This enabled us to view more rosily the ultimate issue.

Kantara soon lay beyond the rolling eternity of sand which all who served through the rigors and privations of the desert campaign call "Hell." It was somewhere in these regions that the Children of Israel wandered for forty years. Forty minutes in the Vimy was quite

DANCING GIRLS OF DELHI

Great excitement prevailed in Delhi when the Vimy arrived, on the afternoon of the same day that Poulet, the French aviator, had departed in the race half around the world.

sufficient for us. We looked down upon that golden sea of desolation, with only here and there a solitary clump of date palms that boasts the name oasis, and we felt very sympathetic toward the Children of Israel. Two things alone stood out clearly in the wilderness—the iron way, which had been thrust forward to carry supplies from Kantara to the fighting front, and the line of water-main beside it.

We were flying at an altitude of 1,500 feet, so that it was possible to pick out all details readily. As we passed over the old battlefield of Romani, I picked out my old camping site and machine-gun nests.

THE AIR LINE ACROSS PALESTINE

El-Arish, Rafah, Gaza—all came into being; then out over the brim of the world of sand. Gaza from the air is as pitiful a sight as it is from the ground. In its loneliness and ruin, an atmosphere of great sadness has descended upon it. On the site of a once-prosperous town stands war's memorial—a necropolis of shattered buildings. The trenches before Gaza and on the hill Ali Muntar looked

A LAUGHING BOY OF DELHI

"I circled above Delhi to allow the people to see our machine" (see text, page 297).

as though they had been but recently vacated.

Next we passed over the Medjdel aërodrome, and as I gazed down at the marks where the hangars had stood, many memories of bygone days came pleasantly back to me. Soon after leaving Medjdel we ran into dense clouds, and on reaching Ramleh heavy rain began to fall. There was an R. A. F. squadron station on the old aërodrome, and I was sorely tempted to land and renew old friendships, for I had been stationed at this aërodrome for five months at the latter end of the war. However,

this was no joy-ride; so I reluctantly passed over this haven of refuge, and then once more out into the bleak world of storm and rain; but I was much cheered by the whole squadron turning out on to their aërodrome and waving up to us.

THE ENGINES' SONG

My past experiences in Palestine rain-storms steeled me for what was to follow, and from Ramleh to the Sea of Galilee the weather was despicable and smote us relentlessly. The torrential rain cut our faces and well-nigh blinded us.

We were soaked through and miserably cold. One thing only comforted me, and that was the merry song of the engines. Whether "in breeze or gale or storm," they heeded not. On through the rain and wrack they bore us, as in the times of warmth and sunshine, singing their deep-throated song—"All goes well!"

Fortunately I knew the country very well, for after passing Nazareth I had to follow the winding course of the valleys, owing to low clouds, until the Jordan was reached.

The River Jordan presented an extraordinary sight. The main stream has eroded a narrow channel between wide banks, down which its waters meander in an aimless way, zigzagging a serpentine course across a forbidding plain of great barrenness and desolation. A narrow green belt, somber in color as age, pursues the river through the Jordan Valley, which for the greater part is an arid waste, speckled with sparse and stunted shrubs. The river enters the Dead Sea at nearly 1,300 feet below the level of the Mediterranean.

TRAVELING IN THE AIR BELOW SEA-LEVEL

The Sea of Galilee is, roughly, 700 feet above Dead Sea level, and, as we were flying 500 feet above the river, most of our journey through the Jordan Valley was done at an elevation several hundred feet below the level of the ocean.

On reaching the Sea of Galilee the weather improved. As we passed over the great lake, where deep-green waters rest in a bowl encompassed by abrupt hills, strange emotions passed over me, for below us lay a hallowed place—a scene of ineffable charm, peace, and sanctity.

I now headed the Vimy northeast for Damascus and climbed up to 5,000 feet. Occasional cloud patches passed below us, but the landscape for the most part was drear and featureless, save for a line of snow-clad summits that lay away to the north, Mount Hermon and the Anti-Lebanon Mountains.

The flight through Palestine had been an ordeal; extreme weariness gripped us all, for we were still soaking wet and very cold.

Then once more joy filled our thankful hearts when our straining eyes picked up Damascus, a miraged streak on the horizon of a desert wilderness. The streak became irregular. It grew into a band assuming height and breadth, minute excrescences, and well-defined contours. Color crept in; details resolved, developed, enlarged; a city arose from out the waste of sands, an oasis, glorious, magical, enchanting—this was Damascus. A city almost ethereal in its beauty, rearing a forest of slender minarets and cupolas, surrounded by dense groves and woods, had sprung into being, as if by magic, from the Syrian desert.

WE LAND IN DAMASCUS

Although one of the world's most ancient cities, age has dealt lightly with Damascus. From the air it appears no older than the blaze of poplars and cypresses that features the gardens and shades the sun-baked mud-houses and mosques. Beyond the city, beautiful gardens and glades extend, gradually dwindling and blending into the desert spaces. To the north and west rise the multicolored foothills of the Anti-Lebanon Mountains, flanked by the higher peaks with radiant snow mantlings.

Damascus invited and offered a haven of rest. Great was our joy on touching the ground; greater still to be welcomed by old comrades, and to be cared for. The Vimy, too, was looked after. Bennett and Shiers attended to their beloved engines, while I overhauled the controls, and my brother Keith filled up with "shell," to be ready for an early start on the morrow.

A NIGHT IN OLD DAMASCUS

After attending to the machine, we drove in another machine—a Ford—into Damascus and took lodgings at the leading hotel, where the fare was excellent and sleep undisturbed by the parasites common to the country. Damascus is wholly Oriental, though in many ways it is adopting Western fashions and customs. Trams run in the city, and though their speed harmonizes with the indolent habits of the Orient, they seem strangely out of place, as also does the electric light, that sheds its beams of searching and misplaced effulgence in the bazaars and

HONOR GUARD OF SIKHS, BOMBAY, INDIA

These upstanding fellows are born fighting men, and are considered the finest and most loyal soldiers among the native military forces of the British Empire. The Sikh religion rejects alike the teachings of Mohammedanism and Hinduism, but acknowledges one God—the God of all mankind.

THE MAN WITH THE BROOM

The old Portuguese geographers called Ceylon the "utmost Indian isle" during the days of their nation's
ascendancy in India: For more than a hundred years it has been a British possession.

Autochrome by Helen Messinger Murdoch

A MAID AND A BRIDE OF KANDY: CEYLON

The little girl, still a pupil in school, wears white. The brown sari of the young married woman is a
treasured heirloom

III

Autochrome by Helen Messinger Murdoch

DOMES AND SPIRES OF THE MOTI MASJID, DELHI, INDIA.

This mosque was built in 1659 by the Mogul emperor and defender of the faith, Aurangzeb. It is constructed of white marble and its minarets are covered with burnished copper.

IV

285

ELEPHANTS BATHING IN THE MAHAVELIGANGA NEAR KANDY : CEYLON

Autochrome by Helen Messinger Murdoch

Nowhere else do the colors of nature challenge the powers of the autochrome more sharply than in Ceylon. There were six elephants in the group, but the mahouts refused to pose them for less than a rupee a head. As two were better than six, the others were kept out of the picture.

WHERE NATURE SEEMS BENT ON OUTDOING ITS OWN RECORD FOR LUXURIANCE
One who has never wandered through a tropical jungle can scarcely believe how little standing room is left where the bamboos and the palms contest for their places in the sun.

287

Autochrome by Helen Messinger Murdoch

A KANDY CHIEF ATTIRED FOR THE KING'S BIRTHDAY LEVEE

To the occidental eye the Kandy chieftain's robes of office may appear effeminate, but no one will gainsay their picturesqueness.

VII

THE TAJ MAHAL, AT AGRA, INDIA, IN THE EARLY MORNING LIGHT

Autochrome by Helen Messinger Murdoch

This alabaster masterpiece is the most beautiful tribute to a woman constructed by man and was built by Shah Jehan in 1629-1650 as a mausoleum to his favorite wife, the Begum Mumtaz-i-Mahal. It represents the supreme achievement of Mohammedan art.

squalid stalls, where shadow, deep shadow, is essential to effect a successful sale.

I looked out of my window before turning in. A myriad spires, misty and intangible, pointed to a heaven brilliant with stars; a faint breeze drifted in from the desert. The atmosphere was laden with mystery and enchantment. I felt contented. The skies promised sunshine, and henceforth the weather would be good!

Conceive my dismay when, on awakening with the morning, I discovered heavy rain falling; still further was I dismayed to find the aërodrome surface rapidly becoming soft, and the wheels of the Vimy sinking in. As there was no sign of the weather clearing up, we greased our tires to assist their passage through the sticky clay, started up the engines, and, to my unspeakable relief, the Vimy moved ahead.

But the take-off was not lacking in excitement. The propellers sucked up water and mud, whirling it in all directions (we happened to be included in one of them), and so we rose into the air, once more to be cut by the lash of the elements. To my intense relief, the storm did not extend more than a score of miles beyond Damascus.

A MUD VILLAGE ON PALMYRA'S RUINS

We were now heading for Tadmur; again the desert extended before us—a rolling expanse of dreary gray sand over which it was some satisfaction to speed at eighty-five miles per hour. Tadmur is a miserable village of mud huts that has sprung up amidst the noble ruins of ancient Palmyra. The modern bazaars are built for shelter among the ancient columns and fragmentary walls of the Temple of the Sun. These magnificent ruins are the bleached skeletons of a glorious past, austere and dignified even in the squalor and meanness that surround them. From Tadmur the route lay east to Abu Kemal, on the Euphrates.

Shortly after leaving Tadmur we observed an encampment consisting of several hundred black goat-hair tents, and gathered around them were vast herds of camels. As we were flying low at the time, our sudden appearance caused a stampede, not only among the beasts, but

also the occupants of the tents. They decamped, evidently terror-stricken. We subsequently learned that the camels were the spoils of a victorious raid. Perhaps the raiders thought we were the Judgment!

FOLLOWING THE EUPHRATES

On reaching Abu Kemal we turned southeast, following down the course of the Euphrates. It was a pleasant change, after the interminable desert, to pursue the lazy course of the great river and to pass again over fertile tracts and numerous villages.

The most remarkable of these villages is Hit, not only on account of the ancient city which lies buried here, but because there are several bitumen springs, from which this valuable commodity oozes in vast quantities. Practically every native who owns a boat on the Euphrates has copied Noah, who was commanded to "pitch the ark within and without with pitch."

On leaving Abu Kemal we encountered strong head winds, which diminished our speed considerably. I was becoming anxious as to whether we could reach Bagdad before dark, as I was not keen to make a night landing there.

The sun was fast sinking in the west, and as we flew over Ramadie it dipped below the horizon. I decided that there would not be time to do the forty miles to Bagdad before dark. We selected a suitable landing ground among some old trenches, close to a cavalry camp, and landed.

We had landed on the old Ramadie battlefield, which was one of the notable sites of the Mesopotamian campaign. Soon after landing the C. O. of the Indian cavalry regiment came out to greet us, and proffered the hospitality of his camp.

We were delighted to learn there was a small supply of aviation petrol here, and we obtained sufficient to carry us through to Basra without having to land at Bagdad. An Indian guard was mounted over the machine, and the Vimy was securely lashed down for the night.

The O. C. of the 10th Indian Lancers and his staff were thoroughly pleased to meet us, and over the excellent dinner

Photograph by Sir Ross Smith

THE WORLD'S FINEST ARCHITECTURAL GEM SEEN FROM THE SKIES

Of all the remembered scenes in the flight from London to Melbourne, Sir Ross Smith accounts the Taj Mahal, at Agra, India, the most wonderful and beautiful (see text, page 301, and Color Plate VII, facing page 289).

that was prepared I told the latest happenings in London—their home. They were a fine, stout-hearted lot of fellows and greatly we appreciated their hospitality. We felt truly sorry for them, stationed in such a remote, isolated place as Ramadie.

A BLOW IN THE NIGHT

About 11 o'clock that night a heavy windstorm swooped down upon us, and my brother and myself rushed out to the machine. The wind had suddenly changed, and was now blowing hard on to the tail of the machine. The Vimy was in imminent danger of being blown over and crushed.

We turned out fifty men from the nearest camp. They hung on to the machine until we started up the engines and swung her head round into the wind. It was a pitch-dark night, and the gale whirled the sand into blinding eddies, cutting our faces and eyes. One very severe gust caught one of the ailerons and snapped the top balance-wires. This allowed all four ailerons to flap about in a very dangerous manner, and it looked as though they would all be wrenched off before we could secure them.

By weight of arms, however, we eventually managed to secure the ailerons before serious damage was done. At last the machine was turned, facing the wind, and in that position successfully weathered the storm. Throughout the rest of the night the guard hung on to the machine and all stood by.

REPAIRING THE DAMAGE

The storm abated by morning. We found that all the aileron control wires were strained or broken. The sand had choked up everything exposed to the weather, and by the time the damage had been repaired and our tanks filled with petrol it was noon.

For the first time since leaving London we had promise of a good flying day with a following wind. This good fortune atoned for our troubles of the night and for our lack of sleep. We were sweeping along at 100 miles an hour, and in less than thirty minutes Bagdad lay below. Glorious old Bagdad! Bagdad today, faded of all its old glory, is a place of poverty and decay, alluring only through name and association. Yet, in spite of its meanness and squalor, the magic city of Haroun-al-Raschid, the hero of the Arabian Nights, of Aladdin, and Sindbad the Sailor, shall remain immortal.

OVER THE GARDEN OF EDEN

It is hard to believe that the land above which we were now speeding was once the garden of the world. Oh, where is thy wealth and prosperity, fair Babylonia? Despoiled by the ravages of the Ottoman Empire, misruled and wasted by the accursed methods of the Turkish Government, it seems incredible that this void of marsh and waste land was once a country of milk and honey, a land of pomp and luxury that led the civilization of the world.

From a height, the aspect of Bagdad is more inviting than from the ground. A maze of narrow streets, wandering through a tesselated plan of flat roofs, of spires and green splashes of cultivation and date palms, of a great muddy brown river, covered with innumerable little round dots, which on closer investigation resolve themselves into circular, tublike boats—all this is Bagdad, and the impression is pleasing and reminiscent of bygone glory.

A LAND OF MEMORIALS

There is but one thoroughfare that stands preëminent today in Bagdad—a wide road which the Turks had cut through the city to make way for the retreat of their routed army before the victorious British under General Maude; and so now may we see the dawn of a new era and fairer days ahead for this outcast land.

Every mile of land and river above which we were passing was a measure of history of valorous effort, mighty deeds, and heroism. The map of Mesopotamia unrolled before us. Here lay the old battlefields of Ctesiphon, Laff, Tubal, the trench systems still being clearly observable.

Kut el Amara, where was enacted the most dramatic and heroic episode of the Mesopotamian campaign, next came into view. For five awful months that little garrison of British men, led and cheered by their beloved general, had held out against the Turk, disease, and the pangs of starvation. The glorious story of the defense of Kut and the surrender is one of those splendid episodes that thrill the heart of every Englishman, and which shall live immortal with the memories of Lucknow, Delhi, Khartoum, Ladysmith, and Mafeking.

IV. CHASING POULET ACROSS PERSIA AND INDIA

In describing Mesopotamia I am inclined to quote the terse, if ineloquent, account of the British Tommy who wrote of it: "A hell of a place, with two big rivers and miles and miles of dam all between them." Yet the possibilities of development are infinite and the potentialities golden—a land of suspended fertility, where animation and prosperity lie for the time dormant—a wondrous garden, where centuries of neglect and rapine have reaped desolation and bar-

Photograph Courtesy Chas. J. Glidden

A RELIGIOUS GATHERING IN THE COURTYARD OF THE PEARL MOSQUE OF AGRA

This, "one of the purest and most elegant buildings of its class to be found anywhere," was built by Shah Jehan, the same ruler who brought into being the Taj Mahal (see page 290). In the center of the court, here almost concealed by the devout multitude, is a marble pool. During the Indian Mutiny the mosque was used as a hospital.

renness. The land is athirst, but the two great rivers, the Tigris and the Euphrates, move sullenly on, ebbing their life out to the sea. Turn back these tides into the veins of irrigation and the land will be replenished, Eden shall be again, and even the valley of the Nile shall be despised to it.

Exulting in the fair weather and following breeze, we swept over the world at 100 miles an hour. Three thousand feet below, the two great rivers conflux and unite in the Shatt-el-Arab, with the miserable village of Kurnah at the junction—a village built of mud, and its humanity of the same color as the turbid streams that bear the soil of Asia Minor away to the Persian Gulf. Clusters of date palms and a scant belt of vegetation fringe the bank, but beyond a half mile or so there is nothing but the dun-colored wilderness, the miraged sky-line, and the blue canopy where the sun rules king.

All this once was the Garden of Eden. Today it is not a delectable site; but who may speak of the morrow? The waters of the Shatt-el-Arab, heavily impregnated with mud, resemble the outflow from a mud geyser, swirling and boiling; they move oozily forward as their caprice inclines, the silt is precipitated, shallows form, mud-banks come into being, grow into islets, and disappear with the next flood.

BASRA, A HIVE OF ACTIVITY

The flight from Bagdad to Basra took just under three hours. The crazy river barge, probing its way through shallows, rips, and mud, generally takes a fortnight! Basra we discovered to be a hive of activity. It was the main shipping port during the Mesopotamian campaign, and a large military base and aërodrome were still in evidence. The aërodrome stretches to the horizon, and with the British camp extends for miles along the eastern bank (see pages 271-273).

We crossed over to the town in one of the characteristic river boats called mabailas—a Viking vessel strangely and crudely carved at prow and stern, and with sails as multi-patched as the garments of the crew. The town is an unlovely place of strange and variodorous perfumes; so after dispatching mails we hastened back to the Vimy.

As there was a Royal Air Force depot here, I decided to delay a day and allow Bennett and Shiers to overhaul and adjust the engines.

A LONG DAY'S FLIGHT

On the morning of November 23 we made a daylight start for Bander Abbas, 650 miles south. Soon after starting, the sun came up from the distant hills; the world threw off its somber gray, and in dawn's fair raiment became beautiful. The delicate shades of pink that flushed the horizon mounted higher and higher until the zenith grew gay; and so another day of the flight had begun.

The sunlight sparkled on our varnished wings, and the polished propellers became halos of shimmering light. Our engines sang away merrily. The Vimy ceased to be a machine and pulsed with life, as if feeling the glory of the morning; my brother scanned the landscape below, plotting off the course on the chart and checking our position from time to time by villages and salient features, remarking how wonderfully accurate the world was created!

Bennett and Shiers had stowed themselves away in the after cockpit and were reclining inside the fuselage with the spare parts, endeavoring to secure well-earned rest from their strenuous efforts of the past few days. As the spare parts crammed all available space, theirs was painful comfort indeed. The dimensions of our front cockpit were of those adequate proportions generally attributed to wedges. The weather continued fine, but for the most part the flight was uninteresting and monotonous.

We passed over Bushire and several coastal villages, but the only really impressive sight was the ruggedness of the coastal belt and the hinterland ranges. Some of the country presents a remarkable sight, and appears as if a mighty harrow had torn down the mountain sides into abysmal furrows. Fantastic-shaped ridges and razorbacks rise precipitously from deep valleys barren of vegetation and desolate of life. Occasionally we passed over small flat plains dotted with abrupt hills and flat tabletops. The whole earth appeared as though some terrific convulsion had swept it and left

ONE OF THE FAMOUS GHATS, OR LANDING PLACES, ON THE BANKS OF THE GANGES AT BENARES, THE HOLY CITY OF THE HINDUS

The flyers passed over this great city, noted alike for its age, its shrines and temples, and for its innumerable host of deformed beggars, who chatter its streets, soliciting alms from rich Hindus who elect to spend their declining years in its holy atmosphere (see page 304).

in its wake this fantastic chaos of scarred mountains and gouged valleys.

DOWN 'THE PERSIAN GULF TO BANDER ABBAS

In striking contrast, the shores of this wild scene are washed by the stagnant waters of the Persian Gulf. The coloration of this phenomenal panorama was equally bewildering. The dead expanse of the Persian Gulf, mingled with the mud of the rivers, was an exquisite shade of green, patched here and there with darker areas, where the wind had caught it up into ripples.

Mountainward, the first impression was that all had been molded in yellow clay. A closer survey showed streakings and strata of infinite shades, of which the rust color of ironstone appeared dominant. At intervals the dry beds of waterways cut well-marked defiles from the high mountains to the sea. They stood out like roadways winding through the maze and seeming as if blasted out by the hand of man.

Throughout this terrible country I scarcely observed a possible landing ground, and had our engines failed us it would have meant either crashing or else an immersion in the Persian Gulf. So it was with no small relief that I brought the Vimy to a safe landing at Bander Abbas, where a hearty welcome was extended to us by the British Consul, the Persian Governor, and a great concourse of interested natives.

Although dog-tired, I could not sleep that night. The coming day's trip, I hoped, would enable us to reach Karachi in a non-stop flight of 730 miles. The distance did not perturb me in the least, but the treacherous country and the isolation from civilization in case of a forced landing, and another long stretch of detestable mountain-scored country was in itself enough to give one a nightmare.

The British Consul had prepared an ostentatious-looking document which we were to carry. It commanded the murderous tribes which infested the country to treat us kindly, in case we were compelled to land among them!

Fortune favored us once more with a following breeze and excellent weather. The country was a repetition of that passed over the previous day, and, with the morning sunlight striking aslant, heavy shadows gave the scene the semblance of a mighty rasp.

The engines were perfectly synchronized, and roared away harmoniously; but it is imperative for the pilot to watch every part of his machine, especially the engines. As I sat there hour after hour, I found myself automatically performing the same cycle of observation over and over again.

My supreme difficulty was to keep my sleep-heavy eyelids from closing and my head from nodding. First of all I would look at my starboard engine and see that the oil-pressure gauge and revolution counter were registering correctly; then listen to hear if the engine was firing evenly. Next, glance over the engine and oil-pipe connections and check off the thermometer which indicated the water temperature in the radiators. The altimeter, air-speed indicator, and petrol flow indicator in turn claimed attention.

I would then look up to the port engine and go over the instruments and engine as before; then over the side to scan the landscape, and ever keep an alert eye for a suitable spot in case of a forced landing.

By the time I had completed this performance it would be time to start all over again. When flying over interesting country the monotony of this ceaseless routine is relieved, but where flying over country such as the present stage the only mental stimulus that buoyed us up was the anticipation of rosier times ahead. Often our thoughts were of Poulet, who was somewhere ahead, and we wondered if ever we would catch up with him.

ENTERING THE AËRIAL GATEWAY TO INDIA AFTER A 730-MILE FLIGHT

Frequently we passed over small villages, and our advent instilled terror into the inhabitants and their animals.

For the last 100 miles we left the coast and flew on a compass bearing direct for Karachi, and so we entered the aërial gateway to India after a non-stop flight of eight and a half hours.

The usual procedure of overhauling the engines and machines and refilling our

A HARBOR SCENE IN CALCUTTA: MORE THAN HALF WAY ON THE LONDON-TO-AUSTRALIA FLIGHT

It was at the race-course outside the city of Calcutta that one of the most terrifying incidents of the trip occurred. Two hawks, flying against the blades of one of the aeroplane's propellers, jeopardized the success of the whole adventure (see text, page 305).

tank with petrol had to be carried out before we could seek rest. This took from three to four hours; and as we had a flight of 750 miles to do in order to reach Delhi next day, it was necessary for us to put 369 gallons of "shell" into the machine. This petrol was in four-gallon tins, which meant that we had to handle, lift up, and filter ninety tins, or well over a ton of petrol (see page 233)

NO ROMANCE IN POURING PETROL

My brother and I generally filled the tanks, while Bennett and Shiers worked on the engines. It was not much fun, after piloting the machine for eight and a half hours in the air, to land with the knowledge that we had to lift a ton of petrol, besides doing innumerable small jobs, before we could go off to rest.

In addition, we had to run the gauntlet of functions and ceremonies, and it was difficult to make folk understand that work had to be done. We deeply appreciated every one's generous kindness, but I fear that on some occasions people must have thought me very discourteous.

The first news that greeted us on landing at Karachi was gratifying. Poulet was at Delhi, only a day's flight in the lead! This was a great surprise, for we fully expected that he would be well on his way to Singapore. From now onward added zest would be given to the flight, for I intended to pursue the chase in keen earnest. Already I considered the race as good as won, for the Vimy was superior both in speed and range.

We had hoped for a good rest at Karachi, but the local Royal Air Force officers had arranged a dinner, and it was not before "the very witching hour" that we turned to bed. Three hours and a half later we were called to continue the flight. This was to be the longest non-stop we had undertaken. Nine hours' flight should land us in Delhi, 750 miles away.

ABOVE THE INDUS

After circling above the aerodrome we turned east, heading straight into the golden sun that was just rising above the horizon. A low ground haze that changed into a golden mist as the sun mounted higher hid the earth from view. Pass-

ing over Hyderabad, the vapors rolled away and we had a grand view of the River Indus. Once more we entered the monotony of the desert. For the next three hours we flew steadily onward, pursuing the railroad track across the dreary Sind Desert.

It was a joy to reach Ajmere, a delightful little city, beautifully situated in a basin of green hills. The country beyond is for the most part flat—a vast verdant carpet irrigated from the great rivers. Practically from the time we had reached the African coast, when on our way to Cairo, the flight had been across deserts or desolate lands. Now the new prospect that opened ahead invited and attracted.

During the afternoon flying conditions became very boisterous, and the turbulent atmosphere tossed the Vimy about like a small vessel in a heavy sea. This I also accepted as a welcome diversion, for the flights of the past few days had cramped me in one position, and now I was kept actively on the move keeping the machine straight and fighting the air-pockets and bumps into which we plunged and fell.

DELHI IN THE DISTANCE

We first noticed Delhi from fifty miles distance—a white streak in a haziness of green plain. Quickly details became apparent, and soon the streak had germinated into a considerable town.

I circled above Delhi to allow the people to see our machine, which had established a record by arriving thirteen days after leaving London—a distance of 5,870 miles. We climbed crampily out of the machine and were welcomed by General McEwan, the Royal Air Force chief in India, and many other old friends.

I regretted that I was quite unable to reply to their kindly expressions, as I did not hear them. The roar of the exhausts for nine consecutive hours' flying had affected my ears so that I was quite deaf.

After several hours my hearing returned, and it was to learn that Poulet had left the same morning for Allahabad. Great excitement prevailed, for one aëroplane had departed and another

THE LOFTY GOLDEN PILE OF THE SHWE DAGÓN, THE CENTER OF BURMESE
RELIGIOUS LIFE IN RANGOON

In no city along the route to Australia was the arrival of the Vimy awaited with keener interest than in Rangoon, the capital of Burma and third seaport of the Indian Empire. When the first news of the departure of the aviators from London appeared in the Rangoon papers a large crowd of natives straightway assembled on the race-course, expecting to see the aëroplane arrive in a few hours. As the time for the actual arrival drew near, people began to congregate, bringing their food and beds, intent upon holding a festival for the duration of the stay of the airmen. The Vimy was the first aëroplane Rangoon had ever seen. An hour after its arrival, Poulet, the French flyer, arrived in his Caudron (see text, page 307).

Photograph by E. D. McDowell

ONE OF THE SHRINES OF THE SHWE DAGÔN PAGODA: RANGOON, BURMA

This, the finest and most venerable place of worship in all Indo-China, attracts pilgrims from Cambodia, Siam, Korea, and Ceylon, as well as from all Burma.

had arrived on the same day, both engaged in a race half-way around the world! After attending to the machine we dined at the R. A. F. mess, thoroughly tired but extremely happy. Half the journey was completed and Poulet was within range.

FROM BASRA TO DELHI

We had left Basra at 6 a. m. on November 23 and arrived at Delhi fifty-four hours later, covering a distance of 2,100 miles. Out of the fifty-four hours we

had spent twenty-five hours ten minutes actually in the air, and in the balance we had overhauled the engines and machine twice, and had by our own efforts filtered two and three-quarter tons of petrol into the machine. I had intended pushing on to Allahabad next day, but on arrival at the aërodrome we were feeling the effects of the past strenuous days so severely that I decided that rest was imperative. We took it—in the form of the proverbial change of work—and, putting in six good hours on the machine, made every-

Photograph by E. B. McDowell

THE BOAT MARKET ON THE ME NAM RIVER, BANGKOK, SIAM

Until modern times Bangkok was built largely on floating pontoons and on piles at the edges of many canals, and it still faintly suggests an oriental Venice.

thing ready for an early morrow start. Toward evening my brother and I drove into the city sight-seeing.

As I had been to Delhi during my flight to Calcutta with General Borton, I played the guide, and an enjoyable ramble through this future capital diverted our thoughts from the Vimy for the moment and enabled us to relax.

Further diversion, with less relaxation, was provided by the native driver of a car we hired. In the language of the realm in which we had been living, he navigated full out and nearly crashed us on several occasions, in his desire to show what a pilot he was. I declare that I "had the wind up" far more often on this bit of journey than during the whole flight. However, the casualties were few and the fatalities nil, and we paid him off at the R. A. F. quarters.

At 4.30 next morning I tumbled stiffly out of bed on the insistence of a Yankee alarm-clock. Oh for another day off! But by the time the others had uncoiled and emerged into the early Indian dawn,

Photograph by E. B. McDowell

WHERE THE BANGKOK HOUSEWIFE BUYS HER VEGETABLES

In 1769 Bangkok was a mere agricultural village on the banks of the Me Nam; today it is a splendid capital with 500,000 inhabitants.

I felt again the keenness of the chase. A friendly R. A. F. pilot came up in a Bristol fighter and flew with us for a few miles along the course of the Jumna.

Half an hour later the oil-gauge surprised us by setting back to zero, and we made an unexpected landing at Muttra, to find that it was happily only a minor trouble—the slipping of the indicator on its spindle. And so into the air once more, and on to Agra—Agra the city of the Taj Mahal.

Of all the remembered scenes, wonderful and beautiful, that of the Taj Mahal remains the most vivid and the most exquisite. There it lay below us, dazzling in the strong sunlight—a vision in marble. Seen from the ground, one's emotions are stirred by the extraordinary delicacy of its workmanship. Viewed from 3,000 feet above, the greater part of its infinite detail is lost, but one sees it as a whole. It lies like a perfectly executed miniature or a matchless white jewel reclining in a setting of Nature's emeralds (page 290).

Photograph by E. D. McDowell

ONE OF THE MANY GROTESQUE STATUES OF THE TEMPLE OF WAT CHANG: BANGKOK

It seems incongruous to think of this sort of art with electric lights, but Bangkok is well illuminated, and the Vimy's mechanics worked all night by electric lamps in this city, grinding the valves of their engines.

We hovered lazily around, exposed our photographic plates, and swung off on our course. In the vastness of space through which we were speeding, the magnificent monument became a toy . . . a mote . . . a memory. New scenes, villages, and towns rose from the unreachable brink ahead, grew into being, passed beneath, then out over the brim of the world behind us.

We were crossing the vast plains of central India, a great flat tessellation of cultivated patches that gave an impres-sion of the earth being covered with green, brown, and golden tiles. These multitinted patches were framed with brimming channels carrying the irriga-tion waters from the great river.

A BULL ATTACKS THE VIMY

Allahabad was reached after four and a half hours, and we eagerly but vainly searched the aërodrome for a glimpse of Poulet. There were several hangars on the aërodrome, however, and we thought that his machine might be under cover.

Photograph by Theodore Macklin

DANCING GIRLS OF SIAM

The Siamese accorded the aviators the warmest hospitality upon their arrival at the Don Muang aërodrome, 12 miles north of Bangkok, the national capital.

On landing we were informed that he had left that same morning for Calcutta.

There is a considerable garrison stationed at Allahabad, and the commandant, fearing that a deformity might overcome us through being cramped hand and foot in the machine, had arranged an active function called a Jazz. Unfortunately, we were unable to test the efficiency of this form of exercise, though we appreciated the thoughtful hospitality of our hosts.

There was great excitement at the aërodrome next morning. While we were taxi-ing to the far end, preparatory to taking off, a fine bull broke on to the ground, and as we swung round to take off he charged head on toward the machine. The position, though ridiculous, was extremely hazardous. No doubt, to quote the celebrated railway engineer, it would have been "bad for the coo," but

a collision would also have been extremely "bad for the Vimy."

I frightened him for the moment by a roar from the engines. Evidently he took the roar for a challenge, and stood in front of the Vimy, pawing the ground and bellowing defiantly. At this point a boy scout rushed out from the crowd to move the monster, and, much to the amusement of ourselves and the crowd, the bull changed his intention and turned on the hero. Our brave toreador retreated to the fence, pursued by the bull.

THE ARRIVAL AT CALCUTTA

We took advantage of the diversion and made a more hurried ascent than usual. What became of the scout I do not know, but as we circled above I noticed that the bull was still in sole possession of the aërodrome.

Once more pursuing the course of the

Photograph by Albert L. Gurloy

A TRAVELING RESTAURANT IN SINGAPORE, MALAY PENINSULA

The successful landing of the Vimy at Singapore was accomplished by a resort to acrobatics. The improvised aerodrome being too small for the landing and "take-off" of a machine as large as the Vimy, one of the mechanics clambered out of the cockpit and slid along the top of the fuselage down to the tail-plane. His weight dropped the tail down quickly, with the result that the machine came to a halt within 100 yards after touching the ground (see page 322).

Jumna as far as Benares, we headed southeast and followed the railroad to Calcutta. Forty miles north of Calcutta we came above the River Hooghly.

Here and there factories and jute mills came into view, with villages clustering around them. The villages grew dense and became the outskirts of a great and expansive city—a mighty congestion of buildings, white, glaring in the sun; green patches and gardens, thoroughfares teeming with people, a vast fleet of shipping, of docks and activities—and Calcutta slipped away beneath us.

Thousands of people had collected on the race-course, at the far side of the city, to witness our arrival, and when we landed it was with great difficulty that

© Photograph by F. O. Koch

A PRIMITIVE OIL MILL USED FOR PRESSING COCONUTS ON THE MALAY PENINSULA

While flying along the Malay Peninsula, the Vimy successfully outrode a monsoonal storm after several narrow escapes from disaster (see text, page 315).

the police kept back the multitude of natives that surged around the machine. A barrier was at last placed around the Vimy, and soon we became the center of a compact mass of peering faces, all struggling to get closer and obtain a better view. The elusive Poulet, we learned, had moved off the same morning for Akyab.

PAST THE HALF-WAY MARK

That night, after the usual overhaul of engines and filling up with petrol, we stayed with friends and slept well. We had crossed India and were now more than half-way to Australia.

Our departure next morning from Calcutta was marked by an incident that to the layman may sound insignificant, but it might easily have spelled disaster to us. Thousands of natives and a great many distinguished white people came down to see the start. The race-course is really too small for a machine as large as our Vimy to maneuver with safety, and I was a trifle nervous about the take-off; but

the surface was good, our engines in fine trim, and she rose like a bird.

A NARROW ESCAPE FROM DESTRUCTION BY HAWKS

Then came our narrow escape. A large number of kite hawks were flying round, alarmed by the size and noise of this new great bird in their midst. When we had cleared the ground by about ten feet two hawks flew across us at an angle; they seemed to become confused and turned straight into us, one striking the wing and the other flying straight into the port propeller. There was a crash as if a stone had hit the blade, and then a scatter of feathers.

It may not sound very dreadful—except for the hawk—but as a matter of fact it was a breathless, not to say a terrifying, moment, for we fully expected to hear the crash of broken propeller blades.

We were at the time flying straight for the high trees, and, had the propeller broken, nothing could have saved us from a terrible crash. However, more hawks were circling about, and in endeavoring

MALAY CHILDREN OF SINGAPORE

When the aviators reached Singapore they found the heat intense. Coming from the cold
of the English winter, they felt it severely (see text, page 322).

to avoid them I almost crashed the machine on the tree-tops. By a very narrow
margin indeed we cleared them, and I
was deeply relieved when we had climbed
to 1,000 feet and were clear of the pesti-
lent birds. I marveled that our propeller
stood the impact, for a very trifling knock
will cause the disruption of a propeller
when running "full out," and so in an
extremely high state of tension. (I have
known so tiny an object as a cigarette
end thrown carelessly into a propeller to
cause the whirling blades to fly to pieces!)
On looking over the machine I noticed
one of the hawk's wings had become
pinned in the rigging, and we secured it

after the day's flight as a souvenir of a
hairbreadth escape.
Calcutta marked the completion of the
second stage of our journey, and from
now onward the route would be much
more difficult and hazardous. We had
had the benefit of R. A. F. aërodromes
and personnel at almost every landing
place, but henceforth we would have to
land on race-courses or very small aërodromes. Also, I knew that the only possible landing places right on to Port Darwin were at stated places hundreds of
miles apart, and that in the event of engine trouble our chances of making a safe
forced landing were very slender.

I had originally intended flying from Calcutta to Rangoon race-course in one flight, but as the next day, November 29, was a Saturday, and I was informed that a race meeting would be held at Rangoon on that day, I decided to stop at Akyàb.

We were now passing above a dreaded span of country, the sundarbans, where engine trouble would have meant the undoing of all our efforts and labor. The mouth of the Ganges here frays out into a network of streams, producing a jigsaw of innumerable islets and swamps. We breathed much more freely after we had reached Chittagong, a place I had reason to remember well, through having spent four days there the previous year, when our ship caught fire and was blown up (see page 230).

From Chittagong we followed the coast - line of Burma, and eventually reached Akyab. My brother peered over the side as we circled above the aërodrome and showed symptoms of great excitement, while Bennett and Shiers waved joyfully from their cockpit and pointed down to the ground. They indicated a small machine near the center of the field. It was Poulet!

V. THROUGH CLOUD AND MONSOON TO SINGORA

Poulet was the first to greet us on landing. He came forward with a cheery smile and outstretched hand—a true sportsman, the hero of a gallant and daring enterprise. I was deeply interested in inspecting Poulet's machine, which was drawn up alongside the Vimy. In proportion, the contrast was reminiscent of an eagle and a sparrow. The Vimy towered above the tiny Caudron, which appeared altogether too frail and quite unsuited for the hazardous task these two courageous fellows had embarked upon. I had a long talk with Poulet and his mechanic, Benoist; they made fun of their adventures and intimated that theirs was a novel and exciting method of touring the globe.

We agreed to fly on together the next day to Rangoon, but when morning arrived, as we still had some work to complete on the machine, Poulet set off, and by the time we were ready he had an

hour's lead. No aëroplane had ever landed at Rangoon before, and naturally I was very keen to win the honor for the Vimy. For the first 100 miles I followed the coast-line southward and did not observe a single landing place in case of necessity. The coast, for the most part, fringes out into vast mangrove swamps, while farther inland the country becomes mountainous, with rice-fields checkering the valleys and every available irrigable area. The hills are densely wooded and very rugged.

Flying east, we crossed a low mountain chain, and on the other side found the Irrawaddy River. We followed down its course as far as Prome. From here the railroad guided us on to Rangoon. I had no difficulty in locating the landing ground—the race-course, a green patch framed by a compact ring of cheering humanity.

THE RACE TO RANGOON

We came to earth midst tempestuous cheering, and were welcomed by the Lieutenant - Governor of Burma, Sir Reginald Craddock, and Lady Craddock. We were told that no race meeting had been so well attended as the present, nor had the betting been so widespread. The multitudes had massed to witness two aëroplanes racing half-way across the globe. To them the race was more than novel; it was a great event in their lives, for few indeed of the vast assemblage had ever seen an aëroplane.

As flying conditions from Akyab had been boisterous, we in our high-powered machine had a great advantage over Poulet, and in spite of the hour's handicap at the outset, we succeeded in reaching Rangoon an hour ahead of him. Poulet's arrival was the signal for another burst of cheering, and he was welcomed no less warmly than ourselves.

The police experienced great difficulty in clearing the race-course that evening, as many of the natives had brought their food and beds, intent on holding a festival for the duration of our stay. I was told that when the first news of our departure from London appeared in the local papers, and the fact that we intended calling at Rangoon became known, a large crowd of natives straightway as-

THE STRINGS AND BAMBOO USED TO SCARE BIRDS FROM THE RICE-FIELDS OF SUMATRA RESEMBLE THE WIRES OF A RADIO STATION

From a primitive watch tower (see upper illustration on page 310) long strings are run to all parts of the field. To these strings are attached light bamboo poles and bits of cloth. Whenever feathered marauders appear the watcher in the tower strikes the pole to which a string is attached, causing the latter to dance and thus frighten the birds from the grain.

sembled on the race-course, expecting to see us arrive in a few hours. Later, when the news of our reaching Akyab was noised abroad, a multitude camped overnight on the race-course, so as to make sure of witnessing our arrival.

A GOOD OMEN

That night we were the guests of Sir Reginald and Lady Craddock, who did everything possible for our comfort and insisted that we should go to bed early. It was the first time such a suggestion had been made to us, and, as we were very weary, we deeply appreciated their kindly consideration.

There is a strange lizard in the East which makes a peculiar noise, like "tuk-too," and it is a popular superstition that if one hears this sound repeated seven times, good luck will follow. That night, just before going to our room, a lizard "tuk-tood" seven times. The omen was good and we slept peacefully.

We had arranged with Poulet to start off together next morning and keep company as far as Bangkok. The Vimy was considerably faster than the Caudron, but by throttling down and maneuvering, it would be possible to keep together. The way to Bangkok lay across high ranges and dense jungle, and the mutual advantage in making the journey together over this unfrequented and practically unknown country, should a forced landing have to be made by one of us, was obvious.

Traffic fills the highways before sunrise in the East, and a considerable portion of it was moving toward the race-course. A great crowd of interested natives swarmed over the aerodrome, and the police and troops were already busily engaged clearing them off prior to our departure. We started up the engines, took leave of our kind friends, and waited for Poulet. Poulet had some difficulty with his machine; and as it was a warm morning and our engines were beginning to get hot, I took off, intending to circle above the aerodrome until Poulet arose on the wing.

The take-off was not without a thrill. As a matter of fact, to this day it is a mystery to me that we ever left the ground. The race-course was much too small for so large a machine as the Vimy and heavily laden as it was. It had barely attained flying speed when a fence loomed up in front of us. The Vimy just scraped over, but ahead were trees and buildings. I acted instinctively. The undercarriage brushed the tree-top, and danger was past. It was over in a breathless moment; but had the machine been but a single foot lower, disaster must have overtaken us. How slender is the cord that holds success from failure!

I circled above the race-course for twenty minutes; but, as Poulet had not yet left the ground, I concluded that he must be experiencing engine trouble, and so reluctantly we had to push off without him.

We flew due east to Moulmein, immortalized in Kipling's famous ballad, "On the Road to Mandalay," and as no aeroplane had ever flown above this land before, Sergeant Shiers, in words worthy of the great poet, said it was fine to be flitting through air that had never smelt a blanky exhaust!

LOST IN CLOUDS ABOVE THE MOUNTAINS

The maps we carried of this country were very poor and sadly lacking in detail, but they indicated that a 7,000-foot mountain range had to be crossed before reaching Bangkok.

After leaving Moulmein we headed southeast over country rapidly becoming mountainous; but, instead of encountering lofty summits, a mighty cloud bank, that seemed to reach to heaven and bar the entire prospect in the direction of our course, extended before us. The monsoon season was now due, and I concluded that this would be one of the initial storms. Somewhere in that dread barrier lay the high peaks over which we must cross, and I admit that I was afraid of the prospect. As time wore on, the storms would grow in frequency and intensity, so I decided to plunge ahead.

The clouds rested down to 4,000 feet, and we were flying just beneath them. Somewhere ahead lay the mountains that had to be crossed, rearing their summits another 3,000 feet higher. Our maps indicated a pass which we tried to find, and so we started off along a deep valley. At first it looked hopeful, but after five min-

A BATTAK WOMAN AND BOY DRIVING BIRDS OUT OF THE RICE-FIELDS NEAR MEDAN,
SUMATRA (SEE ILLUSTRATION ON PAGE 308)

EAR-RINGS WORN BY BATTAK WOMEN NEAR MEDAN, SUMATRA
See "By Motor through Sumatra," in THE GEOGRAPHIC for January, 1920.

A FAMILY GROUP ON THE VERANDA OF A BATTAK HOUSE NEAR MEDAN, SUMATRA

The aviators "bumped across" the Equator off the eastern coast of Sumatra shortly after leaving Singapore.

utes' flying the cliffs narrowed in, and, fearing I might be trapped in a tapering dead end, I turned the Vimy about. There was just sufficient room in which to effect the maneuver.

After a consultation with my brother, we agreed that our safest course was to climb above the cloud-mass or at least to an altitude sufficiently high to clear the mountain tops, and barge our way through the mist. At 9,000 feet we emerged above the first layer; but eastward the clouds appeared to terrace up gradually, and in the distance there ex-tended still another great wall, towering several thousand feet higher.

OUR MACHINE REACHES ITS "CEILING"— 11,000 FEET

Before starting off over this sea of clouds, my brother took observations with the drift indicator, and we found to our dismay that we would have to fight into a twenty-mile-an-hour head wind. He gave me the compass bearing to fly on, and away we went once more, with the world lost to view beneath us. It reminded me of our first day over

FLYING ACROSS JAVA

"With beautiful weather favoring us, we sped rapidly over fertile tracts of this amazing island. Java impressed me as one vast bounteous garden, amid which rise the immense shapely cones of volcanic mountains" (see text, page 325).

Photograph from Sir Ross Smith

France; but the weather was not so cold, so we felt physically more comfortable. The map showed the range to be about fifty miles wide, and after we had flown for half an hour, still another cloud barrier appeared directly ahead.

Our machine had now reached its "ceiling," so there was no alternative but to plunge ahead into the mist. We were then flying at an altitude of 11,000 feet, and were soon engulfed in a dense blanket of mist. As we had left England hurriedly, there had been no time to fit special cloud-navigating instruments, and the only ones we carried for this purpose were the ordinary compass, air-speed indicator, and inclinometer. Any one who has flown through clouds in a big machine, under similar circumstances, will appreciate my feelings at this time. Down below us lay jagged mountain peaks buried by cloud. Ahead, around, and behind, the mist enfolded us in an impenetrable screen, and if I once allowed the machine to get beyond control, a horrible fate would be waiting for us all below.

WHAT FLYING IN CLOUDS MEANS

To those who have not experienced the anxiety of cloud-flying, I will attempt to describe briefly what happens.

The moment one plunges into heavy cloud there is misty blankness; all objects are lost to view; and as time wears on, a helpless feeling grows upon one that all sense of direction is lost. To overcome this predicament, I was provided with the aforementioned instruments, and settled down to try to watch all three at once and maintain their readings correct. In addition it was necessary to glance over the engine and the gauges continually.

At first all went well; but, while turning to check over an engine, I apparently and unconsciously, with the natural movement of my body, pushed one foot, which was on the rudder bar, slightly forward. This turned the machine off its course, and when next I looked at my compass I was ten degrees off course. I then kicked on the opposite rudder to bring the machine back; but, as the Vimy is much more sensitive to respond than the comparatively sluggish compass-needle, I found that I had put on too much rudder. The result was that when the compass-needle started to swing it did so through an angle of forty-five degrees.

In my attempt to correct the course and bring the needle back on to its correct reading, I glanced at the air-speed indicator and found it registering over one hundred miles an hour—twenty-five miles above normal flying speed. This meant that I must have pushed the nose of the machine down. The inclinometer indicated that the machine was not flying laterally correct; in fact, we were flying at an inclined angle of forty degrees.

I realized that the machine was slipping sideways, and that if I did not get matters righted at once, the machine would get out of control and go spinning down to earth.

LONG AND ANXIOUS MOMENTS

It is useless attempting to describe how I acted. A pilot does things instinctively, and presently my instruments told me that we were once more on our course and on an even keel.

All this took but a few seconds; but they were anxious moments, as a single mistake or the losing of one's head would have been fatal. This happened several times, and at the end of what seemed hours I glanced at my watch and found we had only been in the clouds for twelve minutes! Perhaps my nerves were a little ragged, owing to strain and lack of sleep during the past fortnight; but I felt at last that anything would be better than going on under these tense and nerve-racking conditions.

It was now an hour since we first started across the clouds, and both Keith and I concluded that we must surely be across the mountain range. So I decided to take the risk and go lower and "feel."

Shutting off both engines, we glided down, and I held up the machine so that we were going as slowly as possible—only about forty miles an hour.

The sensation was akin to the captain navigating a vessel in uncharted shoaling seas—expecting every moment to feel a bump. Lower and lower we went—ten, nine, eight thousand feet—and then we both anxiously peered over the sides—straining for a glimpse of hidden peaks.

NATIVES DIGGING THE WHEELS OF THE VIMY FROM THE MUD AT SOERABAYA, JAVA

The bamboo-matting runway, constructed to enable the Vimy to "take-off," was composed largely of the coverings of the native huts. Entire villages were stripped of their roofs to provide this material (see pages 316-317 and text, page 327).

As we approached the 7,000-foot level, which I knew to be the height of the range, we huddled together and held on tight, in anticipation of the crash! I noticed a small hole in the cloud, with something dark beneath. It was past in a flash, but instantly I pushed the throttle full open and flew level again. At first I thought it was the top of a dreadful peak, but on further consideration I remembered that in my brief glance the dark patch had looked a long way down.

A GLORIOUS WORLD BURSTS INTO VIEW

Once more I shut off and went lower, and as we had not hit anything by the time we reached 4,000 feet, I concluded that the range had been crossed.

A few minutes more and we burst out into full view of a glorious world, carpeted with trees, 1,500 feet below. The sudden transformation was stunning. It was an unspeakable relief—the end of an hour that was one of the veriest nightmare experiences I have ever passed through.

Before our bewildered gaze there stretched a dark-green forest, only limited by the distant skyline. Here and there the dark green was splashed with patches of bright-colored creeper, and in spite of the fact that there was not the vestige of a possible landing place, it was beautiful and a welcome relief. Later, the Siamese told us that all this country was unexplored.

The country now began to fall away gradually to the east; the hills became less rugged and petered out into undulating, yet heavily wooded, jungle. An hour later and we reached the Mekon River and the haunts of man. Small villages lay scattered along its banks and wide expanses of irrigated lands verdant with rice crops.

Following downstream, we landed at

Don Muang aërodrome, twelve miles north of Bangkok, after a flight that will live long in my memory. Don Muang is the headquarters of the Siamese Flying Corps. They have several hangars, a number of machines, and up-to-date workshops. During my visit to Siam the previous year I had been to Don Muang, so that on landing I found myself among friends. We were met by the British Consul General, Mr. T. H. Lyle, with whom I had stayed on my previous visit and who now rendered us valuable and appreciated assistance.

The Siamese also displayed the warmest hospitality, and the commandant very kindly placed his own bungalow at our disposal. It was found necessary to re-grind the valves on two of the cylinders of the starboard engine; and, as this was a lengthy job, Bennett and Shiers worked all night to complete it, so that we might keep to our usual scheduled starting time. An electric lamp was rigged up over the engine, and all the flying ants and insects in Siam collected around it, which greatly added to the discomfort and hindrance of the work.

My original plan was to fly from Bangkok to Singapore, roughly 1,000 miles, in one flight; but as I learned there was a good aërodrome at Singora, about halfway, with 500 gallons of petrol depoted there, and as I was anxious to reserve the machine as much as possible, I decided to land at the latter place.

We left Bangkok in good weather, and were escorted for the first fifty miles by four Siamese machines. For the first hour the flying conditions were ideal, with a good following wind helping us; then ahead again lay our old enemies, the clouds. At this time we were flying along the coast, so did not deem it necessary to climb above them. The clouds became lower and heavier and soon we found ourselves only 1,000 feet above the sea.

CAUGHT IN A MONSOONAL STORM

Ahead we saw the rain, and I dreaded what was to come. While we were over the sea, with the land on our right, there was comparatively little chance of our crashing into anything. This was fortunate, for in a few moments we were soaked through, our goggles became saturated, and all vision for more than a few hundred yards or so was obliterated. The rain came down literally like a sheet of water, and, as we had to remove our goggles and maintain a constant lookout ahead, we were almost blinded by the rain lashing our unprotected eyes.

At this time we were doing ninety miles per hour, and as the torrential rain dashed against us and the machine it pattered and smote like hail. Narrowing my eyes down to slits, I peered out ahead as long as I could endure it; that was but a few minutes. I then tapped Keith to keep the watch while I rested my eyes; then, when he could see no more, I would "carry on" again. So it went on for the best part of three hours. Fortunately, this heavy rain was not continuous, but the squalls which we went through at frequent intervals generally took ten minutes to pass.

RACING THE STORM

Still another difficulty presented itself. As long as we were flying south, the strong wind helped us; but as we had to follow the coast-line in detail, and there were many bays and headlands, we frequently found ourselves fighting right into the teeth of the gale to get out of a bay or weather a headland.

I was afraid to go inland, as the rain only allowed us limited visibility. Once we almost crashed on to a hill, which suddenly loomed up through the rain ahead. I just had time by a hair's breadth to pull the machine around in a climbing turn and go farther out to sea. I have never experienced worse flying conditions, and had it been at all possible to land, I gladly would have done so.

All the flat stretches along the coast were paddy-fields under water. We were wet and miserable, and the thought oftentimes came over me of what an ignominious end it would be if we had engine trouble and were forced to land in a paddy-field of mud and water. I wondered at our marvelous engines—through the snows of France, the blaze of the tropics, and through these terrible rains, they still roared merrily on.

An hour before reaching Singora we passed through and outstripped the storm. As the clouds were still low, we kept our

HAULING THE AËROPLANE ONTO THE BAMBOO TRACK: SOERABAYA, JAVA

Photograph by Sir Ross Smith

At first the pathway of mats was merely laid on the ground. When the engines were started the "slip-stream" from the propellers whisked up the sheets and threw them into the tail-plane, causing the fabric to tear, a tire to puncture, and the machine to run once more into the bog (see text, page 328).

HAULING THE AEROPLANE UPON THE BAMBOO TRACK: SOERABAYA, JAVA

At first the pathway of matting was not wide enough. When the wheels laid on the track, and threw them into the telephone, wiring the failing to ... a fire to minutes, and the machine was hauled from the other side.

Photograph by Sir Ross Smith

PARALLEL BAMBOO ROADS CONSTRUCTED OF MATTING AS RUNWAYS FOR THE VIMY, AT THE AËRODROME IN SOERABAYA, JAVA

The necessity for the construction of this roadway grew out of the fact that the Vimy became bogged in the mud of the aërodrome. Two hundred coolies were required to haul the machine out of the mire (see text, page 327).

317

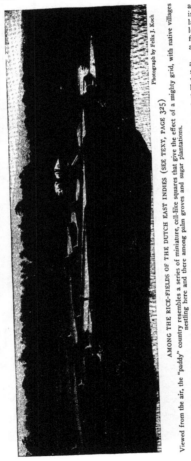

AMONG THE RICE-FIELDS OF THE DUTCH EAST INDIES (SEE TEXT, PAGE 325)

Viewed from the air, the "paddy" country resembles a series of miniature, cell-like squares that give the effect of a mighty grid, with native villages nestling here and there among palm groves and sugar plantations.

Photograph by Felix J. Koch

altitude down to 1,000 feet, passing here and there scattered villages, scaring the water buffaloes, which would career off, flashing across the paddy-fields as fast as their bulk would allow.

At last we reached Singora, and a glance at the aërodrome showed that at least half of it was under water. There was, however, a narrow strip along the center which appeared more or less dry, but I would have to make a landing across wind. I came down low to examine this strip, and to my utter dismay noticed that it was covered with small tree-stumps!

A wide and anxious circling around the aërodrome showed me there was no other spot on which to land; so there was nothing for it but to attempt to make a landing on this narrow strip of stump-studded ground.

As we touched and ran along, I expected every moment to feel a jolt and the under-carriage wrenched off, or else the machine thrown on to her nose; but by the merciful guidance of Providence we miraculously came to rest safely.

The only damage sustained was to our tail-skid, which had caught in a stump and been wrenched off. I walked back along our tracks and found that in several instances our wheels had missed by a few inches stumps a foot to eighteen inches high.

THE DEVIL IN THE MACHINE

The whole native population assembled to see us. None had ever seen an aëroplane before, and at first they would not venture near. Three Englishmen live at Singora, and one of them had imposed upon the simple native minds that the devil was going to arrive in a flying chariot to take charge over all the convicts there. When, however,

Photograph from Henry Ruschin

TEMPLE GROUNDS IN THE CAPITAL OF BALI, DUTCH EAST INDIES

The Vimy skirted the coast of Bali at an elevation of 2,000 feet. Bali is the first island to the east of Java.

they saw that four ordinary humans climbed out of the machine, they quickly surged around us. I noticed that they were staring, arguing, and pointing at us in a peculiar way; but it was not until I heard of our friend's joke that I understood the full significance of their interest in us.

Several of them walked in front of the machine, flapping their arms and performing birdlike evolutions. We concluded that they were solving the mystery of flight and demonstrating how the Vimy flapped its wings to rise from the ground. My brother, unobserved, climbed into the cockpit and, seizing the control column, vigorously moved it to and fro, which caused the ailerons* and elevators to flap about.

There was a wild scamper in all directions. We learned afterwards that the natives imagined that we were flapping our wings preparatory to starting off!

A PETROL SHORTAGE

My first inquiry was as to the quantity of petrol available. I discovered that

the supposed 500 gallons was only 500 liters, depoted there for Poulet. This meant we would be compelled to remain here until I could get sufficient petrol from Penang or Bangkok to take us on to Singapore. I accordingly sent off an urgent wire to the Asiatic Petroleum Company at Penang, asking them to send me 200 gallons of aviation petrol as speedily as possible.

I also wired the resident councilor at Penang, asking him to assist, in the event of there being difficulty in getting this quantity of petrol shipped at such short notice. I next requested the Governor of Singora to have part of the aërodrome cleared of stumps to enable us to take off.

Our machine was left standing on the strip of high ground and we pegged her down securely for the night.

Our next contract was to mend the tail-skid. An examination showed that the fitting which attaches it to the fuselage had broken off. This meant at least six hours' work, provided we could find the

*Hinged portions on the ends of the wings used for banking when turning.

Photograph by Sir Ross Smith

THE BRITISH AIRMEN AROUSED INTENSE INTEREST AMONG THE NATIVES OF THE ISLAND OF SUMBAWA WHEN THE VIMY LANDED AT BIMA

The Sultan of Sumbawa and the leading Dutch official accorded the voyageurs every consideration, and the Bima aërodrome was in excellent condition, clearly marked with a white ring and encircled by a water-retaining wall. (see text, page 329).

necessary materials. One of our English friends took us to a local Chinaman, a jack of all trades and the master of a promising heap of scrap-iron. Bennett unearthed a piece of steel shafting which, provided a lathe was available to turn it down to shape and size, fitted our purpose.

We then proceeded to a near-by rice mill which was just whistling off for the night. There we found a good lathe, but of primitive motive power. Four coolies turned a large pulley-wheel, and their power was transmitted by belt to the lathe.

Bennett got to work at once by the light of a kerosene lamp. After an hour's hard work, little impression was made on the steel, and our four-coolie-power engine "konked out."

Four more coolies were secured, but after half an hour they went on strike and demanded more money. I gave them the increase, but fifteen minutes later they went on strike again. This time we called the foreman from the rice mill. There was a different kind of strike, and so the work proceeded.

VI. How We Made Our Homing Flight

By 10 p. m. Bennett had completed the job, so we returned to the bungalow which had been placed at our disposal by H. R. H. Prince Yugula.

Just before midnight we were awakened by the sound of a torrential downpour—the storm which we had passed through during the day had reached Singora. The wind increased to a gale, and, fearing that the machine might be in danger, we all turned out and kept watch. Fortunately, we had pegged her nose into the wind, but during the heavy squalls the Vimy so strained at her lashings that several times I feared she would be swept away and crushed.

We stood by all night, obtaining what little shelter we could from the wings, and at every squall rushed out and held on to the planes. Needless to say, we were drenched to the skin, and when the wind eased down, shortly after daybreak,

we felt tired and miserable, with no dry clothes to put on.

Ten inches of rain had fallen during the night, and the whole of the aërodrome excepting the ridge on which the machine was standing resembled a lake. Luckily, the ground was sandy, and after the rain ceased the water drained off rapidly. Squalls continued throughout the day, but Bennett and Shiers, after rigging a tarpaulin shelter, were able to work on their engines.

After breakfast in the bungalow we returned to the machine and found that the government had sent down 200 convicts from the local jail to clear away the stumps; and so we set them to work to clear a strip about 400 yards long and fifty yards wide across the aërodrome.

The day's rest from flying was a delightful relaxation; in fact, an imperative necessity, for my brother's and my own eyes were almost too painful for vision, after the previous day's battling with the storm.

Late that afternoon our petrol arrived from Penang, but it was raining too heavily to risk putting it into the machine. We were greatly indebted to Captain Owen Hughes, an ex-Royal Air Force officer, for bringing up the petrol and also for his prompt attention in arranging for its transport.

A TRICKY TAKE-OFF

After a much-needed night's rest, we were down at the aërodrome at daylight, and after putting the 200 gallons of petrol into the tanks, started up the engines. Getting the machine into the air was a questionable problem, but, as our time for reaching Australia was fast closing in, we decided to make the attempt.

Three large patches of water extended across the aërodrome at intervals of about fifty yards. This water was, on the average, six inches deep; but, as the aërodrome was sandy, our wheels did not sink appreciably into it. A clear run of fifty yards allowed the machine to gather fair headway. Then she struck the water, which almost pulled us up; a race across another fifty yards of hard ground, and by the time we had passed through the second patch of water the machine was moving very little faster than at the beginning.

The third patch of ground was a little longer, and when we reached the third pool we were traveling at about thirty miles per hour.

The sudden impact with the water almost threw the Vimy on to her nose, and water was sucked up and whirled in every direction by the propellers. Our flying speed had to be gained on the seventy yards of dry ground which now remained; beyond that extended scrub and gorse bushes.

The Vimy bounded forward as soon as she left the water, and just managed to get sufficient lift on her wings to clear a ditch and scrape over the scrub.

I had been informed that the weather would be much better on the western coast of the peninsula, so we followed the railway line across to that side. As the clouds hung only a few hundred feet above the railroad, we were compelled to descend to a perilously low altitude, which was rendered the more hazardous by huge limestone outcrops, rising four to five hundred feet, scattered over the country.

A FASCINATING LAND

Along the western shores we found the weather much improved; the clouds were higher, and occasional bursts of sunshine threw weird light and shadow effects across the paddy-fields and scattered villages. We still maintained a low altitude, which added greatly to the interest of the flight and also gave us a splendid opportunity of studying intimately this remarkable and productive country.

Near Kaular Lumpar we entered the tin country and observed many tin dredges in full operation. Lower still we flew across the rubber plantations, cheered by the planters and waving back. Then, passing above Malacca, we reached Singapore in the afternoon, after one of the most interesting stages of the journey.

I had been dreading the landing and take-off at Singapore, as the improvised aërodrome, the race-course, was altogether too small for our large machine.

I glided the Vimy down at as low a speed as possible, and just before we touched the ground Bennett clambered out of the cockpit and slid along the top of the fuselage down to the tail-plane. His weight dropped the tail down quickly, with the result that the machine pulled up in about one hundred yards after touching the ground.

The next day, December 4, was my birthday, and to reach Australia within the specified thirty-day time limit meant that we had to arrive in Darwin on the 12th, eight days from now, and four more landings to make after leaving Singapore. Thus it will be seen we still had four days in hand. I therefore decided to remain the whole of next day at Singapore and work on our machine.

AN "EARLY" DANCE

We now had, roughly, 2,500 miles to complete, and in all that distance I knew of only five places at which a landing could be made; the rest of the country was either mountain, jungle, or swamp; so it behooved us to look well to our machine, for a single engine trouble and a forced landing away from any of these aërodromes would have ended all.

The heat at Singapore was intense and, coming from the cold of the English winter, we felt it severely. After a heavy day on the machine, we were asked that night to a dance at the Tanglin Club, but physical weariness compelled us to refuse.

My host, in a very persuasive manner, did his utmost to induce me to go, assuring me the dance would be over very early. However, when we arrived at our machine, at daylight next morning, and were getting ready to start off, my quondam host of the night before and some of his party arrived, all still wearing evening dress. They had just come from the early dance!

As I have mentioned previously, the ground was much too small for an aërodrome, and the heavy rain which had fallen overnight made it very heavy.

My brother and I paced over and examined the ground and discussed the best way to take off, but we were both very dubious as to whether we could get the machine into the air or would pile her up on the adjacent houses in the attempt. I taxied into the position, so as

Photograph from Sir Ross Smith

"WE DRANK SUCCESS TO THE LAST STAGE OF THE FLIGHT IN THE VINTAGE OF
TIMOR—COCONUT MILK"

The aërodrome on the island of Timor, the last landing place before reaching Australia, had been completed by the Dutch officials the day before the arrival of the Vimy (see p. 335).

to give the maximum amount of run, and then opened the throttle full out.

We gathered way slowly, and I watched the fence around the course come rapidly nearer and nearer, and still we were not off the ground. It was a tense and anxious moment. When fifty yards from the rails, I pulled my control-lever back; the trusty Vimy rose to the occasion and just cleared the rails. There were still houses and trees to be negotiated, and I set the Vimy climbing at an alarmingly steep angle.

Another breathless moment passed, and the wheels of the under-carriage just brushed the tree-tops. It was a great triumph for the Vimy. She achieved the seemingly impossible, and to this day I regard our escape from disaster during this perilous take-off as providential and one of the finest maneuvers of the whole voyage.

HEADING FOR JAVA

After a wide sweep above Singapore, we headed for the open sea and Java.

Photograph by Sir Ross Smith

AN AUSTRALIAN WARSHIP PATROLLING THE SEAS BETWEEN TIMOR AND PORT DARWIN

"A faint smoke haze resolved into the smoke-plume of a fighting ship—the *Sydney*—and we knew that, whatever might befall, we had a friend at hand." It was the *Sydney* which sank the famous German commerce raider, *Emden* (see page 335).

Passing down to the Sumatran coast, we ran into characteristic doldrum weather—isolated patches of dark thunder-storm clouds, from which the rain teemed down in heavy, murky columns.

Occasional forks of lightning seared the clouds, throwing up into relief their immense bulk and shedding a flickering gleam over the calm sea, where almost stagnation expanded. Occasionally a light zephyr came out of the east, but almost in the course of a few minutes the puff had boxed the compass and died away.

BUMPING OVER "THE LINE"

The spectacle of these local storms was extremely uncanny, and by navigating accordingly it was easy to avoid them. On reaching the coast of Sumatra we encountered a light head wind and flying conditions became very bumpy. One immense vacuum into which we fell made us hold tight and wonder.

"That's the Equator," ejaculated my brother, and, sure enough, by dead reckoning, we had bumped across the line into the Southern Hemisphere.

Our entrance into the Southern Hemisphere was welcomed by improved weather, but the landscape below—dense jungle inland, fringed along the seashore by belts of mangrove swamps and the blue tropical sea—often kindled in my mind thoughts of utter helplessness in case of engine trouble.

There developed in me a strange admiration—almost reverence—for the super-mechanism that hummed away rhythmically, that had now covered 10,000 miles without an overhaul, and at the opposite side of the globe was still singing a hymn of praise to the makers, as it had done when the bleak wintry snows had carpeted the aërodrome at Hounslow and northern France. How far away this all seemed!

These were times, indeed, for musing, as we sped along above this tranquil tropical landscape, home only a few days away—an achievement!

Numerous small islets—emeralds in a setting of turquoise—passed below us. There were yearnings to land and explore their mangrove-fringed bays and foreshores, but the nearest landing was our destination, Batavia.

Soon the large island of Muntok came below, and in the strait separating the mainland we passed a vessel. Subsequently we learned she was equipped with wireless and had transmitted news of our arrival on to Batavia.

I had originally intended to hug the coast of Sumatra on to Java; but as it was all dense mangrove swamp, with no sign of a possible landing-place, I reasoned that we might just as well fly over the sea. My brother computed the compass course, and so we headed direct for Batavia.

THE LANDING AT KALEDJAT

The hazy contours of the mountains marking the western end of Java soon began to show up to starboard, and ahead a scene of rare enchantment began to resolve itself upon the bosom of the tropical sea.

The sea was a glorious mirror almost as rippleless as the canopy above, and scattered broadcast lay a thousand isles, each one beautiful, and all combined to make one of the most beautiful sights I have ever looked down upon. Many of the islands are heavily grown with palms extending to the very water's edge; others, sparsely cultivated, fringed with a narrow ribbon of beach; but around each is a setting of an exquisite shade of green, marking a sand-girt shallow; then deep-blue and depth.

Myriads of tiny white fisher-sails passed through the channels, gleaning their harvest from the sea.

Reluctantly we turned from this glimpse of fairy-land, and, bearing for the Garden Island of the East, soon reached Batavia, the city of canals and beautiful avenues.

Following the railway line, we landed at the Dutch Flying School at Kaledjat. The Dutch had sent an escort of four machines to welcome us; but, although they passed within about 500 feet of the Vimy, they missed us.

The distance of 650 miles from Singapore we had covered in just nine hours. Hearty greeting was extended to us by His Excellency Count Van Limburg Stirum, the Governor-General of the Netherlands Indies, and a large number of leading officials.

After a well-enjoyed meal, we set to work on the machine. The petrol available was very heavy, and it took us eight hours to filter 350 gallons through the chamois leather strainer into the tanks. As the next stage to Soerabaya was only a short lap, we did not leave Kaledjat before 7.30.

THE BOUNTEOUS GARDEN OF JAVA

With beautiful weather favoring us, we sped rapidly over fertile tracts of this amazing island, charmed by the unsurpassable beauty that unfolded below. Java impressed me as one vast bounteous garden, amid which rise the immense, shapely cones of volcanic mountains.

Perhaps one of the most striking sights was the "paddy" country. From our height, the whole expanse of the land appeared to be inundated by irrigation water—all contained in miniature, cell-like squares, that gave the effect of a mighty grid, stretching away to the moun-

Photograph by Captain Frank Hurley

THE CROWD AT PORT DARWIN INSPECTING THE VIMY AFTER THE COMPLETION OF ITS HISTORIC FLIGHT HALF WAY AROUND THE WORLD

Through every possible rigor the Vimy had come, and not once, from the time of its departure from Hounslow, near London, had it been under shelter.

tains on our right. Even there the irrigation did not cease, but climbed up the mountain sides in a system of stairlike terraces.

Here and there native villages nestled beneath the shelter of the palm groves or among the verdant green of sugar plantations. Always in the background, subdued by tropic haze, rose the chain of peaks, practically all quiescent, and far away to the left that faint blue line which marked the Pacific horizon.

Nearing Soerabaya, flying became very bumpy, and it was no small relief when the town, like a magic carpet of multicolored fabric, spread beneath us. Heading the Vimy down, we made a low circle above the town, to the infinite amazement of the teeming native population that swarmed out into the streets, petrified, evidently, by the visitation.

As the aërodrome was small, I decided to land on the north side, so that, if we overshot its length, I could, perhaps, swing round to the left. This maneuver, however, I discovered to be unnecessary. We made a good landing and were easing off to rest when the machine seemed to drag, and from past experience I knew at once the Vimy was becoming bogged.

IN THE MUD

Opening up the starboard engine, we began to swing slowly, but the port wheels immediately sank into the mud and we tilted on to our fore-skid. At once I shut off both engines and the Vimy gradually eased back to her normal position.

The natives and people, who had been kept back by the Dutch soldiers, rushed the ground, and their weight on the sun-dried crust soon broke it up, and mud began to ooze through. In a very short while the Vimy subsided to her axles and was surrounded by a pond of semi-liquid mud.

The proposition literally was a decidedly sticky one. It was midday, broiling hot, and the tenacity of the mud reminded me forcibly of that clinging tendency familiar to our black-soil plains. Moreover, only four days of our prescribed time remained in which we must make Port Darwin.

The engineer of the Harbor Board arrived, and together we discussed the situation. He collected a horde of coolies and a large quantity of bamboo matting, and so we set to work to dig out the wheels.

After some hard work we got the matting almost under the wheels, started up the engine, and, aided by the coolies and Dutch soldiers, the Vimy was hauled from the bog. I then stopped the engines, tied ropes to the under-carriage, and the machine was pulled on to a pathway of mats (see page 316).

After a couple of hours the machine was safe out of the morass, and the ground on which we stood felt quite solid; so I thought we had landed on the only soft spot on the aërodrome, and decided to taxi to the opposite end under our own engine-power.

I was soon disillusioned, for, after moving but ten yards, down went the wheels again. More digging, tugging, and pushing, and we, apprehensive all the while as to whether the coolies would drag off the under-carriage, finally had to lay down a pathway of bamboo mats and have the machine hauled by 200 coolie-power.

I should say here that our hearty thanks are due to the harbor engineer and the officials—indeed, to all who lent a hand—for without them it would have been impossible for us to have completed the flight on the scheduled time.

MAKING A RUNWAY OF NATIVE ROOFS

My brother and I had decided that it would be impossible to get the Vimy into the air in the usual way, so we consulted with our invaluable friend, the engineer, and he agreed to collect bamboo matting from far and wide, so that we might construct a mat-paved roadway.

I observed that this matting formed the principal covering of the native huts, and subsequently learned that entire villages in the immediate vicinity were stripped bare to provide us with the necessary materials.

The British consul invited us to a "quiet" dinner that evening, but when we arrived at the café every British resident in Soerabaya had gathered there to welcome us.

Photograph by Walter Burke

GAWKY PETS IN AUSTRALIA

On the left is an "old man" kangaroo, standing nearly seven feet high when in action. The claws on his hind legs are so powerful that he can rip open an attacking dog with one stroke. The animal on the ground is a female of the species. During their enforced stay in a dried-up swamp one night, the aviators' camp was visited by several of these animals.

Next morning saw us at the aërodrome by daylight, and a gladsome sight met our eyes. Natives were streaming in from every direction bearing sheets of bamboo matting—they were literally carrying their houses on their backs—and already a great pile of it lay by the Vimy (see pages 314-317).

At first a pathway of mats was merely laid down, but in our keen anxiety to get off we had overlooked the "slip-stream" from the propellers. The engines were opened up and we were just gathering speed nicely when some of the sheets were whisked up and blown into the tail-plane. The fabric was torn. a tire punctured. and the Vimy ran off and bogged badly.

Once more we had to extricate the wheels and reconstruct the roadway. This time we pegged down and interlaced the mats.

More matting arrived on a motor-lory, so we lengthened the road, said a second

WHITE-ANT BEDS OF NORTHERN TERRITORY, AUSTRALIA

When a split propeller blade necessitated a forced landing in northern Australia, some of these large ant-hills provided the material for an improvised roadway to enable the Vimy to "take-off" after repairs had been made.

good-bye, and, just twenty-four hours after our arrival at Soerabaya, made a sensational take-off, with the mats flying in all directions. We circled low over the town and anchorage, so as to give the engines time to settle down to normal running, and then we headed on a direct compass course for Bima.

From the point of view of a prospective forced landing, the 400-mile flight to Bima was impossible. Not a single flat occurred on which we might have landed. Scenically, this lap was glorious. We skirted the coast of Bali and Lombok, keeping only 2,000 feet above the sea.

The Bima aërodrome we found in excellent condition, clearly marked with a white ring and encircled by a water-retaining wall. The natives scampered in all directions and would not venture near until they saw us walking about the machine.

The local sultan and the Dutch officer met us and proffered the hospitality of a native bungalow a couple of miles from the machine. Here we aroused intense interest; eyes taking little furtive glimpses at us peered through every crack and gap.

During that night some gentleman tried to force my window. I waited quietly until he had raised the sash half-way; then a shot from my Very light pistol put him to a screaming and, I have no doubt, a terror-stricken flight.

The natives had recovered from their shyness by next morning, and on our arrival were swarming around the machine with presents of coconuts sufficient to start a plantation; evidently they thought the Vimy a very thirsty sort of bird.

We took a cargo of nuts on board, as the water was unsuited for drinking, and, setting off in dazzling sunshine, once more pursued our course above scenes of tropical enchantment and alluring charm.

AT TIMOR

After following the north coast of Flores to Reo, we crossed over to the south side of the island and ran into isolated rainstorms.

We observed a volcano in active eruption on the eastern horizon, but, as the weather indicated a change for the worse, we could not afford to make a deviation. We flew on as far as Pandar, and then swung off direct for Timor.

Photograph by Captain Frank Hurley

ENJOYING MORNING TEA IN AUSTRALIA

Even in the smallest hamlets the aviators were always the center of an eager group of spectators, as the Vimy continued its triumphal flight.

THE GREAT GORGE OF THE BLUE MOUNTAINS, NEW SOUTH WALES, AUSTRALIA (SEE PAGE 337)

On the voyage across Australia Sir Ross Smith was accompanied by Captain Frank Hurley, the Antarctic explorer and war photographer, who was commissioned to illustrate this leg of the journey.

SIR ROSS SMITH ARRIVES IN AUSTRALIA'S FAMOUS SEAPORT OF SYDNEY

Photograph by Captain Frank Hurley

A tumultuous welcome awaited the aviators upon their arrival in the capital city of New South Wales.

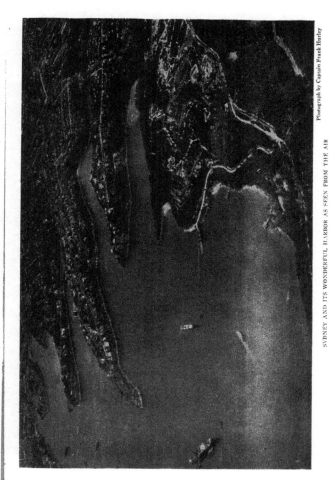

Photograph by Captain Frank Hurley

SYDNEY AND ITS WONDERFUL HARBOR AS SEEN FROM THE AIR

"Like a mighty fern leaf, ramified and studded with islets, this glorious waterway unfolded below. The city and its environs, massed along the waterfront and extending into the hinterlands, flanked by the Blue Mountains, composed a spectacle of exquisite charm and beauty (see text, page 337).

STUDIES IN EXPRESSION WHEN YOUNG AUSTRALIA IS INTERESTED

Photograph by Janet M. Cummings

We had by this time acquired such confidence in our engines that it mattered little what lay below us—sea or land.

The thick volcanic smoke soon obscured the land and all distant vision, but we eventually picked up the Timor coast a few hundred yards from our calculated position. Ten miles inland we came down on the aërodrome at Atamboea, our last landing ground before Port Darwin.

The Dutch officials, who welcomed us, had thoughtfully arranged our petrol and oil supply close at hand, saving us a good deal of valuable time, which we were able to devote to a thorough overhaul.

THE NIGHT BEFORE THE FINAL "TAKE OFF"

Tomorrow would be the great day whereupon reposed the destiny of our hopes, labors, and ideals.

This was one of the aërodromes specially made by the Governor-General of the Netherlands Indies for the Australian flight, and had been completed only the day before our arrival. A guard of Dutch soldiers kept watch over the machine while we proceeded with their officers to camp, some six miles away.

It is hardly necessary to say that none of us overslept. We were too excited at the prospect of the morrow. We felt sure that if it dawned fine and hot, our homing was assured; but as we stepped out, before sunrise, into the still, sluggish air, we realized that our hopes of an early start were small. A heavy haze lay over the sea and the coast, obscuring everything; so we decided to await its clearing.

We were at the aërodrome before sunup to discover that a great swarm of natives were even earlier risers than ourselves. Most had come afoot, but many had ridden their ponies, and they clustered on and around the fence, behind and beside the Vimy, like swarming bees. We had hauled the machine well back and raised the tail over the fence in order to take advantage of every foot of the short run.

ONE OF OUR CLOSEST SHAVES

Our start off was brightened by one of those incidents that usually make material for comic papers. The propellers were just "kicking" over, like two great fans, and those natives sitting on the fence in the line of the slip-stream were enjoying the cool breeze and looking pleased with themselves. When I opened up the engines and both propellers swung into action, the sudden blast of air sent these particular spectators toppling back into the crowd, where ponies and natives made a glorious mixup, at which we all laughed heartily.

Soon after 8 the fog began to thin, and by 8.35, to be exact, I opened up the engines and just managed to scrape out of the 'drome. Scrape is exactly the word, for the branch-tops of the gum-tree rasped along the bottom of the machine as we rose. It was indeed one of the closest shaves of the trip.

In front of us rose a chain of high hills, and, as the atmosphere was hot and we climbed very slowly, we made a detour to avoid them. Still flying low, we approached the coast and pulled ourselves together for the final lap—the jump across the sheet of blue Indian Ocean that lay between us and Port Darwin.

Keith took all possible bearings, noted wind direction, and made numerous calculations of ground speeds. Then we set compass course for Darwin, and with a "Here goes!" we were out over the sea. All our hearts were beating a little quicker; even our fine old engines seemed to throb a trifle faster.

SIGHTING THE "SYDNEY"

Our watches registered 11.48 when Keith nodded ahead, and, dead on the line of our flight, we made out a faint smoke haze that soon resolved itself into the smoke-plume of a fighting-ship. It was the *Sydney*, and we knew now that, whatever might befall, we had a friend at hand.

We swooped low, and exactly at twelve minutes past noon passed over the vessel, seeing plainly the upturned faces of the sailors and their waving hands. It was a cheer of welcome quite different from anything that we had experienced on the long journey. Perhaps it is not to be wondered at that the result of our snapshot was blurred through the shaking of the camera (see page 324).

Photograph by Captain Frank Hurley

THE £10,000 PRIZE PRESENTED BY THE COMMONWEALTH OF AUSTRALIA TO SIR ROSS
SMITH FOR THE FIRST LONDON-TO-AUSTRALIA AIR VOYAGE

The commander of the expedition decided that, as all four voyageurs had participated
equally in the perils and the labors of the enterprise, they should all share alike in its
financial rewards. Each man received a fourth of the prize money.

We took the opportunity of snatching a speed test, and found that we were averaging seventy-five miles an hour.

THE FIRST GLIMPSE OF HOME

An hour later both of us saw ahead and to port what appeared to be haze, but which we hoped was land, though neither dared express his hopes. They were justified, however, ten minutes later, and hailing Bennett and Shiers, we pointed joyfully to Bathurst Island lighthouse.

It was just 2.6 p. m. when, as our diary prosaically notes, we "observed Australia." At three o'clock we not only observed it, but rested firmly upon it, for, having circled over Darwin and come low enough to observe the crowds and the landing-place, we settled on Terra Australis on December 10th, 27 days 20 hours after taking off from Hounslow.

Two zealous customs and health officials were anxious to examine us, but so were about 2,000 just ordinary citizens, and the odds of 1,000 to 1 were rather long for those departmental men, and our welcome was not delayed.

The hardships and perils of the past month were forgotten in the excitement of the present. We shook hands with one another, our hearts swelling with those emotions invoked by achievement and the glamour of the moment. It was, and will be, perhaps, the supreme hour of our lives.

Almost reverently we looked over the Vimy, and unspoken admiration crept over us as we paid a silent tribute to those in far-off England for their sterling and honest craftsmanship. The successful issue of the venture, in a great degree, was due to them, and surely they merited and deserved a large proportion of the praise.

Through every possible climatic rigor the Vimy had passed, and practically without any attention. Not once, from the time we took our departure from Hounslow, had she ever been under shelter. And now, as I looked over her, aglow with pride, the Vimy loomed up as the zenith of man's inventive and constructional genius. I could find neither fault nor flaw in the construction, and, given a few days overhaul on the engines, the Vimy would have been quite capable of turning round and flying back to England.

OUR WELCOME HOME

These reflections were of brief duration, for the crowd, having satisfied its curiosity over the machine, directed it to us The Administrator of the Northern Territory and the Mayor of Darwin were given barely time to make an official welcome when the assemblage, brimming with enthusiasm, lifted us shoulder high and conveyed us to the *jail!*

This sinister objective, for the moment, gave us qualms, for we fully expected a charge of exceeding the speed limit to be preferred against us. That drastic apprehension resolved itself into being dumped on a tree-stump, historic or otherwise, in the garden, while raucous howls of "Speech! Speech!" came from the hospitable multitude.

After the exchange of much hot air on both sides, we returned to the Vimy, made all snug, and lashed her down for the night.

OVERWHELMED WITH TELEGRAMS

During our stay at Darwin we were the guests of Mr. Staniford Smith, at Government House. And now we were to be bewildered by an amazing array of cables and telegrams. They arrived in great fifteen-minute shoals from every corner of the globe.

What had gone wrong? Surely every one had gone mad—or had we? Why all this fuss and excitement? In our race across the globe we had not read a newspaper, and, beyond the local natural attention evinced at our numerous landing-grounds, we knew nothing of the interest the rest of the world was taking in the flight.

Great indeed was our astonishment when, on turning up back files of newspapers, we read of our exploits, recorded with a degree of detail that must have taxed the imaginative resources of editorial staffs to gray hairs.

The rush, strain, and anxiety were over; henceforward the conclusion of our flight across Australia could be undertaken leisurely; and, what was more to the point, we could afford to wait for the best possible flying conditions.

Much of the flight over featureless wastes would have been drear and monotonous, but it was Australia and that was compensation enough. Moreover, we had the occasional diversion of passing over the small out-back towns, where many of the inhabitants rushed into the streets and stood looking up, waving and cheering.

The sublimest spectacle of the entire flight from Hounslow to our journey's end was to burst upon us when we arrived over Sydney and its wonderful harbor.

Like a mighty fern-leaf, ramifying and studded with islets, this glorious waterway unfolded below. The city and its environs, massed along the waterfront and extending into the hinterlands, flanked by the Blue Mountains, compose a spectacle of exquisite charm and beauty.*

We headed up the coast and, turning through the entrance, entered the port.

Planing down to 600 feet, we flew above a myriad ferry-boats and vessels, from the whistles of which little white jets of steam spurted up, screaming a welcome; then across the roof-tops, where crowded waving and cheering humanity, and over the streets below, where little specks paused to look up and join in the greeting. It was a great day—a time that comes once in a lifetime.

THE PRIZE IS DIVIDED EQUALLY

Not the least pleasant incident upon our arrival finally in Melbourne was the paying over of the £10,000 prize by the Prime Minister, the Right Hon. W. M. Hughes, on behalf of the Commonwealth Government. As all participated equally in the perils and labors of the enterprise, the prize was divided into four equal shares.

In Melbourne I formally handed the Vimy over to the Commonwealth Government on behalf of Messrs. Vickers Ltd., who generously presented the machine to the Commonwealth as an historic relic of the first aërial flight from London to Australia. At the request of the authorities, I flew the machine on to Adelaide, my native city, and thus realized to the full my ambition and dream of flying from London to my own home.

It would be hard indeed to comprehend the feelings that surged through me as I landed the Vimy on the sod of my native city—the recognition of familiar faces; the greeting of well-known voices; the

* See "Lonely Australia, the Unique Continent," by Herbert E. Gregory, in the NATIONAL GEOGRAPHIC MAGAZINE for December, 1916.

Photograph by U. S. Army Air Service

LOOKING DOWN ON THE DECKS OF THE U. S. BATTLESHIP "OHIO" FROM AN AIRPLANE

A single air bomb containing from 1,000 to 1,400 pounds of high explosive and dropped from an airplane would wreak more destruction on this great floating fortress than 25 large projectiles from an enemy battleship (see text, page 341).

hand-clasp of innumerable friends; but, greatest of all, the reunion with my parents after five long years.

Our heartfelt thanks are due to the officers and mechanics of the Royal Air Force; to the Dutch authorities for constructing aërodromes and other assistance, and for the coöperation of numerous friends, whose willing and generous

help laid the paving-stones over which Fortune piloted me.

My brother Keith shares equally any worthiness that the effort might merit, as also do my two master mechanics, Sergeants Bennett and Shiers, whose loyalty and devotion to duty have done much to bind closer the outposts of the Empire through the trails of the skies.

AMERICA IN THE AIR

The Future of Airplane and Airship, Economically and as Factors in National Defense

By BRIGADIER-GENERAL WILLIAM MITCHELL

ASSISTANT CHIEF OF AIR SERVICE, FORMERLY COMMANDING AVIATION, FIRST CORPS, FIRST ARMY, AND GROUP OF ARMIES, A. E. F.

THE flying-machine, dreamed of for centuries, became a reality with the development of the gasoline engine. Before that all sorts of appliances had been tried, ending with an actual flying-machine, developed by Professor Langley, of the Smithsonian Institution.

A steam-engine furnished the motive power for Langley's creation, and it actually flew alone, but it did not succeed in carrying a man to and from an airdrome until a gasoline engine was fitted to it years after Langley's death.

Although the Wright Brothers had experimented for several years, their first public demonstrations took place in 1908. So, really, the practical application of aviation has been within the last twelve years.

Similarly, lighter-than-air craft—that is, dirigible balloons, or "airships," as we call them today—had been experimented with for a long time, but they also had to wait for the gasoline engine as a propelling force.

Much as we would like to see the greatest application of aëronautics to civil and commercial uses, this will come gradually and not at once. It took many years for the railways to supplant the stage-coach and for the motor cars to do away with the horse-drawn vehicles.

At present over 90 per cent of all aëronautical appliances are used exclusively as elements of national defense by the countries owning them.

An airplane* is one of the most complicated instruments in all its parts, and changes more rapidly as the knowledge of its properties expands than any other creation which has been known heretofore.

THE AIRPLANE AS A FIGHTING MACHINE INTRODUCED IN WORLD WAR

Up to the time of the World War, all military power was exerted either on the land or on the water, and offensive and defensive equipment was made so as to withstand attack in a single dimension. With the coming of the fighting airplane all of these notions had to be modified, and a third element, acting both over the ground and over the sea, considering no frontiers such as rivers or mountains, deserts or coast-lines, and whose only limit was the amount of fuel in its tanks, had to be considered.

Slow as the old services, both army and navy, were to recognize the power of this new arm, it was forced upon them to such

---

*In the preceding article Sir Ross Smith's use of the British form *aëroplane* has been retained, but *airplane* is the official name for a heavier-than-air machine in America.—Editor.

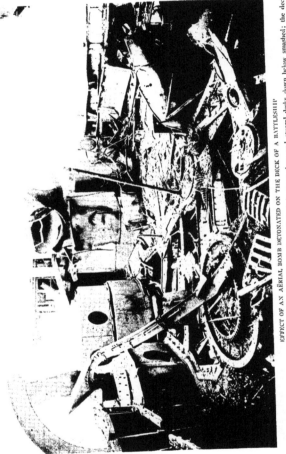

EFFECT OF AN AËRIAL BOMB DETONATED ON THE DECK OF A BATTLESHIP

It will be noticed that the turret is cracked and demolished; the whole deck is swept clean, and several decks down below smashed; the deck above blown off. The damage extended to practically all parts of the U. S. S. *Indiana*, whose days of usefulness as a fighting craft were over, and on which the tests were made.

A direct hit by an aërial bomb on a battleship will break every electric light globe on the ship and throw her into absolute darkness below-decks; will disrupt telephone, radio, and interior communication systems; fill with noxious gases the fire-rooms, engine-rooms, and all compartments practically all over the ship; will disrupt ammunition hoists, dislodge and jam turrets, ventilated by force draft; cause shell-shock to the personnel practically all over the ship; or any one standing on deck; will cause dish upper decks, kill all personnel on upper decks, anti-aircraft gun crews, fire-control parties in the tops, and sink or disable the battleship. If the bomb is exploded in the fire to break out, exploding all the anti-aircraft ammunition on upper decks, and sink or disable the battleship. water alongside, the battleship will either sink or be permanently disabled (see pages 341 and 347).

340

an extent that, after the second year of the war, no movement was tried on the Western Front without a most thorough aërial preparation, and from that time until the end of the war the offensive air service, or that which gets out and fights the opposing air service and then his ground troops, constantly increased.

Since the war the lessons gained have been carefully studied, and their application to future requirements has been accurately estimated.

The branch of aviation which has been developed for bringing the enemy's air force to combat and forcing it to fight is called pursuit aviation. It is equipped with the fastest, most maneuverable, and most heavily armed airplanes that it is possible to obtain.*

Pursuit airplanes in the possession of the principal powers at present are of a speed of from one hundred and sixty to one hundred and seventy miles an hour, can climb to a height of 20,000 feet in twenty minutes, and are equipped with from two to four machine-guns.

With an air force, team-work is more essential than is the case in any other military organization, because the space in the air is so vast and separation from one's companions is so easy that the utmost care has to be exercised to prevent distribution and thereby allow an enemy the advantage of concentration against an isolated detachment.

The degree of success of these operations depends on the training of the pilots and their commanders and is a matter of years and not of months. There is no movement, combination, or method of flying which must not be known in its every phase by the navigating personnel of pursuit aviation.

This is the branch that has to be depended upon to win the air battles, and. at the conclusion of the World War it constituted more than 60 per cent of all the offensive aviation.

AN AIR FORCE IS THE ONLY DEFENSE AGAINST SIMILAR FORCE

The only defense against an air force is another air force. Anti-aircraft guns or any defenses against aircraft from the

* See the NATIONAL GEOGRAPHIC's "Aviation" number January, 1918.

ground have comparatively little effect. Only about one-tenth of 1 per cent of the airplanes going over the line in the American air service during the war were, shot down by anti-aircraft weapons. While they are necessary, they are really auxiliaries of an air force and can do nothing decisive by themselves.

Although the war probably advanced aviation more than would fifty years of peace, still a great deal has been done since the war in the development of appliances really thought out during the war.

This has been particularly true of the second great branch of aviation, which is known as bombardment. This branch carries heavy missiles and drops them, or projects them, at the targets that they are designed to attack.

THE AIR BOMB AS AN ENGINE OF DESTRUCTION

Loaded with the high explosives of today, the modern air bomb will cave in the whole fronts of buildings, shatter armor, and demolish all sorts of military objects, including the destruction of life by concussion alone. For instance, the whole water-front at Halifax was destroyed by an explosion in the harbor.

Heretofore projectiles from large cannon have been designed to pierce the armor of battleships, and then cause their effect by driving the fragments through the bulkheads and into the various parts of the ship. Twenty-five such shots went clean through the German flagship *Derfflinger* in the Battle of Jutland, but, aside from killing about 200 of the personnel, never destroyed the speed of this ship. These twenty-five shots altogether had no more than about 1,000 pounds of explosive in them.

But one of our present air bombs, which weighs one ton and contains from 1,000 to 1,400 pounds of explosive, dropped on her from an airplane, would have wrecked this ship to such an extent as to put her completely out of action and end her usefulness as a war vessel (p. 347).

Our cities in this country are particularly subject to the destructive effects of bombardment on account of the inflammable character of the constructions and in many cases the difficulty encountered

THE WHOLE CENTER PART OF THE "INDIANA" WAS SMASHED TO PIECES BY ONE
AËRIAL BOMB WHICH WAS PLACED ON DECK AND DETONATED

ANOTHER VIEW SHOWING DEMOLITION CAUSED BY THE SAME BOMB (SEE PAGES
340, 343, 347)

THE DETONATION OF A 600-POUND DEMOLITION BOMB

in getting people out of them in the event of such an attack.

NEW YORK CITY AS A TARGET

New York is a notable example. It is situated on a narrow peninsula, and so marked out by two rivers on each side of it that it cannot be mistaken from aircraft, either by day or by night. Communications off of this peninsula for people trying to get away from a bombardment attack are very bad, so that the population could not get away, and the conflagration which would be incident to it would be more serious than anything that has ever occurred in any city.

Bombardment from airplanes is not confined to a coast-line, because there is no coast-line in the air. The airplanes can fly inland to the limit of their gas, if unopposed, and deliver their loads of bombs. Airplanes can carry now from one to three tons of bombs, and a group of 100 airplanes is able to carry 100 tons, as distinguished from the groups that we

had during the war, when one group of four American squadrons could carry only 3½ tons to the trip.

The initiation of gas attacks by the Germans has centered attention on the effect that gas would have if dropped from aircraft. It was not used in Europe from aircraft, because neither side desired to start it, as it would have led to a great deal of useless loss of life on both sides, and not much would have been accomplished, on account of the equality of aviation with both belligerents. If, however, one side had a decided air supremacy and had destroyed the air force of its enemy, and it desired to adopt these barbarous methods of warfare, it might believe that its ends would be furthered by the use of this matériel.

ALL NEW YORK WOULD HAVE TO WEAR GAS MASKS

In an area the size of New York, if two tons of crying gas were dropped every eight days, the whole population

© Eastman Kodak Company

A MOSAIC MAP OF ROCHESTER, N. Y., A CITY OF 300,000 INHABITANTS, MADE FROM
AN ELEVATION OF 10,000 FEET IN A SINGLE FLIGHT

The map is made from prints of eighty separate negatives, in the form of a single strip
of film ten inches wide and seventy-five feet long, all taken in the course of a single flight of
one hour and twenty minutes. Such an accomplishment suggests the great value of this form
of aërial photography in map-making over otherwise inaccessible territory.

would have to wear gas masks and goggles in order to protect itself. Two hostile airplanes could accomplish this effect. If it were desired to use mustard gas, 70 tons of this very poisonous element would be necessary.

The greatest precautions would have to be taken against such an attack. As a matter of fact, it would cause the evacuation of the whole city, paralyze its means of transportation, and result in its virtual extinction as a port. If phosgene gas were used, 200 tons would be required to be dropped every eight days. This is a very deadly gas and will have almost immediate effect, and will kill every man, woman, and child not carefully protected against it.

Gases even more deadly than this are being experimented with and considered by all the nations. ' While we all hope that they will never be used, still precautions must be taken against their use; and the best precaution against any such danger is an adequate force of aviation to shoot the enemy out of the air.

In the defense of an area such as New York, all elements for its protection, such as search-lights for use at night, anti-aircraft artillery, machine-guns, and barrage balloons, must be used as auxiliaries under the command of the air force. Barrage balloons act as aërial barbed-wire fences, and at present can be raised to a height of about 20,000 feet. They are attached to the ground by thin cables, which will cut an airplane wing, break its propeller, and otherwise damage a ship. The barrages, however, have to be protected by airplanes; otherwise the enemy will shoot them down. The anti-aircraft defenses of London were particularly efficient during the last war and were arranged as indicated above. Until their completion, London was subject to innumerable raids from both German airplanes and airships. Once the necessity for co-ordinated action in the air was realized, the effect was immediate.

The third great branch of offensive aviation, which was being developed just as the war ceased and which has received a great deal of development since that time, is what we call attack aviation—that is, the branch which utilizes machine-guns and cannon for shooting at objects on the earth or the water.

RADIO PROVIDES CONSTANT COMMUNICA-
TION WITH THE AIR SQUADRON

The airplanes are armored over all of their vulnerable parts, so as to resist fire from rifles or machine-guns—in fact, they are really flying tanks—and one of their first objectives of attack will be actual tanks on the ground. These airplanes carry small cannon, which can fire at the rate of over 100 shots a minute, and from six to ten machine-guns, each of which can fire from 500 to a 1,000 rounds of ammunition per minute.

When one of these airplanes attacks its object, it looks almost as if the ground were being plowed up, from the intensity of the fire. They fly at very low altitudes and surprise the troops, motor trains, railroad trains, or whatever they desire to attack.

Nowadays air forces can be handled by radio telegraph and to some extent by radio telephone; so that a means of communication exists between airplanes while in the air, which gives very much added power of combination to all aërial operations.

PHOTOGRAPHS MADE FROM 28,000-FEET
ALTITUDES

In addition to the three principal branches of aviation that I have mentioned above, which are just as different in their methods, armament, and the training required for their personnel as are the infantry, cavalry, and field artillery, there is a fourth branch of aviation, called "observation," which is necessary for reconnoitering and scouting and photographing the country.

The photograph forms the most accurate representation of anything that is possible, and with the equipment that we now have, photographs can be taken from altitudes as high as 28,000 feet. '

Military maps of whole areas are made by aërial photography, far into the enemy's country. His railroad lines, his roads, his depots for ammunition and stores of all sorts, and all of his cantonments or concentration points for troops can be photographed and an estimate made of the size of the body of troops which is occupying it.

Observation aviation also adjusts or regulates the fire of artillery. In fact, the artillery is virtually handled from the air. The target is reported by the airplane and the direction and range given to the gunners.

In the World War the observation airplanes had to keep in touch with the troops themselves when all other means of communication failed. Although the telephone could be kept with the troops almost always, yet in the heat of a great battle the wires were severed and the troops lost complete touch with those behind them, even by runners and by all the devices known; so the airplane had to fly right through the barrages of artillery fire, and even infantry and machine-gun fire, to get to their troops and find out where they were; also to see

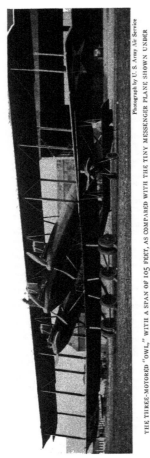

Photograph by U. S. Army Air Service

THE THREE-MOTORED "OWL," WITH A SPAN OF 105 FEET, AS COMPARED WITH THE TINY MESSENGER PLANE SHOWN UNDER IT, AT THE RIGHT

that more ammunition was brought up to them, or even drop ammunition and food to them in some cases.

DIFFICULTIES IN SIGNALING FROM THE AIR

The airplanes, when they reached the point where they thought the troops were, would fire a certain kind of a rocket, and the troops would answer by putting out a panel, or a piece of cloth of a certain dimension and color, on the ground, indicative of the unit to which they belonged.

It took a long time to get this branch of the Observation Air Service and the troops working together, because the troops all thought that when the airplane shot the rocket it indicated its position to the enemy and called their attention to that locality; and then when they showed their panel, that the enemy knew exactly where they were.

Observation aviation is the branch above all others where coördination with the ground troops is essential, and on account of the many and diversified appliances, such as radio, photography, and signal devices of all kinds, it is very difficult to maintain in an entirely efficient condition.

THE AIRPLANE'S SPEED ADVANTAGE OVER THE BATTLESHIP

During the war almost all aviation was used over the land, because that was where it would have the maximum effect.

After the Armistice, however, and particularly after the British airship R-34 had flown across the Atlantic from England to New York and back again, and the transatlantic flight had been made by airplanes, attention was drawn to the possibility of using aircraft for protecting the sea communications of a nation in very much the same way that navies are today; so that the first problem presented is what effect aircraft would have against heavily armored ships—warships.

There never was any question that, even with small aërial bombs, unarmored vessels can be sunk and disposed of at will—that is, within the radius of operation of the airplane. Airplanes have from five to eight times the speed of battleships, and therefore can always catch up to them when and where they wish, and consequently have the power of initiative.

As we lost so few airplanes from anti-aircraft fire during the war, it is entirely safe to say that we will lose very few as the result of any action from the ships themselves.

The defense of the vessels with anti-aircraft artillery and machine-guns will not be as efficient as the same defense on land would be, because these guns are placed on a movable platform when on the ship.

The effect of a bomb on an armored battleship is terrific. Not only does it cause great material damage, but it shatters all the navigating appliances, kills a great many of the personnel, knocks out all lighting systems, and stands a good chance of blowing the structure completely to pieces (see pp. 340 and 342).

The probability of hitting the battleship is very much greater than is the case with gunfire at a range of over 20,000 yards. In fact, it is estimated that at 40,000 yards the great guns will only make 1 per cent of hits against a battleship, and, as the life of the largest guns is only about 200 rounds, they will only make about two hits during the whole time they are in existence. Even to do this, an organization of airplanes has to be furnished to observe the fire and tell the gun crew where their shots are falling.

These airplanes, if of a bombardment type, would make many more hits than do the cannon. In fact, good bombardment airplanes will make from 30 to 40 per cent of hits, at least, which will affect in varying degrees the ability of a battleship to exist.

BATTLESHIPS MAY BECOME LIKE ARMORED KNIGHTS OF OLD

Before the coming of an air force the development of war vessels on the water had been in increasing armor, increasing gun-power, and increasing speed. The increase in armor, of course, increased the weight and diminished the speed of these leviathans.

Compared to an airplane, these great vessels are very much like the knights in the middle ages, encased in their heavy armor, in which they could scarcely move, as compared to the light-armored

foot soldier, equipped with a musket. If the weapons which the air force now has for attacking battleships are so efficient, and these weapons consist only of bombs that were developed for use on land, one can expect remarkable results when this problem is studied and armament devised especially for the attack of shipping.

As the airplanes undoubtedly will be able in future wars to control the surface of the water, an air force will be the key to the command of the sea.

Submarines undoubtedly will have a great value; but, on account of their slow speed and expense, they will gradually give way to an air force, as the latter develops in its radius of action and power.

The present battleship, with its accessories, costs about $45,000,000, and for this amount of money about 1,000 bombardment airplanes can be constructed, each one of which can carry a bomb sufficient to sink a battleship. The airplanes require a personnel of only two or three men, whereas the battleships require 800 or over.

These are only some of the comparisons that interest the nations today, in culling the lessons from the last war.

As the airplane engines increase in reliability, the forced landings will become less and less, so that airplanes can act over water with almost as much security as they can act over land.

Furthermore, vessels of very high speed can be equipped to form movable airdromes, or airplane carriers, as they are called; and as these are not weighted down with armor or heavy guns, and as everything can be stream-lined on them in a way that has never been done with battleships, a corresponding increase of speed is possible.

The same airplanes that are used over the land can be used over the water, both as a means of coast defense from land bases and from airplane carriers.

A NON-STOP FLIGHT OF 4,000 MILES

So far, only airplanes, or heavier-than-air craft, have been discussed. Lighter-than-air craft, or airships, as we call them today, have taken on an increased interest because of their economical application to commerce. The rigid airship as con-

A VIEW OF BATTERY PARK, LOWER MANHATTAN, FROM THE AIR

Photograph by U. S. Army Air Service

With its incomparable towers and giant office buildings, the skyline of New York presents the most impressive of man-created panoramas.

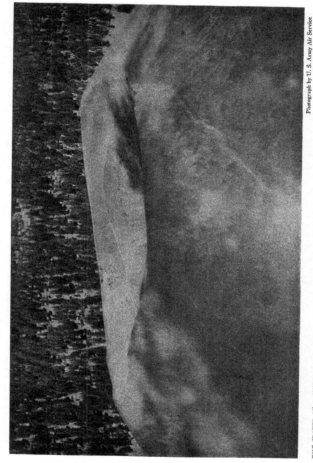

THE CRATER OF A CINDER CONE, NEAR MOUNT LASSAN, TAKEN FROM THE AIR BY THE U. S. ARMY AIR SERVICE FOREST FIRE PATROL,

Photograph by U. S. Army Air Service

Each day sees a new application of aircraft to the arts of peace, and, while the World War advanced the science of aviation more rapidly than the most optimistic prophet could have foreseen, commerce and industry are now reaping the rewards. The airplane is proving a potent factor in checking the ravages of forest fires in the West.

A SEAPLANE AT SOUTHPORT, NORTH CAROLINA, EN ROUTE FROM NEW YORK TO KEY WEST

In order to escape a heavy storm, this passenger plane made harbor at the tiny village of Southport, known for its shrimp fisheries. The photograph was taken at dawn. Recently a squadron of twelve U. S. Navy seaplanes flew from San Diego, Calif., to Panama, and despite a heavy squall in the Gulf of Tehuantepec, only one forced landing is reported. Never before had so large a squadron of any navy made a flight of nearly 500 miles without a stop.

Photograph by John Oliver La Gorce

structed in Germany has proved its great usefulness, both in war and peace.

Originally devised for reconnaissance and observation over the North Sea, these great airships were used for bombardment even against England and France, and did a great deal of work over Russia during the war. In fact, one made a non-stop flight of over 4,000 miles, to German East Africa and back again to Europe. On account of their ability to remain at very high altitudes, comparatively few were damaged or destroyed by airplanes.

At the end of the war the Germans had one airship, the L-72, that was designed to attack New York. This ship had a fuel capacity of about 17,000 miles, and was designed to be able to maintain itself at a height of 30,000 feet, with crews provided with oxygen apparatus and the engines arranged to be kept warm at this great height, which it was then impossible for airplanes to reach.

Had she appeared over New York and bombarded that city, there would have been absolutely no defense against her whatever, and this airship could have come down to any height desired.

The French now have this airship in a hangar in the southern part of France and are allowing their engineers to obtain all the technical data from her with a view to building new ones.

When a nation or organization has developed airships of this kind and worked for years on them, it is a very difficult matter for any other organizations to catch up with them and be able to construct as good ones for many years. This is on account of the engineering and structural difficulties encountered.

AN AIRSHIP CARRYING 20 TONS CAN TRAVEL 90 MILES AN HOUR

In a military way, a nation needs airships for reconnaissance at a great distance, over the land or over the water. They are needed for attacking hostile airships, for dropping explosives against targets on land or water. A large airship can carry 200 fully equipped infantrymen and drop them off in parachutes if necessary; also, it can act as a means of transport for army units or other air units.

As a large airship can transport 20 tons

or more at a speed of 70 to 90 miles an hour, more points can be reached than would be possible in other ways. In fact, in most places in the United States today, with the roads that we have, it would be impossible to supply an army of any size over them; and as railroads exist only in certain places, airships will be the natural means of supply and should be developed accordingly.

For commercial purposes the airship offers very interesting possibilities. There is no dust or smoke or unpleasant experience in traveling by them. The degree of safety, with a proper ground organization of airship stations, is very great. In fact, the Germans have carried over 200,000 passengers without a fatality.

For communication across the Pacific, or particularly to South America, airships will be a very efficient means of travel.

When we turn to airplanes and commercial possibilities we are immediately concerned with the high cost of the necessary ground organization. Due to their limited gas capacity and the fact that their engines must be running at all times to stay in the air, being heavier-than-air, every provision has to be made for guiding them properly across the country and having airdromes, or landing places, at convenient intervals. An organization of this kind we call airways, and without it no real commercial development of aviation is possible.

These airways should be established by the government; should join the principal centers of population by well-marked routes through the country, and provide airdromes at 200-mile intervals, where facilities for repair, fuel, and proper attendants on the airplanes can be obtained. Between these places emergency landing places should be distinctly marked, so that a landing could be safely made if trouble occurred.

Of course, this seems impracticable from the standpoint of cost; but when one considers that automobiles have to have roads everywhere they go and gas stations from which they can get oil and gasoline, one concludes that the establishment of airways through the country would be not nearly as expensive and would tremendously facilitate rapid communication from one part of the country to another.

IT COSTS $60 AN HOUR TO OPERATE AN AIRPLANE TODAY

The cost of operation of airplanes is very great, the average.at this time being about $60 an hour, and is about 20 to 50 per cent more for seaplanes. In fact, seaplanes, or flying-boats, are the most expensive aircraft to operate.

Of course, seaplanes have the advantage of being able to land in many places not provided with airdromes. This is particularly so in Canada, where a very great percentage of the surface of the country is covered with water and lakes. In fact, seaplanes can fly all over Canada and almost always have a landing place within gliding distance. On the other hand, seaplanes are worthless in the winter time, when the lakes and rivers are frozen or have floating ice on them.

It is estimated that carrying passengers costs almost 70 cents per mile.

The great advantage about airplanes is their speed; and it has been found that the higher one is able to get, the greater the speed that can be obtained, on account of the lessened resistance of the air. At 30,000 feet, speeds of over 200 miles an hour have been obtained, with ordinary equipment, merely adapted for that purpose and not constructed specifically for it.

The way that these high altitudes are obtained is by the use of a turbo-booster in connection with the gasoline engine. The turbo-booster is a turbine which is actuated by the exhaust from the engine, this in turn works an air compressor, or air pump, that delivers compressed air to the carbureter of the engine, thereby keeping up a proper mixture, which otherwise would be lost on account of the rarefaction of the air at high altitudes.

By the use of this device greater altitudes will be obtained and the power of an engine kept up in a remarkable manner. For instance, by using the turbo-booster with a 400 horse-power engine, this engine will deliver as great horse-power at 30,000 feet as a 1,000 horse-power engine would without the turbo-booster.

SPEEDS OF 300 TO 400 MILES PER HOUR ARE POSSIBLE

We believe that, with the development of equipment of this kind and the construction of special airplanes to be used at very high altitudes, speeds of 300 miles an hour and over can be obtained.

At these altitudes, also, there are regular winds blowing, similar, in a way, to the trade winds, some of which have a speed of over 100 miles an hour; so that, by combining the speed of our airplanes with the speed of the wind, we will get velocities of from 300 to 400 miles an hour in many cases. This will enable very rapid communication between America and Europe, San Francisco and New York, or other places of corresponding distances. This class of transportation we believe is not very far away.

The United States has the best climate, all the natural resources for the construction of airplanes, the manufacturing ability, the engineering ability, and its men make the best pilots in the world.

From the standpoint of national defense, we can get greater security, dollar for dollar, from an air force than we can from any other military element.

VOL. XXXIX, No. 4   WASHINGTON   APRIL, 1921

MODERN PERSIA AND ITS CAPITAL

And an Account of an Ascent of Mount Demavend, the Persian Olympus

By F. L. Bird

For five years American college instructor in Teheran

ALONG series of catastrophes has failed to bring to a final anticlimax the Persian chapter in the world's historic record. The crushing impacts of Greek, Roman, Arab, Mongol, and Russian armies could never quite remove the epic charm from the story of Persia and her people; and now another world upheaval renews a waning interest in the trend of her affairs.

Babylon, Assyria, and Chaldea rose to power in rapid sequence, served their day of world dominion, weakened, and quickly disappeared. Persia, following in their footsteps, elevated southwestern Asia to still higher eminence as the center of civilization and empire; struggled with Greece for European hegemony; disintegrated; but maintained its entity, through a diverse ebb and flood of fortune, down to the present day.

Darius would fail to recognize as his mighty empire the narrow limits of modern Iran, its borders now far withdrawn from the waters of the Oxus and the Indus, from the shores of the Mediterranean and the widespread Mesopotamian plains; but the nucleus still is there in territory, race, language, and customs.

PERSIA IS THREE TIMES THE SIZE OF FRANCE

Persia of today includes within a territory still three times the size of France ancient Media, mountainous Parthia, and the province of Fars, whence sprang her first great dynasty. Such monuments to the glory of the great kings as the ruined capitals of Susa, Persepolis, and Ekbatana still stand on Persian soil (see map, page 418).

The majority of the present inhabitants, although tinged with the blood of Greek, Arab, Turk, and Mongol conquerors, are the lineal descendants of the original Iranian, or Aryan, population, and speak a language which has for its basic element the ancient Persian tongue.

The Mohammedanism of their Arab conquerors penetrated to the foundations of Persian life, and the ravages of Turk and Mongol often threatened their very existence, yet their national characteristics and culture have time and again triumphed over their oppressors.

Time after time, as the centuries passed, Persia has drawn together her scattered provinces and with surprising virility, renewed now and again by the infusion of foreign blood, has forced back the contracting circle of encroaching enemies; but during the last century the increasing power of her neighbors, combined with her own decay, has definitely turned the scale against her and she has drawn behind her last barriers—the mountains and deserts which doubly guard the western portion of the Iranian

Photograph by Roland Gorbold

A PERSIAN GENDARME

The police service of Teheran was entrusted to Swedish organizers in 1913. The gendarmerie numbers 8,400 men. Before the World War, Russian officers trained the Persian army.

plateau, a lone remainder of her inheritance (see map, page 418).

Sultan Ahmed Shah, the one hundred fifty-sixth "king of kings," sits on the tottering Persian throne, while the future of his kingdom rests in the hands of outside powers.

A VAST, MOUNTAIN-RIBBED DESERT PLATEAU

Modern Persia, with the exception of the prosperous northwest province of Turkish-speaking Azerbaijan and the semi-tropical region between the Elburz Mountains and the Caspian Sea, can be characterized as a vast, mountain-ribbed desert plateau, studded here and there with oases which most frequently form ribbons of fertile green fringing the desert at the bases of sterile mountain slopes from whose snow-clad summits comes the life-giving moisture.

The encircling mountain walls shut out the rain from the central table-land. Rivers with sources but no mouths flow half the year and lose themselves in the parched desert wastes.

The density of population is less than that of Texas, and more than half the country is an uninhabited Sahara, some of it unexplored. Much of the remainder is suitable only for sheep-grazing part of the year, thus forcing upon a fourth of her ten million people a semi-nomadic existence between the high, well-watered mountain valleys in the summer and the warm plains in the brief winter season.

Some of these tribes, like the Kurds, rarely leave their mountain homes, where they exist independently of central government control. Others, like the Ghashgais and Bakhtiaris, sometimes by coercion and sometimes through necessity of political alliance, are vassals of the state, although they pay allegiance only to their chiefs, who arrange with regal authority for their followers the matters of taxes and military service.

Cities are naturally few and small, there being but two or three of more than 100,000 inhabitants. The lower mountain valleys and the oases are the centers for both town and agricultural population, and the wonderful fertility of these scattered areas, snatched from the blighting grasp of the desert, forms the basis

PERSIAN WOMEN IN INDOOR COSTUME

This modified ballet attire was introduced from Europe in the latter part of the nineteenth century. The story is told that Naser-ed-din Shah, upon attending the opera in Paris during a European visit, was much attracted by the ballet and ordered that the entire front row be purchased at once for his harem, which already contained several scores of carefully selected beauties. Disappointed in this desire, he had to be content with the adoption of the ballet skirt by the ladies of his harem, whence the new fashion rapidly spread.

Photograph by Roland Gorbold

A BEAUTIFUL TRIBESWOMAN OF BAKHTIARI
LAND, WEST OF ISPAHAN

She is a pure Iranian type, rare in these days.
Her costume is a negligée worn only in the
home. These tribeswomen are not secluded;
they ride and shoot like the men, and, of course,
wear a more practical costume, including balloon-like trousers, when in the saddle.

for the startling contrasts in the climate
of this unusual country.

Water is the chief concern of the Persian peasant. Wherever he can divert
the flow of a mountain stream or build a
crude canal from a well or spring, a small
portion of the desert becomes a paradise
and he prospers. Certain of these regions are said to be among the most fertile in the world, producing in abundance
not only the finest of wheat and barley,
but grapes, apricots, peaches, nectarines,
pomegranates, figs, and melons which are
unsurpassed among the fruits of the
Temperate Zone. Cotton and tobacco
thrive, and roses, as well as other flowers,
gloriously deserve the frequent association of their names with that of Persia.

A LAND OF CONTRASTS—ROSE GARDENS AND DESERT

It is the desert contrast that has made
the Persian poets sing of rose gardens
and of nightingales. Dwelling in a land
of barrenness, where the cooling shade of
trees, the refreshing greenness of vegetation, the life-giving productiveness of the
soil itself are possible only by struggling
years of human toil; where only high
mud walls guard tiny groves of elms,
chinars, and poplars, so carefully reared
along priceless flower - edged watercourses, from the encroaching. waste
without, is it to be wondered at that they
cherish these artificial beauty spots and
idealize them as typical of heaven itself?

The day is at hand, as one of the by-
products of the war, when Persia is to
begin to learn from British experts, not
only how to reclaim more desert land by
building better aqueducts and by throwing barrages across mountain gorges to
store the surplus of the spring freshets,
but how to establish closer communication with the outside world and to develop her great potential resources.

Lacking in the energy, initiative, and
coöperative spirit necessary to develop
their country themselves, the Persians
have suffered from the jealous rivalry of
their neighbors, and from a seclusion
forced by nature, but belied by their central geographical location, in all the recent history-making disturbances in the
Near and Middle East.

In spite of her position as a veritable

Asiatic Belgium, Persia is strangely cut off from world intercourse by those same natural barriers which so affect her climate.

PERSIA'S FIRST HIGHWAY BUILT IN 1900

At the opening of this century not a single highway suitable for wheeled conveyances pierced the mountains to the plateau. Handfuls of foreign officials and infrequent venturesome travelers made their toilsome way by caravan over tortuous passes to the Persian capital or to other Persian cities, and the Persians themselves for the most part stayed at home. But about 1900 a government-subsidized Russian company opened a post-road, as a military-commercial venture, which climbed from the Persian port of Enzali, on the Caspian Sea, to the capital city, Teheran.

Five years ago three or four post-carriage routes and a narrow-gauge railway running five and a half miles from Teheran to a suburban shrine were the only competitors of the picturesque but slow-moving caravan.

Then came the pressure of the World War. Russia rapidly substituted a military railroad for the old carriage route from her Transcaucasian boundary to Tabriz, the provincial capital of Persian Azerbaijan; but again the mountains intervening between this projecting northwestern corner of Persia, with its Tatar population, and the Persian-speaking portion of the country, have prevented this from being of more than local advantage.

HOW THE WORLD WAR REDISCOVERED PERSIA

It was a more famous road, however, over which Persia's neutrality became a mere expression. Almost from the dawn of history a great international highway has held its threadlike course through the plains of Mesopotamia from Babylon, Ctesiphon, and Bagdad to the western scarp of the Zagros Mountains, spiraled up this mountain stairway, and continued its way down trenchlike mountain valleys, over wind-swept passes, under the great Behistun rock, which still bears the triumphant inscriptions recorded by Darius and his successors, and through

Photograph by Roland Gorbold

A LUR TRIBESMAN FROM THE MOUNTAINS OF WESTERN PERSIA

These nomadic inhabitants of Luristan are the lineal descendants of the old Iranian stock of the time of Darius and Xerxes. This young man's tribe is indicated by his figured sash, which has been carefully adjusted from a twenty-yard length of cloth.

THE LAGOON OF ENZALI, PERSIA'S CHIEF SEAPORT ON THE CASPIAN SEA

The first glimpse which the Western traveler gets of Persia on approaching Enzali from Baku is an impressionistic one—thatched or red-tiled roofs of the low-lying town, a wealth of wide-branching trees, and in the distance the dusky, cloud-mantled mountain range which bars entrance to the desert hinterland, the real Persia.

NATIVE CRAFT IN A PERSIAN HARBOR

This is Piri Bazaar, at the upper end of the lagoon leading inland from the Persian port of Enzali, on the Caspian Sea. Here shipments bound for the interior receive their final transfer to caravan, cart, or motor lorry.

ALONG THE HIGHWAY FROM ENZALI TO RESHT

"Near fragrant orange groves, past lily-padded lagoons, and through flower-carpeted jungles alive with an endless variety of semi-tropical song-birds and water-fowl, the traveler proceeds from Enzali to Resht" (see text, page 365). The journey is usually made in a Persianized Russian drosky.

Photographs by John B. Jackson

A PERSIAN PEASANT'S HOME NEAR RESHT

Water is the chief concern of the Persian peasant. Wherever he can divert the flow of a mountain stream a small portion of the desert becomes a paradise and he prospers (see text, page 356).

Photograph from Faye Fisher

THE SHOPS AT RESHT, ON THE ROUTE FROM THE SEAPORT OF ENZALI TO THE PERSIAN CAPITAL. Resht is a town of 60,000 inhabitants, with sodden roofs, narrow alleys, and crumbling walls. These queer little shops have their fronts open to the street. The shop-keeper sits where he can reach everything without rising.

great cities on whose sites stand modern Persian towns, across the Iranian plateau.

Making its way eastward, where the foothills of the Elburz Mountains meet the northern edge of the great central desert of Persia, it passes by way of Meshed, Merv, and Bokhara into central Asia and on to the borders of China. In time of peace it has served as a thoroughfare of commerce between Mesopotamia and China; in time of war it has directed the march of armies.

The war chariots of Cyrus and Xerxes rumbled over it, Alexander led his cohorts along it to Asiatic conquest, Persian liberty and religion fell before the Moslem Arabs who surged through this gateway in the seventh century. Still later it gave too ready passage to the devastating Mongol hordes of Hulagu and Genghis Khan; then, as Persia's power declined and ocean trade routes opened up, this highway accompanied its builders into sad dilapidation.

But once more, with the outbreak of the World War, remote Persia came within a scheme of world conquest; and German, Turkish, Russian, and British armies fought along this crumbling highway, where Turkish ox-carts in retreat outdistanced the motor trucks of the Russians.

Before the war was over a new, well-macadamized road existed, built by the British Royal Engineers—a road which is but the forerunner of a railway, already in operation to the Persian frontier near Kasr-i-Shirin, that will connect Teheran, and eventually all of Persia, with the Bagdad Railway at Bagdad.

So the war has rediscovered Persia and brought new prominence to the affairs of its capital.

A CITY AS OLD AS PERSIAN HISTORY

No one knows how long there has been a city where this modern capital stands. It has not always been called Teheran, and it has not always been in quite the same spot; but a city has existed in this locality as far back as Persian history records, for so suitable a location could not well be overlooked.

The present city stands 3,810 feet above sealevel, at the foot of the inner slopes of the Elburz Mountains, which

rise at this point nearly 13,000 feet. To the southeast is the great, lifeless desert, shaped like a huge hour-glass, 900 miles in length, from the foothills of the Elburz range, in the north, almost to the Indian Ocean, in the south, and ranging in width from 300 to 100 miles.

This untraversable wilderness determines this point as the junction of the great trunk route from Mesopotamia into the East, the north and south road through central Persia, and the old caravan trail westward through Kazvin and Tabriz to the Black Sea.

TEHERAN HAS ARIZONA CLIMATE

Passes through the Elburz from the Caspian Sea converge upon Teheran from the east and west; and water, whose presence is of such supreme importance in the location of a Persian city, is here in abundance. The annual rainfall is only ten inches, but in the mountains which overshadow the city are springs, wells, and rushing streams; and not far to the northwest and northeast the Karaj and Jajrud rivers burst from their mountain gorges to irrigate the rich surrounding plains which produce the city's food supply.

The district, which lies in about the latitude of Cape Hatteras, has a temperate, healthful climate which is invigorating and pleasant during nine months of the year. The three summer months are excessively hot and dry, but if one wishes the luxury of a summer resort, it is necessary only to load one's belongings on a string of donkeys or a springless cart and move six or eight miles to one of the cool mountain villages, where the six or seven hundred members of the foreign colony and many of the Persians take refuge from the heat.

Teheran weather is similar to that of Arizona, but several degrees cooler, both in summer and winter. The nights are always cool, the sun shines nearly every day of the year, the winter is brief and moderate, and the long spring and autumn are like those seasons in delightful southern California.

Although occupying an ancient site, Teheran is a very modern city. It has been the capital of Persia only a little more than a century, and has been an

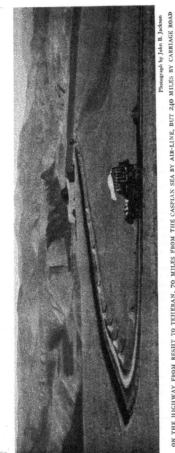

Photograph by John B. Jackson

ON THE HIGHWAY FROM RESHT TO TEHERAN, 70 MILES FROM THE CASPIAN SEA BY AIR-LINE, BUT 240 MILES BY CARRIAGE ROAD

The trip over the mountains is made at a cost of 50 cents a mile, but the varied scenery and the thrills of travel in the "débris of a coupé of a long-forgotten era" is worth the price (see text, page 367).

important metropolis for a much shorter time than that. Rhages, or Rei, its predecessor in this district, was a populous city of ancient Media, thrived in the middle ages, is said to have had a population of one and a half million, and to have been the largest city east of Babylon, but found itself too centrally located for its own permanence a n d continued prosperity, when the Mongols swept through western Asia.

When the inhuman Agha Mohammed Khan, after a bitter civil war, in 1793 founded t h e Turkish dynasty which now rules over Persia, he did not dare to establish the seat of government at a spot so far removed from the pasture lands of the Kajars as Shiraz, the former capital. So Teheran, which not only commanded the highways of the plateau, but also the entrance to the Elburz passes leading to this tribe's original possessions on the southeastern shores of the Caspian, became the capital of the K a j a r kings.

At that time the new capital, which had been wiped out by the Afghans in 1723, consisted of not more than three thousand houses of sundried brick. A European traveler who visited Teheran in 1796 wrote that, "In spite of Agha Mohammed's efforts to induce people to settle and merchants and manufacturers to establish themselves there, t h e

Photograph by J. W. Cook

TWO AMERICAN WOMEN TRAVELING A LA MODE IN PERSIA

These two-passenger vehicles, called *kajavehs*, resemble chicken-coops balanced on the back of a diminutive horse or donkey (see text, page 371).

population does not amount to more than 15,000 souls, including a garrison of 3,000."

But once the prestige of the new dynasty was established, this mud-walled hamlet grew with amazing rapidity, considering the decadence of its surroundings; and today, with its 300,000 inhabitants and its foreign colony representing at least a score of nationalities, it is not only the metropolis of Persia, but a city of considerable international importance.

The substitution of internal peace for anarchy in the country was bound to repopulate this productive district, but the shifting of trade routes westward, through the rise of Russian commerce, largely restricted Teheran's commercial importance to that of a local distributing center. It grew, therefore, like Washington or Petrograd, because it was the capital; because the Kajars were ruling with a firm, steadying hand, and because Persia was being forced into the widening commercial and political plans of the great powers.

"THE FOOT OF THE THRONE"

The Persians commonly refer to Teheran by its title, "The Foot of the Throne." As the Kajar sway extended,

all the chiefs of the royal tribe, all the great nobles, wealthy land-owners, and famous generals, and all the hungry office-seekers — in fact, all those who wished to bask in the sunlight of royal favor—crowded into Teheran and set up their lavish establishments. With them came their obsequious hordes of parasitical clients, and after them the poets, scholars, buffoons, and quack scientists, a shabby band of flattering pensioners who by nimbleness of wit would live by royal or noble patronage.

The leading merchants and architects and the most highly skilled artisans of the country found ready employment, and the bazaars began to resound with the metallic tapping of the silversmith from Shiraz and the coppersmith of Ispahan, to be scented with the pungent-smelling product of the Hamadan tanner and to overflow with the wares, not only of Persia, but of Manchester, Birmingham, and Moscow.

Naser-cd-din Shah, who ruled justly and well for 47 years, till in 1896 he was removed by the hand of a fanatical assassin, traveled extensively in Europe and introduced many modern western institutions to beautify his capital, such as broad streets, substantial buildings, frock

Photograph from F. L. Bird

THE "MORNING EXPRESS" LEAVING TEHERAN ON PERSIA'S FIRST FIVE MILES'
OF RAILWAY (SEE TEXT, PAGE 387)

In 1916 a Russian company opened a 67-mile railroad from Julfa, on the Perso-Russian
frontier, to Tabriz, and the British Government is now coöperating with Persia in the construction of railways, but for more than a quarter of a century this little narrow-gauge line
running from the southern end of Teheran across the hot plains to the village of Shah
Abdul Azim, the seat of a golden-domed shrine, was Persia's only railway.

coats, ballet skirts for the feminine population, and a dozen or more foreign
legations and consulates.

During his long, peaceful reign the
city outgrew the old mud walls which
had inclosed it within a four-mile circuit.
They were torn down, the moat filled in
to provide more building sites, and a
larger and more extensive earthen em-
bankment, pierced by twelve great gates
and surrounded by a huge dry ditch, was
thrown up, giving the city its present
size and contour, that of an irregular
octagon more than twelve miles in circumference.

The new area quickly filled, and now
the city has outgrown this latest boundary and residences are springing up outside the wall, which soon will disappear,
like other obsolete oriental institutions,
to provide for necessary modern growth.

For very obvious reasons, travel agencies do not feature Teheran in their world
tours. Nevertheless it is usually quite
accessible, if one will tolerate a little discomfort while getting there. At least
nine-tenths of all travelers to Teheran

use the Caspian port of Baku in Transcaucasia as a jumping-off place.

With good luck, in peace times, one
can reach Baku over the uncertain route
via Constantinople and the Black Sea
to Batum, and thence by rail through
Tiflis, the capital of the new Republic
of Georgia.

THE FIRST GLIMPSE OF PERSIA SUGGESTS
SUMATRA

Baku itself is the capital of Azerbaijan,
another of these precocious backyard
republics, which, by the way, is not identical with the neighboring Persian province
of the same name. Its importance as the
center of a great oil district has given it,
in spite of its polyglot population of
Tatars and Armenians and the fifty-seven
other varieties of the Caucasus, a familiar
and prosperous western appearance. But
upon embarking on the little Russian
steamer which navigates the Caspian between Baku and the leading Persian seaport of Enzali, one takes passage to another world as surely as though crossing
the Styx.

Photograph by John B. Jackson

THE NEWER, OR NORTHERN, PORTION OF TEHERAN BOASTS A TRAMWAY

Wide, well-graded streets, electric lights, motion-picture theaters, European shops, and semi-Western architecture distinguish this section of the Persian capital from the southern section, with its great bazaars, narrow, twisting alleys, and blind-fronted adobe house walls (see text, page 371).

The first appearance of Persia is disconcerting, because it does not look like Persia. It agrees very well. with what one might expect of Mindoro or Sumatra, but the standard requirements for the "Land of the Lion and the Sun" are conspicuous by their absence.

Soon after the uncertain haze to the south has resolved itself into shore-lines, comes one's first impressionistic glimpse—the thatched or red-tiled roofs of the low-lying town; then a wealth of wide-branching trees, the outposts of a dark, enveloping mass of jungle; and behind this, and rising swiftly to unbelievable height, the dusky, cloud-mantled mountain range which bars entrance to the desert hinterland, the real Persia.

If the exotic luxuriance of vegetation and the careless primitiveness of the thatched huts and rustic booths of the inhabitants disturb your preconceived visions of the country, you will find them fading with shocking suddenness at your first introduction to its population, when the boat ties up at the pier and an ill-smelling rabble of ragged, half-naked villains swarms on board to wrangle about getting your luggage ashore.

A courteous, frock-coated Persian official, conventionally crowned with what appears to be a cross-section of an opera hat, passes you through the ceremonies of the custom-house, and in a brief space of time you are rolling inland in a Persianized Russian drosky, near fragrant orange groves, past lily-padded lagoons, and through flower-carpeted jungles alive with an endless variety of semi-tropical song-birds and waterfowl.

Arrival at the city of Resht after a twenty-mile ride of the rarest kaleidoscopic loveliness is certainly a transition from where every prospect pleases to where only man is vile.

The sixty inches of annual rainfall, which have made the surrounding country a Garden of Eden, have conspired with man's inventive genius to turn this town of 60,000 inhabitants, with its sodden roofs, narrow, slimy alleys, and crumbling walls, into an odorous, undrained mudhole, a veritable Slough of Despond to any one with such lofty illusions of Persia as those of a certain disgusted American traveler who had gone all the way to Arnold's "majestic Oxus stream" only to find it muddy.

A 240-MILE MOUNTAIN TRIP TO ADVANCE 70 MILES

The trip over the mountains, with its ever-changing variety of unusual sensa-

THE PERSIAN MERCHANT SEEKS THE BUYER IN THE HOME

The striped articles on the business man's shoulder are bath towels, which consist of large cotton squares. It is a common sight to see men with these towels wrapped around them coming from the public baths.

tions, even at fifty cents a mile, is worth the price.

Teheran is only seventy miles south of the Caspian, but the road must climb and twist for 240 miles in order to arrive there.

At the post-house your means of transportation awaits you. You clamber gingerly into the débris of what may have been in a long-forgotten era a very elegant and commodious coupé, but which now, with your variegated assortment of luggage lashed to every available projection, approximates more closely an itinerant peddler's van than anything else.

A dark-visaged bandit, whom you have been regarding with suspicion, pours a pail or two of water on the warping wheels and axles for lubricating purposes, clambers to the driver's box, leers back from under his huge, pot-shaped felt hat, grunts to the four gaunt ponies harnessed abreast, and you clatter off with a jangle of bells along the well-built Russian road to Teheran.

Theoretically, if one travels day and night, the trip requires a day and a half; actually, about twice that long. At sixteen-mile intervals there are exasperating delays, capable of abbreviation by the judicious use of baksheesh, in the changing of horses and drivers, to say nothing of the additional halts while the driver has his tea or pipe of opium at a roadside tea-house or ties up a broken spring or dilapidated harness with a bit of rag. But, unless the whole equipage rolls over a cliff while the *kismet*-trusting coachman takes a nap, you are almost certain to turn up eventually at the walls of Teheran.

THE HOME OF THE HYRCANIAN TIGER

From the oppressive humidity of the region of the rice-fields, the road gradually ascends to the shade of the deep forest belt, whose labyrinth of close-growing trees and interwoven giant creepers forms a dark, silent lane, with impenetrable green walls, into which the carriage intrudes with its ceaseless rumble of heavy-tired wheels and the constant jingle of the pony bells.

Perhaps a passing band of hunters, with a freshly killed wild boar or leopard,

serves to break the solitude and to remind the traveler that from this very jungle, which extends without a break for hundreds of miles along the northern slopes of the Elburz, came the fierce Hyrcanian tigers used by the Roman emperors for their spectacular contests of the arena.

Sometimes the road follows the Sefid Rud, or White River, the only stream to break its way through to the Caspian from the plateau; or leaves it to follow the brink of a canyon.

The abruptness of the ascent increases; the forest area is left behind; thriving vineyards, bearing luscious seedless and skinless fruit, cover the hillsides; occasional clumps of olive trees appear, and frequent groups of tattered peasants stare curiously at the infidel invader in the passing carriage.

The aspect of the country now changes rapidly. All signs of habitation, except a few wretched dugouts, disappear, and the old coach climbs heavily, over barren rocks, to the bleak summit of the pass, 7,000 feet above the sea.

TRAVELING BY THE LIGHT OF THE PERSIAN MOON

The journey by daylight is novel, by night it is weird. The dubious accommodations at the post-houses render the extreme night cold and the uncertainties of the dark road the lesser of two evils; so perchance the traveler finds himself bowling along the upper reaches of the pass in the soft light of the Persian moon, which smoothes the jagged outlines of the surrounding crags, works fairy magic with the snow patches on the neighboring peaks, and reveals the silver flood of near-by dark-walled gorges.

The night wears on and the moon slips down behind a distant ridge, leaving the cool, gray stars to light one on the way. With only the monotonous roll of heavy wheels to break the perfect stillness of the night, a dreamy, drowsy feeling creeps over one, when out of the darkness and far away there comes a faint suggestion of strange, uncanny music; and as the night breeze freshens, it bears a deep and rhythmic ringing, which slowly grows in volume until the mellow donging of a hundred swinging bells pro-

Photograph from Faye Fisher

THE "HOKEY POKEY" PEDDLER OF THE WEST HAS HIS COUNTERPART IN THE
HOT-SOUP VENDER OF PERSIA

These peripatetic caterers to hungry youths and adults do a thriving business on the streets
of Persian cities morning, noon, and evening.

A PERSIAN WATER-CARRIER AND HIS SHEEPSKIN JUG ON HIS ROUNDS TO SERVE
THE THIRSTY

The entrances to many Persian doorways are rather attractive with the mud bricks joined by
blue cement. The quaint door-knockers are admired by all foreigners.

Photograph from Faye Fisher

AN OPEN-AIR RESTAURANT IN PERSIA

These public cooks are preparing mutton, which is cut in small pieces, skewered on a long iron pin, and broiled over hot beds of charcoal. It is very delicious. The flat sheets on the ground at the right, resembling bits of paper or cloth, are pieces of native bread which comes in strips only a half inch thick but two and a half feet in length. A piece is torn off and the mutton *en brochette* when well done is taken off the pin and eaten with the bread.

claims the near approach of a camel caravan.

Then comes the *mush, mush* of padded feet; shadowy, ungainly forms loom out of the darkness, and camel after camel shuffles past, bearing a slumbering driver swaying aloft in the folds of his rough felt mantle.

After a rapid descent through barren gullies comes a sweeping view of actual Persia.

Broad, brown, rolling plains extend beyond the limit of vision, even in the clear, thin air of the plateau, and the naked southern scarp of the mountains shows not a vestige of green. At lower levels irrigating ditches, which seem to flow uphill, sluggishly follow the curving hillsides; orchards and mud-walled gardens begin to appear; and before long the turquoise domes and crenellated walls of the city of Kazvin come in sight.

Passing through a gaudily tiled gateway (see Color Plate VI), the route leads along a wide avenue shaded by beautiful plane trees to a pretentious and much-ornamented building, which is nothing more nor less than the post-house and hotel, where one may actually occupy a spring bed or eat a passable attempt at a European dinner.

A large portion of the last ninety miles between Kazvin and Teheran is a monotony of drab, stone-covered waste, of which the road itself is an almost indistinguishable part.

The route is level and parallels the great northern mountain rim of the plateau. The only sign of approach to an important city is the increasing traffic on the road, not only of the leisurely mule and camel caravans, but of primitive prairie schooners, with wild-looking, shaggy-hatted drivers, and bare-legged villagers driving strings of heavily laden little donkeys to market.

Even an occasional Persian family, evidently moving to the metropolis, jogs along, the head of the family astride an undersized mule, which is enveloped in capacious saddle-bags bulging with copper kitchen utensils; the good wife, fetchingly attired in lavender hose and balloon-like trousers, perched on a pile of bedding under which labors the counterpart of her husband's mount, and the numerous offspring distributed between two huge chicken-coops slung on the flanks of a diminutive, moth-eaten donkey.

THE SUDDEN TRANSITION FROM DESERT TO CITY

So sudden is the transition from desert to city that before one realizes that the journey is at an end he finds himself clattering across the stone causeway over the moat toward the most surprising of gateways, a great multicolored façade overlaid with a gay mosaic of glistening tiles and topped with numerous minarets ornamented in the same fashion. Over the iron-bound gate itself is wrought in glazing of many colors some stirring scene from Persian mythology, and the flanking walls are niched with tiled and arched recesses.

Teheran is one of those numerous cities between the Near and the Far East which calls for a modification of Kipling's oft-quoted line; for here East and West have met, but have not mixed.

WITHIN THE WALLS OF TEHERAN

Undoubtedly the strongest impressions for a stranger to the Orient when first entering the city are those made by the wide, shaded avenues, with their bordering high mud walls inclosing beautiful gardens and palatial residences, as contrasted with the noisy, primitive street life itself.

For the most part the buildings, the homes of the middle class, are of one- or two-storied, flat-roofed, adobe construction, many of them plastered dingy white or pale blue or pink and with projecting balconies. At intervals rows of slim poplars project above the street walls, and through a stately gateway one gets a glimpse of conventionally ordered flowers and shrubbery, spraying fountains, and the brick or stucco residence of some Persian prince or noble.

Outside, along the streets there is the hubbub of an overgrown Persian village, increased by the clatter of carts and droskies, and the raucous shouts of impudent hucksters, mixed with the importunate chattering of repulsive beggars.

In the city, as in the country, the ugly wall is a symbolic dividing line between present-day development and yesterday's primitiveness.

The northern portion of the city, built up largely during the last generation, is quite different from the southern, or older, section. This newer part, the product of western influence, has many wide, well-graded streets, some of them lined with elms and plane trees; and it boasts of a tramway, electric lights, motion-picture theaters, hotels and restaurants, European shops, and numerous respectable buildings of semi-Western architecture.

In this area are located some of the government buildings, the foreign legations, the homes of the foreign residents and of most of the wealthier Persians, numerous foreign business and philanthropic institutions, including the large American mission schools and hospital, and also the small Armenian and Parsee quarters.

Photograph from Faye Fisher

THE ALMOST BLIND LEADING THE REALLY BLIND IN PERSIA

There are many blind persons in Persia, owing partly to the intense light rays of the sun. Tradition gives the following origin for the wearing of veils by Mohammedan women: One day when the Prophet was seated with his favorite wife, Ayesha, a passing Arab admired her, expressed a wish to purchase her, and offered a camel in exchange. This experience so angered Mohammed that the custom of requiring women to wear veils resulted.

PERSIAN LADIES LEAVING A PUBLIC BATH-HOUSE PRECEDED BY A DOMESTIC SERVANT

Every Friday is "bath day" in Persia, and a bath is obligatory before the faithful can worship. Frequently there is a public bath attached to the mosque.

Photographs by Lt.-Col. Alfred Heinicke

PERSIAN WOMEN IN CHADARS

Both Christian and Mohammedan women wear the *yashmak* (veil) out of doors, but the *chadar* (chuddar), or enveloping garment, is peculiar to the followers of Mohammed.

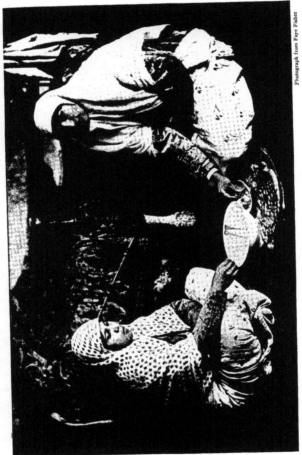

A REPAST IN THE WOMEN'S QUARTERS

Persian women have their faces unveiled only in their quarters, where no men trespass. The meal consists generally of a thick soup and bread. A piece of bread is broken off and used as a scoop, for very few people use silverware. The kalian, or water-pipe, is smoked by both men and women (see also Color Plate VII). It is so arranged that the smoke goes first through the water, which purifies it before it is taken into the mouth. The tops to these pipes are frequently studded with turquoise.

374

The southern part of Teheran is an undisturbed bit of old Persia—the great bazaars, the narrow, twisting, dirty alleys with their filthy gutters, the blind-fronted adobe house walls, and the unkempt, sickly people, who stare or hoot at the foreign interloper, all representative of that major portion of Persian city life as yet untouched by Western ways.

ARTILLERY SQUARE IS THE HEART OF TEHERAN

The city centers around a large public plaza, the Maidan-e-Toop Khaneh, or Artillery Square, which has been developed into a public park, where a number of antiquated cannon, the spoils of former conquests, are exhibited.

Fronting the eastern end of this square is the headquarters of the British-managed Imperial Bank of Persia, a striking building of gayly adorned Perso-European architecture. Facing the square on the other three sides are arched and balconied military barracks.

Six important avenues lead, through brilliantly tessellated, arched gateways, from this inclosure. From the northeastern corner Khiaban-e-Lalehzar, or Tulip Field Avenue, the chief business street, runs north, past the post-office and custom-house. From the corresponding northwestern corner the Khiaban-e-Alacd-Dowleh, along which are located many of the important legations and the two European hotels, parallels this street.

Long tramway streets run east and west; from the southeastern corner a busy thoroughfare leads to the bazaars, and through the southwestern gateway passes the broad, tree-arched Khiaban-e-Almasieh, the Avenue of Diamonds, to the royal palace.

The palace, with the treasury, foreign office, royal college, telegraph department, and various other government buildings, is located within the old, mud-walled citadel.

One is struck by the abundance of clear, flowing water in the well-kept palace gardens, and although the buildings themselves are architecturally and structurally disappointing, they are substantially built of brick and exceedingly interesting because of their bizarre and fantastic exterior and interior decoration.

The royal museum is well worth a visit, for in a somewhat amusing conglomeration of trinkets, ranging all the way from an American company's sewing-machine advertisements to a collection of mechanical clocks, there are many rare treasures, among them being the sword of Tamerlane, the famous jeweled globe, and either the original or a replica of the jewel-studded Peacock Throne supposed to have been taken by Nader Shah in the sack of Delhi.

Teheran's handicap as a modern city is felt in her lack of fine historic institutions. There are no mosques or religious colleges of any distinguished antiquity or holiness, although modern ones are numerous. The finest is the Masjid-e-Sepahsalar, which was built by a former prime minister. It stands in the northern part of the city, near the Baharistan Palace, at one time the residence of this same official, but since the granting of the Constitution in 1906 occupied by the Persian Parliament.

TEHERAN HAS ITS AVIATION FIELD

A somewhat unusual point of interest is the great Maidan-e-Mashk, or Drill Square, a forty-acre military parade ground in the midst of the city, not far north of the central square, which is one of the largest inclosures of its kind in the world. At present it is used chiefly as a race-course, and by the young Persians, who are enthusiastically adopting this Western game, as a football field. It is also proving an admirable flying field for recently introduced airplanes.

There is a splendid, unobstructed view of the great mountain range north of the city from this large field, as well as of the mighty snow-clad cone of Demavend, which, off to the northeast, holds its solitary position nearly four miles upward in the clear blue heavens (see text, pages 393 to 400).

However unalterable the laws of the ancient Medes and Persians may have been, it would be incorrect to speak of present-day Persia as unchanging. The traveler who reaches this conclusion after noting habits and customs handed down from Achæmenian times has failed to consider that the passing Persian civilization never reached the submerged masses

RELIGIOUS PROCESSION ON THE 10TH OF MOHARRAM, THE GREATEST ANNIVERSARY
OF THE PERSIAN YEAR

Moharram, the first month of the Mohammedan year, is for the Persian Shia Mohamme-
dans a month of mourning. The procession shown in the illustration commemorates the
death of Hosein, son of Ali and Fatima and grandson of Mohammed, who was barbarously
slain while attempting to gain the caliphate. This anniversary is observed with a vast amount
of mourning and a sort of Persian passion play. Crude floats depict the scenes of the
tragedy, and effigies on biers represent the torso and gory head of the murdered Hosein.

of the people, and that the new and more
penetrating Western civilization has had
time, as yet, merely to touch the surface.

The streets and bazaars of Teheran are
picturesque examples of all the stages in
the transformation which is now taking
place.

WHERE MOTOR CAR MEETS CAMEL

A luxurious motor car dodges a camel
or two and a drove of donkeys laden with
charcoal or street refuse, and draws up
at the main entrance of the Hotel de
Paris on Ala-ed-Dowleh Avenue. The
distinguished occupants descend and
make an unceremonious break for the
doorway; but before they can reach its
protective shelter they must run the
gauntlet of a swarm of indescribably
filthy professional beggars, who claw at
their garments and wail for alms in the
name of the Prophet or the Holy Virgin.
Or perhaps a fawning creature, clutch-
ing at the bridle of a patient ass draped
with a dubious collection of Persian rugs,
waylays the party and calls upon the
Sahib to note the quality of his rare assortment of antique carpets.

The corner, on this avenue, at the
southeastern end of the stately British
Legation garden, is also a favorite haunt
of the proletariat; for the location has
two indispensable requirements for comfortable outdoor Persian existence—an
abundance of shade, where a perspiring
pedestrian can squat to munch a refreshing cucumber on a scorching midsummer
afternoon, and a warm south wall, where
even in crisp January a tormented citizen
can pause and leisurely remove his upper
garment to pursue the elusive ceremonies
of the chase.

A GROUP OF MOURNERS OBSERVING THE RELIGIOUS FESTIVAL OF MOHARRAM
BY FLAGELLATION

The men who participate in the rites in honor of the assassinated Hosein (see illustration on opposite page) work themselves up to such a pitch of frenzy that they go through the streets shrieking and striking their heads with long knives.

At such vantage points a mendicant dervish, sketchily garbed in a tattered crazy-quilt, is usually on hand to croak "Ya Hakk" at the passerby (see Plate VIII).

THE POMPOUS PERSIAN GROCER AND HIS STOCK

The narrow cross-street from this interesting corner to the northern end of Khiaban-e-Lalehzar, one square to the east, passes a number of typical native groceries. These are merely large stalls set in the street wall, with almost the entire stock in trade exhibited at the broad entrance—long cones of sugar and strings of very-much evaporated figs suspended overhead, and matches, soap, and trays of rice, dried beans and fruits, raisins and walnuts displayed on the broad sloping counter, where the passer-by can bargain with the fat proprietor without entering.

The green grocer, also, has on display his entire assortment of lettuce, spinach, onions, tomatoes, pomegranates, apples, oranges, peaches, grapes, and long, yellow melons—in fact, a very wide variety of seasonable vegetables and fruits—which he has grouped with natural art in a beautiful harmony of color in pleasing contrast to the dingy street. True, they are open to the flies and street dust, but this fails to annoy the average patron.

The pompous grocer himself is an imposing type in his flapping, capacious trousers and skirted robe, belted at the waist with a voluminous sash or shawl, green in color if by good fortune he is a descendant of the Prophet. He anoints his beard and finger nails with henna, and if he should lift his large, egg-shaped, black felt hat he would reveal a modish haircut that has left smoothly shaven a five-inch path straight back over the top of his head.

Khiaban-e-Lalehzar is Teheran's Fifth Avenue and the pride of all the inhabitants. In the evening this short street is thronged with male promenaders.

Photograph by E. K. De Witt

A STREET CROWD ON A RELIGIOUS HOLIDAY: PERSIA

This is not only an interesting study of Turko-Persian racial types, but also an enter-
taining exhibition of Persian headgear so useful in identifying the residence and class of the
wearer, the rough felt dome of the peasant or artisan, the black pill-box of the merchant or
student, the skull cap of the porter, the white lamb's wool of the police officer, and the
cushion-like turban of the ecclesiastic in the right foreground being but a few among a strange
variety.

Fastidious, self-important Persian gentle-
men of leisure, garbed in frock coat or
flowing mantle, saunter along, jostling
humbly dressed tradesmen or peasants,
and an occasional Westernized Armenian
family elbows through the crowd. But
the hurrying European intruder usually
takes to the street, where the faithful
modern police force has had better success in training the drosky drivers to
keep to the right than in regulating the
confusion on the sidewalks.

Persian women are conspicuous for
their absence, and if a few brazen ones
do appear they suggest nothing quite so
much as black shrouds tottering along on
high-heeled slippers; even their faces are
concealed by black horse-hair blinders.

The variety of architecture along this
avenue is more striking than its quality.
Modern shops, with show-windows displaying actual European creations or
their ludicrous imitations, alternate with
junk-shops and second-hand stores,
where every conceivable commodity can
be unearthed, all the way from rusty
opera hats to astronomical telescopes. It
is the accepted custom for homeward-
bound foreigners to dispose of their dis-
carded effects, at a profit, to these enterprising traders; so it is not unusual to
see the familiar last season's wardrobe of
some legation-circle society leader dang-
ling from a shop door as a ghostly reminder of the departed, later, no doubt,
to adorn some brown-eyed harem beauty.

STREET LIFE SUGGESTS A TRAVELING
CARNIVAL

The precursors of the popcorn wagon
and the peanut-vender are there too. The
man pushing the red and yellow perambu-

Photograph by E. K. De Witt

A SHIA MOHAMMEDAN SHRINE

The mural decorations, done in vivid color, represent scenes in Hosein's ill-fated attempt to gain the caliphate (see also illustration on page 376).

lator has rose-flavored ice cream to sell, and the gentleman industriously fanning the little charcoal brazier is dispensing another delicacy, hot-boiled potatoes, or possibly succulent slices of huge sugar beet.

All of this, with a wandering magician performing his amazing feats at one corner, and at the next, perhaps, a professional story-teller in the center of an entranced crowd, conveys the impression rather that a traveling carnival has come to town than that this is the customary life along the most prominent avenue of an important capital.

The bazaars possess a never-failing interest for the Westerner. Here a large part of the city's trade is carried on in what might be described as one immense, primitive department store. Under low, vaulted brick and mud roofs covering many acres of territory, the leading Persian merchants and craftsmen not only sell their wares, but manufacture many products as well.

More than twenty-five miles of narrow, arched passageways wind and twist past literally thousands of small shops, which are merely alcoves, from six to twenty feet square, set in the flanking walls. Here and there archways in the wall open to caravanserais, which are large courtyards surrounded by arcades and ware-rooms, where caravans can be loaded and unloaded and the goods safely stored.

Round holes, which appear at regular intervals in the tops of the continuous series of domes forming the roof of the passages, let in dusty bars of dim light on the busy interior.

On a busy afternoon this labyrinth of half-lighted tunnels is crammed with a hurrying, shoving, noisy mob in which donkeys, camels, horsemen, and pedestrians mingle in a confused mass; and when a reckless carriage driver tries to force his way along, with shouts of "Khabar dar, khabar dar!" (Take notice, take notice!), there is a mad scramble of the crowd to flatten itself against the walls.

The dealers in different types of commodities have grouped themselves roughly

Photograph from Faye Fisher

THE LORD CHIEF OF THE PARSEES, OR FIRE-WORSHIPERS, OF PERSIA

The cashmere coat is both costly and picturesque. Lest they pollute the earth by burial, the Parsees dispose of their dead by placing the bodies in a Tower of Silence, which is situated on a hill far from human habitation, but accessible to the "corpse-eating dogs and birds."

in separate quarters, but each merchant has his own little hole in the wall, where he squats on the elevated floor beside his small show-case, or, if his shop be more pretentious, exhibits his goods from behind a counter.

The customer makes his purchases standing in the public thoroughfare; and the process is a complex one, for often prices cannot be agreed upon even after protracted bargaining.

Individual initiative and skilled hand production still prevail in Persian industry, and the sections of the bazaar occupied by the master craftsmen, who execute the delicate gold and silver filigree-work, the unique engraved copper ware, or other native products are exceedingly interesting.

PERSIA IS NINETY-EIGHT PER CENT MOSLEM

Since Teheran is the capital and has drawn its rapidly growing population from throughout the whole country, it affords an easy opportunity to acquire a general idea of the religious groups to be found in Persia.

Nearly 97 per cent of the population of Teheran and more than 98 per cent of that of all Persia is Moslem. In Teheran there are about 5,000 Jews and 4,000 Armenians. Nearly all, however, of the fewer than 100,000 nominally Christian population of Persia live in the western part of the province of Azerbaijan.

A remnant of the old Zoroastrians, or Fire Worshipers, of pure Iranian stock, still exists within the confines of Persia, and 400 of these ten or eleven thousand who have remained faithful to the ancient Persian religion reside in the capital. They are distinguished as being better business men and more honest than the Mohammedan Persians, and their women have greater freedom.

Especially among the military class are found many representatives of the two million people of predominating Tatar blood in northwestern Persia.

BY THEIR HEADGEAR YE SHALL KNOW THEM

Every city, town, or district of any importance is sure to have sent enterprising citizens to the capital, and there are picturesque representatives of numerous tribes as well. The readiest means of distinguishing the latter is by their distinctive headgear.

The Kurds, of whom there are 600,000 in the country, wear hats which look for all the world like huge, inverted black coffee-pots bound round with gay silk handkerchiefs. The Bakhtiaris, from the mountains in the direction of the British oil fields, in southwestern Persia, whose chiefs maintain a numerous retinue in Teheran, wear white felt preserving kettles. In fact, a dissertation on masculine Persian headdress (women are not allowed to wear hats) would give a ready key not only to recognition of the different races of Persia, but even of the different classes of society, since hats are rarely removed except when the owner sleeps, and vary in appearance and dimensions from the huge, pillow-like turban of the *mollah* or lady's woolly muff of the Persian Cossack to the round, brimless felt or lambskin cap worn by the middle and upper class urban residents.

The tribesmen especially have splendid, powerful physiques. While the Persians of the peasant and working classes are of medium height and sparsely but solidly built, a large portion, in particular of the city population, has degenerated through poverty, vice, and the use of opium. And the urban upper class, because of a very sedentary, idle, overfed life, inclines to obesity, the present shah being a good example. Large brown eyes, dark complexions, and straight black hair are dominant characteristics of the entire Persian population.

MOST OF TEHERAN LIVES IN APARTMENT-HOUSES

The great mass of the Teheran population lives in apartment-houses. This may seem strange or impossible, considering the forced seclusion of women; but it is a natural requirement of city life, and the poor women have to move about as unobtrusively as possible.

The typical apartment-house is of one-story, mud-brick construction, built around a court, in the center of which is usually a tank or pool of water.

The rooms, or apartments, all open upon the central court and in the ma-

Photograph by John B. Jackson

THE CURTAIN BEFORE THE THRONE IN THE PALACE COURT-YARD AT TEHERAN

The alabaster throne of the Shah (see opposite page) is under open skies.

jority of cases are lighted only by the entrance or an additional small latticed window. The single street entrance is a tunnel-like passageway into the court-yard. A family may occupy one whole side, but more commonly just a single mud-floored, mud-walled room, which rents for perhaps one dollar a month; and only this trifle makes a large hole in the monthly income of an unskilled laborer who receives but thirty cents a day.

Even in these lowly dwellings, however, the Persian's artistic sense and love of natural beauty assert themselves, for almost always there are potted plants and a tiny, carefully tended flower bed in the sunny area of the court.

IN THE AIRY PALACES OF THE RICH

What a contrast are the rich, airy palaces of the grandees, their white columns and porticos gleaming invitingly through the luxuriant green foliage of stately gardens. At a distance they convey the general impression of magnificence and wealth but close at hand most of them recall the imposing but crumbling St. Louis or San Francisco Exposi-

tion edifices as they were at the end of their term of service. There is much that is tawdry about them, and a great part of the really skillful and artistic workmanship in their architectural adornment is wrought in fragile plaster, which soon deteriorates.

The impressive structure which occupies the dominant position in the great garden is the *berun*, the abode of the male members of the family. In the rear of the premises, where the average American usually has his garage, is located the unpretentious *anderun*, or harem, the humble retreat which shelters and segregates the numerous feminine adjuncts of the household.

There are several national institutions, as yet unmentioned, which are exceedingly essential to Persian life—the bakery, the public bath, the tea-house, and perhaps the ice factory.

THE BAKER WHO OVERCHARGES IS BAKED IN PERSIA

Wheat bread is the most important, almost the only, food of the Persian masses on the plateau. Rice is the staff of life in the Caspian Sea region and a favorite

THE SHAH OF PERSIA'S MAGNIFICENT ALABASTER THRONE

From this famous dais Persia's ruler holds his New Year's reception (which takes place in March). The ministers of all the countries having diplomatic relations with Persia are present on this occasion, wearing their court costumes, and it is a brilliant assemblage. The Shah sits in a jeweled armchair on the throne and the court poet (in his official regalia, which consists of a long coat of beautiful cashmere) reads his greetings. The Shah's band plays throughout the celebration.

delicacy throughout the country when prepared in the form of a Persian *pilau*, but it is second in importance to bread as a staple article of diet.

Bread is prepared in a number of ways, but the most approved variety in Teheran must be baked in the large ovens of the public bakeries. The dough is spread on huge mounds of red-hot pebbles, comes out deliciously crisp, in thin sheets thirty inches long, and is displayed on sloping counters at the street entrances to the shops (see illustrations, pages 449-450).

The method of government price control of this important factor in the cost of living is gruesomely effective when put into operation. The path of the profiteering baker is precarious, for he is sometimes thrust into his own oven and nicely browned.

The evidence of one's eyes might not rate the public bath in Persia as an important institution, but it is indispensable; for by religious law it is encumbent upon the devout Moslem to bathe at least

once in ten days. The fact that in the cheaper baths there is a common pool the water of which remains unchanged for months at a time would seem to militate against the sanitary value of the performance, but the high temperature to which the water is raised no doubt has a more or less valuable sterilizing effect.

The street entrances to the baths are entertainingly marked by lines of varicolored bath-cloths, groups of semi-nude attendants, and mural paintings resembling in spirit and color the comic section illustrations of American Sunday newspapers. The fuel employed in heating the baths—dung collected from the streets and dried in cakes—is but one example of the many ingenious economies practiced by the resourceful Persians.

THE TEA-HOUSE IS THE PERSIAN CLUB

The tea-house is the democratic Persian's political and social club, a splendid institution for which we have no adequate equivalent in America. It is every-

PERSIANS ASSEMBLED FOR THE SHAH'S NEW YEAR'S DAY RECEPTION

The City Palace of the Shah, gaily decorated with tiled designs, is in the heart of Teheran. It is one of many palaces of the Persian potentate and is used chiefly for state functions.

PERSIANS ASSEMBLED FOR THE NEW YEAR'S DAY RECEPTION
It is one of many palaces of the Persian potentate.

Photograph by F. L. Bird

SMALL TRIBAL WANDERERS OF PERSIA

This little group belonged to an encampment of nomads in a mountain valley near Teheran. While big sisters are making butter in the hairy goatskin churn and mothers are doing the day's baking in their earthen ovens, it has fallen to the lot of these small lassies to keep the camp babies out of mischief. Large brown eyes, dark complexions, and straight, black hair are characteristic of the entire Persian population.

Photograph by E. K. De Witt

DERVISHES OR MOHAMMEDAN MENDICANT FRIARS

These holy men are members of one of the Mohammedan brotherhoods of dervishes. Having forsaken all their worldly goods, they are dependent for their livelihood on the liberality of their co-religionists. The small axe, the calabash receptacle, the gnarled staff, and the small hand-woven bag carried by the pair are distinctive insignia of these picturesque wanderers.

385

Photograph by J. W. Cook

A FUNERAL PROCESSION IN HAMADAN

The closely veiled Persian woman in the foreground has so far forgotten her modesty in her curiosity to view the bier of the deceased Persian dignitary as to lower a fold of her domino and thus reveal a portion of her countenance to public gaze. Hamadan is the ancient Ekbatana, the home of Esther and Mordecai.

where—in the city, in the village, even along the desert caravan trail. Here the harassed business man or weary traveler can refresh his careworn soul with a glass of tea, a leisurely cigarette or water-pipe, and a bit of light gossip or exchange of current news with fellow-beings of kindred spirit.

The tea-house may be in external appearance anything from an adobe hut with a few crude benches to the glorified cafés of Lalehzar, but it always possesses those unfailing essentials, a big, brass Russian samovar, an adequate collection of little tea-glasses, bright-colored saucers, and filigree spoons, a bubbling hubble-bubble or two for public use, and a

genial atmosphere of camaraderie reminiscent, perhaps, of the obsolete American bar-room.

The ice factory is merely a mud wall, two stories high instead of one, throwing a cold shadow on a puddle at its northern side in winter. The stored product is neither crystal nor clean, but it serves to cool many a refreshing glass of Persian sherbet during the thirst-compelling dryness of the withering summer's heat.

PRIMITIVE ENGINEERING PROVIDES TEHERAN WITH WATER

The problem of food and water supply for a city nearly as large as Washington, D. C., without railway or steamship communication, without modern machinery of any kind, without even such primary essentials as farm wagons and cast-iron water pipes, has been solved very ingeniously by the people of Teheran. The water system especially is a marvel of primitive ingenuity.

The unusual topography of the plains about Teheran always arouses the traveler's curiosity. Row after row of earthen craters, which look like series of gigantic ant-hills or, perhaps, entrances to the subterranean abodes of mammoth moles, lead toward the mountains. These mark the courses of the underground aqueducts, which bring the mountain well or spring water a distance of from five to ten miles to the city.

After the source of supply is located, a party of *moqannis*, or professional well-diggers, sinks a line of narrow shafts, at intervals of about one hundred yards and often to a depth of more than one hundred feet, all the way from high mountain levels to the interior of the city, and joins the bottoms of these shafts along a stratum of impervious soil which will retain the water.

These primitive engineers join their tunnels underground without even the use of a compass and accomplish the whole tremendous task with merely a crude windlass, a bit of rope, and a few small picks, shovels, and canvas sacks. Thirty or more of these burrowed channels convey the entire water supply to Teheran, including that used for irrigating purposes.

It is the delivery end of the system which is deficient. The water is run about the city in open ditches, collected in . pools, or impounded in huge underground reservoirs, from which it is transported in large skins, by water-carriers, to private houses.

The mysterious little brooks that magically appear and vanish along the city streets are a refreshing sight when there has been no rain for months, and they afford a ready supply for the street-sprinkler with his big dipper, the thirsty populace, and the busy laundress who wishes to rinse out a few garments. But the dangerous and disgusting pollution results in much otherwise avoidable illness.

The mean flow of water the year round is estimated at nearly one million gallons per hour, which if properly utilized would be abundant; but, with the winter supply too large and the summer supply too small, the distribution unequal, and the wastage in open ditches and by leakage so great, there are portions of the city which in the dry season receive no water whatever.

Photograph by J. W. Cook

A PERSIAN MULLAH OR TEACHER: NOTE HIS HENNA-DYED BEARD AND NAILS

PERSIA'S FIRST RAILWAY, 5½ MILES LONG

An account of Teheran would be incomplete without some mention of the first Persian railway and its route. This abbreviated narrow-gauge line runs from the southern end of the city, five and a half miles across the hot plains, past the ruins of ancient Rei, to the village of Shah Abdul Azim, the seat of a famous golden-domed shrine which attracts great crowds of excursionists on every holiday.

Absurd as its antiquated equipment may appear, this road has one important advantage over many great American railway systems in being able to earn a generous return on its capital investment.

Of the famous old city of Rhages, or Rei, founded, according to tradition, in the fourth millennium before Christ, capital or metropolis of many dynasties, advanced base of Alexander the Great in his campaign against Darius III, and the birthplace of the mother of Zoroaster and of Haroun-al-Raschid, all that remains are a few ruined walls still massive in their decay. An occasional cultivated field or garden dots the site, and here and there inquisitive treasure-hunters have excavated the old house walls.

A GROUP OF PERSIAN SCHOOL BOYS AT HAMADAN

While modern schools are increasing, the old *Makhtab Khaneh*, with the Arabic Koran as the text-book in the primer class, is still a national institution (see illustration on opposite page). More than 98 per cent of the people of Persia are followers of Mohammed.

Photographs by J. W. Cook

A GROUP OF ARMENIAN AND PERSIAN SCHOOL GIRLS

Probably 95 per cent of the women of Persia are illiterate, but in recent years foreign educators, and since the war the Persian Department of Education, have done much to promote the cause of education for Persian girls. It has only been within the last few years that girls have been allowed to attend the mission schools. Formerly they were married at the age of ten of twelve.

Photograph from Faye Fisher

A PERSIAN SCHOOL FOR BOYS

Schools such as these are scattered through the shopping districts of Persian cities. They resemble shops with one side all open to the street. The teacher sits at one end with a long stick which he uses on the boys if they stop studying aloud for a moment.

A noted spring gushes from under the broken ramparts. Chashmah-i-Ali, or The Fountain of Ali, it is called, in honor of Ali, the son-in-law of Mohammed. Above it is a modern rock carving, and near by a palimpset replacing a Sassanian bas-relief, both of them commemorating the reign of Fath Ali Shah, who ruled in Persia a hundred years ago. Neither is of historic value except as a reminder of the conceited king who presumptuously obliterated the ancient and valuable inscription to provide for the record of his own supposedly greater glory.

TEHERAN'S TOWER OF SILENCE

To the northeast of Rei, on a bare, shelving hillside, the Zoroastrian Tower of Silence stands, visible from all the surrounding country.

On this circular, whitewashed tower, which is perhaps fifty feet in diameter and thirty feet high, the modern Zoroastrians, according to ancient religious custom, expose the bodies of their dead to the vultures and the weather. Gruesome as this strange cemetery may seem, it is a better tribute to the still glowing embers of old Persian national life and customs than the mouldering walls of forgotten Rei.

In the case of Teheran, as of a great many other things Persian, distance lends enchantment to the view. The shaded avenue from a northern gate leads mountainward through bare, rolling foothills, past a deserted palace of a former shah, and ascends through cultivated fields, brown walled villages, and past the cool summer gardens of the legations, the Persian aristocracy, and the royal family, to the very base of the great inner wall of the Elburz.

Then a narrow bridle-path climbs sky-

A PERSIAN KHAN WITH HIS BODYGUARD

On account of the numerous bands of marauding bandits which infest the highways, a traveler in Persia does not venture forth without a guard of armed retainers.

Photographs by E. K. De Witt

PERSIAN HUNTERS WITH THE FRUITS OF THEIR CHASE, A WILD BOAR

The jungles in the semi-tropical Caspian Sea region fairly teem with wild boars, which render themselves doubly unclean and obnoxious to the Moslem peasants by their never-ceasing inroads on the rice fields. These doughty hunters have not only proudly demonstrated their skill with their antiquated flint-lock muskets, but are very evidently satisfied at having disposed of one more hated marauder.

Photograph by J. W. Cook

A SUMMER ENCAMPMENT OF PERSIAN NOMADS

Nomadic tribal life in Persia still exists even in the environs of the capital. Fully a fourth of the population of the country still dwell in their home-woven black tents. live on the product of their flocks and herds, and wander with the change of seasons from the warm plains in winter to the cold mountain valleys in the summer time.

ward over slides of shale, around boulder-strewn promontories, into an unexpected hollow, green with stunted willows about a spring, and up a breathless zigzag along a snow-fed rivulet bordered with purple iris, to the snow-streaked crest 12,600 feet above the sea. From there the view to the south commands a sweeping area of the plateau and, to the north, of the second great range in the triple wall which bars Persia from the sea.

Far down the desolate southern slopes, crevice-like gorges open, ground by the tread of the ages in the forbidding gray rock of the mountain side; and down these deepening canyons flow silent streams of greenest foliage, concealing foaming torrents and splashing water-falls, and spreading, as they emerge from their narrow confines, to hide the crudeness of mud-walled hamlets.

On the open slopes of the low-lying foothills irrigated patches of wheat and barley stand out like the squares of a checker-board, or well-watered villages

gleam like emeralds in their yellow desert setting.

A dusty haze overhangs Teheran on the northern edge of the level plains, and beyond the sunlight falls on the ghostly burial tower and the gilded dome of Shah Abdul Azim. From the right the Karaj River bends in a threadlike, silver bow to the glimmering salt marshes far to the south, and distant mountains inclose the scene in a giant, bowl-like hollow.

A COMPLETE PANORAMA OF DEVELOPING CIVILIZATION

Within a twenty-five mile radius of Teheran a complete panorama of developing civilization unfolds, perfectly illustrating the whole shifting scale of human existence from the primitive to the modern stage. In the city itself are the conflicting institutions of modern society and Mohammedanism, along with an underlying stratum of unchanged primitive customs.

A PERSIAN WOMAN APPARELED FOR A PILGRIMAGE

The elaborately embroidered saddle-bag is a *khorjon*, in which both clothes and food are carried for the journey. The white veil over her face is the *yashmak*.

aloof from the modern world, dominated by bigotry and conservatism, and not yet ameliorated by medical science or enlightened education.

From their centralized abodes the peasants go each day to till the surrounding fields with tools and methods similar to those employed by their forebears. And within this same limited area rove rough, untutored nomads, self-dependent, prosperous in terms of flocks and herds, desiring no better shelter than the black wool tenting woven by their virile, unsecluded wives and daughters, but unconcerned with the affairs of state.

From the more progressive centers of Teheran and other important Persian cities waves of reflected enlightenment are moving in slowly widening concentric circles to reduce the divergence in present-day Persian life.

The rise of the present Kajar Dynasty

The railroad and the motor car have not yet won the competition with the camel and the donkey; and while modern schools are increasing in number and quality, the old *Makhtab Khaneh*, with the Arabic Koran as the text-book in the primer class, is still a national institution.

While Teheran is the seat of an experiment in representative government, most of the surrounding villages are a part of an oriental feudal system, as the property of the crown or the wealthy land-owning nobles. In these dreary, insanitary, adobe villages, still centers the Persian peasant life, an existence entirely

was a final attempt to restore a conservative oriental despotism in the midst of a swiftly progressing modern world, and under the long, benevolent rule of Naser-ed-din Shah, who maintained order and obedience throughout the length and breadth of his kingdom, this anachronism gave strange promise of success; but the failure of his successors disclosed the futility of competing with antiquated equipment in the modern economic and political struggle.

The progressive element among the Persians realized their weakness, and in 1906 demanded and secured a constitutional form of government; but they were illy prepared to combat internal anarchy and reaction or to resist the aggression of Russian imperialism.

International intrigue, with their country as a helpless tool, has prevented a fair test of the ability of the Persians, by their own unaided efforts, to restore the Persian state.

But now a new factor has appeared, as though Aladdin had rubbed his magic lamp, portending a rapid change in Persia's status—oil, which has brought wealth and progress to many an unpromising region. This, as well as undeveloped stores of copper, lead, iron, and other products, has attracted the foreign capitalist; and in these days of

dollar diplomacy there are bound to be railways and valuable concessions for this most-sought-for of all present-day commodities of commerce.

An overland rail route to India may soon become a reality, and plans are already complete for a railroad to Teheran.

The old Persia is swiftly passing. The new Persia is bound to be economically prosperous. The unsolved problem is the future of the Persian national life. The end of another generation will reveal either a Persian resiliency and moral strength sufficient to establish securely a modern state or merely another failing experiment in the decanting of new wine into old bottles.

AN ASCENT OF MOUNT DEMAVEND, THE PERSIAN OLYMPUS

ANY ONE who has carefully examined a small map of Asia has probably noticed as one of the few designated features of the portion occupied by Persia the rather striking name, Mount Demavend. The emphasis given to this spot by the absence of many other defined locations throughout the country is quite in keeping with its size, magnificence, and importance.

Demavend, which outrivals in proportions any mountain in Europe, is the highest peak in southwestern Asia, for it rises to an altitude of nearly 20,000 feet above sealevel. Whether viewed through the mists from the Caspian Sea or in the clear, thin air of the Iranian plateau, its snow-ribbed volcanic cone is a vision of surpassing splendor. But to realize its full grandeur one must see it from the crest of a neighboring range, where the eye can take in with a single sweep the unbroken rise from base to summit, or from some point a hundred miles southward across the desert, where it still dominates the hazy horizon long after the rugged outlines of its surrounding ranges have dissolved in the distance.

THE PARADISE OF ZOROASTER

Is it to be wondered at that such a superb landmark should hold a prominent place from the earliest times in the

legend and the superstition of the Iranian peoples?

As Mount Olympus in Greece was the home of the gods, so the paradise of Zoroaster was the summit of Demavend in Persia. Many legends have developed from its mysterious, fear-inspiring grandeur, important among which is that of the monster tyrant Zohak, who, halted in his worldwide tyranny, was chained, Prometheus like, upon this peak. This tale, which is found in the sacred writings of the Fire Worshipers and in Persian classical poetry, is still cherished in the folk-lore of the inhabitants; and so strong is the inherited superstition that even today the venturesome traveler is warned not to attempt the ascent of the mountain, because "the devils will get you."

Not only has this great mountain held a lofty place in mythology, but it has cast its far-reaching shadow over many epoch-making events in history. Almost at its very base (in the Median metropolis of Rhages) was born the mother of Zoroaster. It marks the eastern limit of the raids of the Assyrians before the rise to power of the great kings of Persia, and its frowning eastern face overlooked the mountain home of the rising Parthian Empire. Alexander the Great paused beneath it in his pursuit of Darius III

Photograph by E. K. De Witt

A PERSIAN PLOWMAN

In striking contrast to the sun-scorched desert, with its drifting sand and rock-strewn wastes and toiling, thirsty caravans, is this oasis-like garden, a paradise, as the Persians often call it, set down in the midst of barren wilderness. The slender, stately poplars and dense foliage in the background indicate the abundance of the watercourses, and the good natured smile of the young peasant plowman and the placid contentment of his beasts carry some idea of the productiveness of the fertile soil he is scratching with his crudely fashioned plow.

and sent expeditions through the neighboring passes to subdue the impregnable regions of Hyrcania. Following in his footsteps came Antiochus the Great against the Parthians, and westward along this same route Genghis Khan, Hulagu Khan, and Tamerlane led their ravaging Mongol hordes.

A GUIDEPOST FOR GENERATIONS

One can imagine that even Alexander or Tamerlane, sweeping on to world conquest, must have felt his insignificance as he faced this unconquerable barrier.

Rising not far from a great international highway, Demavend has served as a gigantic guidepost for scores of generations of daring merchants, who, long before Columbus, exchanged the wares of the West and the East by means of their slow-crawling caravans; and its lonely grandeur has gripped the imagination of intrepid adventurers of all ages. Within its shadow a score of great dynasties have risen and fallen, and today it stands as one of the few remaining glories of the Persian Empire.

The Elburz Mountain range, of which Demavend is an outstanding member, is a unit in the great mountain system that stretches from southern Europe to central Asia, and, with regard to Persia, is the great dividing line between the northern limits of the Iranian plateau and the Caspian depression—a 12,000-foot wall separating a basin 81 feet below sealevel from a table-land averaging 4,000 feet in altitude. Beginning near Ardabil, in Azerbaijan, it extends southeastward and eastward more than 500 miles along the southern shores of the Caspian and into Khorasan.

This great mountain wall gives northern Persia the anomaly of two almost contiguous but quite different climates. The moisture of the Caspian basin is excluded from the interior, resulting, on the northern side, in a semi-tropical climate, with an annual rainfall of over 50 inches and luxuriant orange groves and crops of rice and cotton, and, on the mountain sides themselves, dense forests of hardwood, while the southern escarpment is barren and supplies scarcely enough water for the narrow but fertile desert fringe at its base, with its crops of grain and fruits of the Temperate Zone.

A QUIESCENT VOLCANO

Demavend itself is about 45 miles northeast of Teheran, in the central of three parallel chains. It towers high above these flanking mountains, whose summits do not exceed two-thirds its elevation, the only mountain peak among endless series of ridges. Its conical form and seemingly even slope of about 45 degrees from top to bottom at once indicate its volcanic nature.

Although there is no record of an eruption in historic times, the volcano can be called quiescent rather than extinct, for about the base are numerous hot springs, and at the summit one finds evidence of volcanic heat at no great distance beneath the surface.

While from a distance its cone seems to taper almost to a point, it really terminates in a bowl-like crater about one hundred yards in diameter, which is almost entirely filled with snow. The internal heat is sufficient to melt the snow about the summit, and thus uncover to view masses of basalt and limestone rocks and huge deposits of sulphur.

Strange to say, the exact altitude of Demavend still remains uncertain. Numerous measurements have been made, ranging from 18,000 to over 22,000 feet, an average of the most reliable giving an altitude of about 19,000, though the single measurement commonly accepted is 18,-464 feet.

THE ASCENT

For a mountain of this size, the ascent cannot be considered especially difficult, there being few obstacles other than the cold, the rarity of the atmosphere, and fatigue.

Because of the superstitious awe with which the inhabitants regard the mountain, very few of them have tried to scale it, and it requires the inducement of a month's salary to secure a guide for the climb, if one can be found at all. The first European to make the ascent was William T. Thomson, in 1837. Since then it has been accomplished by several Europeans, by three Americans, and in 1914 by seven Persian boy scouts.

Late summer, with its settled weather and minimum of snow, is, of course, the best time of the year for the ascent. Although Teheran, the nearest large city, is the logical point of departure, the precipitous nature of the western scarp necessitates a circuitous approach. A three-day trip takes one across the first range of mountains by the Afcheh Pass, with an elevation of 9,000 feet; then, between the two ranges, down the well-watered Lar Valley, which during its brief summer season supports the flocks and herds of nomadic tent-dwellers, who pass their winters in the plain villages south of Teheran, and skirting the southern base of the mountain itself, to the village of Rena, above the canyon east of Demavend, where the Heraz River cuts through on its way to the Caspian.

This mountain village, which has an elevation of about 6,000 feet, makes an excellent base, for from this point a well-defined trail winds upward 7,000 feet, to where a few shepherds pasture their

A TURKEY HERDER-SALESMAN AND HIS FLOCK ON THE STREETS OF TEHERAN

The fowls serve in the capacity of garbage collectors while foraging, for they feed on refuse flung from the grimy front doors of the shops. Note the Persian woman in street dress at the right.

Photographs by J. W. Cook

PERSIAN SHEEP IN THE FOOTHILLS OF THE ELBURZ MOUNTAINS

The barrenness of the hills suggests the explanation for the development of the fat-tailed variety of sheep, the tail furnishing food storage for the dry season when pasturage is scant.

Photograph by E. K. De Witt

PERSIAN GYPSIES

Ragged but carefree, they have the same characteristics as gypsies everywhere, and among the Persians, with all their superstitions, the gypsies are especially in demand as fortune-tellers.

flocks on the green, moist areas immediately below the snow-fields.

This part of the ascent, made either on horse or mule back or afoot, requires the better part of a day, during the early hours of which it is necessary to grope one's way through heavy cloud banks. It is a glorious moment, however, when a sudden movement of the clouds clears the sky and reveals the summit, its great golden cap of sulphur glowing in the sunlight, seemingly so near in the dry, clear atmosphere that one is deceived into thinking that the climb is almost over.

The camping facilities at the 13,000-foot level of the snow-line are adequate in some respects and lacking in others. Water is there in abundance, and camp-fires ready for the match all too copiously stud the landscape in the form of clumps of dried camel's thorn (see Color Plate V), each much larger than a bushel basket.

MOUNTAIN PEAKS RESEMBLE ISLANDS IN A FAIRY SEA

The inadequacy of the sleeping quarters, however, impresses one when, being forced to turn in on an artificially and precariously constructed rock shelf, he

Photograph from Faye Fisher

PERSIAN BOYS ON A HIKING EXPEDITION AMONG THE ELBURZ MOUNTAINS

The Persians are great lovers of nature and the Boy Scout movement has many enthusiastic adherents among the youthful subjects of the Shah, seven of their number having made
the difficult ascent of Mount Demavend in the summer of 1914 (see text, page 395).

MOUNT DEMAVEND HAS SERVED AS A GIGANTIC GUIDE-POST FOR MANY GENERATIONS

"Its lonely grandeur has gripped the imagination of intrepid adventurers of all ages. Within its shadow a score of great dynasties have risen and fallen" (see text, page 394).

Photographs by F. L. Bird

ON THE SUMMIT OF MOUNT DEMAVEND, PERSIA'S MOUNT OLYMPUS

The superintendent of the American school at Teheran is seated in the foreground at the left. He and the author, who made the photograph, are two of the three Americans who have accomplished the ascent. Mount Demavend is variously estimated at from 18,000 to more than 22,000 feet in height (see text, page 395).

dare not go to sleep for fear of rolling off upon one of the patent camp-fires.

The ascent from this temporary camp to the crater requires about ten hours of actual climbing; so, however one arranges it, part of the trip up or down must be made at night. Although with nightfall the cold becomes extremely severe, there is the compensation of enjoying the wonderfully luminous moonlight of Persia under very unusual conditions. Soft, mellow, yet throwing almost the radiance of day on every peak and crag and snow-field, it holds one entranced by a scene more glorious than befalls the lot of most mortals to behold; for, far below, the feathery clouds roll and swirl like a soft ethereal ocean, dashing their gentle billows against the lesser mountain peaks that raise their black summits like islands above this fairy sea.

Considering the aridity of the region, it is not surprising that there are no great glaciers to be traversed; but the remains of glacial formations, in particular one immense chasm-like groove, at the head of which is a huge immovable ice-mass, suggest that at one time the country enjoyed a more salubrious climate. The angle of the incline varies only from 40 to 55 degrees, thus rendering the ascent as monotonous and tiring as that of an endless ladder.

NATURE BUILDS LADDERS OF SNOW

The steeper portions of the snow-fields present a most peculiar formation. The snow, or coarsely crystallized ice, instead of offering a smooth surface, is molded into tier upon tier of narrow, tapering cones, averaging two feet or more in height, which evidently are produced by the daily routine of thawing and freezing. A rather uneven melting process starts little trickles of water which seek out miniature channels, and the night's freeze establishes these slight elevations and depressions.

The next day's thaw finds small channels waiting, which are deepened and connected and the elevations accentuated thereby. After this process continues for two or three months, the results are these strange stalagmitic arrangements, which serve admirably as natural ladders for the more difficult sections of the climb.

The sliding, shifting fields of powdered pumice not far below the summit seriously obstruct progress, for the climbing here is similar to what one would encounter in trying to ascend a steep roof covered with two feet of loose snow. Although it requires one and one-half hours for this portion of the ascent, the descent over the same ground is easily made in four minutes and with comparative safety.

ON THE SUMMIT OF THE GREAT LANDMARK

The expanse of the great golden sulphur cap, the edge of which is reached a hundred yards below the rim of the crater, is startling. Thousands of tons of sulphur are exposed to view and the fumes which permeate the air are almost nauseating.

The rocky rim of the bowl-like crater, about 300 yards in circumference, is practically level for a width of five yards, and then slopes gradually inward to the snow which fills the crater itself. The only signs of present volcanic activity are the gaseous fumes issuing from small fissures in the sulphur area.

The lofty isolation of the great peak makes it an admirable observation point. On a clear day the country spreads out in every direction like a giant relief map on which a comprehensive view of the whole complex arrangement of mountain chains and drainage systems replaces the restricted vision of one on the plains. Close at hand the great inter-mountain valleys and far away to the south the green desert fringe and the vast desert itself are visible, while to the north hover the mists and vapors that rise over the Mazandaran jungles and the Caspian Sea.

A feeling of utter insignificance comes over one with the realization that he is at last on the summit of the great landmark which has borne the scrutiny of heroes of many ages of history and is at present the focus for the gaze of the camel-driver of the Persian Desert and the sailor cruising on the Caspian, of the peasant in his rice-fields on the Caspian shores and the village or city-bred dweller on the plateau, of the Turkoman tent-dweller on the transcaspian steppes and his Aryan brother, the shepherd nomad of the Iranian tableland

Photograph by Harold F. Weston

THE PORTALS OF PERSEPOLIS, THROUGH WHICH THE ARMIES OF DARIUS AND XERXES WENT FORTH TO CONQUER ASIA MINOR

Here is shown the northern entrance to the Hall of a Hundred Columns where Darius held court 2400

SHIRAZ, PERSIA'S MECCA FOR BEGGARS, POETS AND PHILOSOPHERS

Business activities have ceased for the hot noon hours at this khan or alcove, off the main bazaar, and an old beggar is keeping watch. Here merchandise is unloaded from the caravans and weighed on primitive scales. The arcades around the court are decorated with colored tiles.

THE BREAD AND COOKED RICE MERCHANT IN A PERSIAN BAZAAR

Commerce takes its tranquil way in the marts of Shiraz where the multiplicity of odors rivals the variety of wares.

A GIFT LAMB FROM A PERSIAN POTENTATE

In the absence of her husband, the wife of Ambdullah Khan of Dehbid presented the western traveler with this token of hospitality in lieu of personal entertainment. In the Orient a man never travels except on a pilgrimage or for some special gain. The Persians could not understand a tourist's motives and insisted on assuming that the traveler was on some important mission of a political nature.

PERSIAN PEASANTS WHOSE BURDENS ARE BALES OF CAMEL'S THORN

The black wool caps worn by these men are strikingly similar to the balloon-shaped head-coverings worn in ancient Sassanian days. Camel's thorn is one of the most useful plants to be found in arid regions. It provides excellent forage for camels, is much used for fuel in central and southwest Asia and, being leguminous, it enriches the soil.

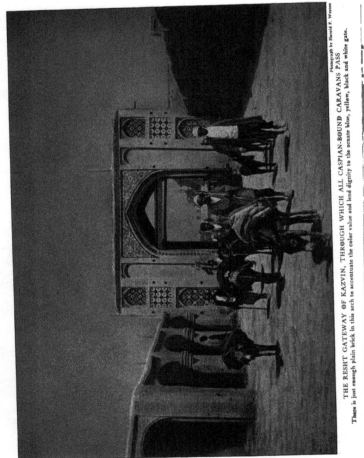

Photograph by Harold F. Weston

THE RESHT GATEWAY OF KAZVIN, THROUGH WHICH ALL CASPIAN-BOUND CARAVANS PASS

There is just enough plain brick in this arch to accentuate the color value and lend dignity to the ornate blue, yellow, black and white gate.

SMOKING THE KALIAN, THE PERSIAN PIPE OF PEACE, AT GABARABAD

Tobacco, dry and powdered, is placed in a charcoal-filled bowl over the water jug. Each smoker puffs deeply once or twice, then passes the carved wooden mouthpiece to his nearest companion as he spits lustily on the earthen floor.

Photograph by Harold F. Weston

VIII

409

A DROVE OF DONKEYS RETURNING FROM PASTURE AMONG THE PERSIAN HILLS

As only one meagre crop of grain is reaped each year, these poor beasts do not get much barley ; they depend largely upon the grass of the highlands for their forage.

Photograph by Alfred Heinicas

IX

"GOD WILLED IT" IS THE ENGLISH EQUIVALENT OF YEZDIKHAST, THE MOST PICTURESQUE
TOWN IN CENTRAL PERSIA

The only access to this peculiarly situated town is from its northern end by a single drawbridge which spans the deep breach between the eminence and the
former river bank. It has been compared to a petrified ship left stranded beside a river bed that has been dry for ages.

411

THE MESJED KOUCHEK (LITTLE MOSQUE) AT KAZVIN, PERSIA

The magnificent tile design of this, the most beautiful mosque in Kazvin, if not in all northern Persia, is sadly in need of repair. The dark object projecting to the right from the dome is a bush that has taken root and grown in a large fissure.

Photograph by Harold F. Weston

THE HEAVY MASONRY BRIDGES OF PERSIA SEEM TO HAVE BEEN DESIGNED TO WITHSTAND THE DELUGE

Between the mountain ranges of Kurdistan, the broad valleys which are from ten to thirty miles wide are watered by streams which frequently disappear in the summer months. Although built hundreds of years ago, this bridge, which is only a foot thick at the top of the arch, sustains the weight of loaded two-ton motor lorries.

441

SASSANIAN SCULPTURES IN THE LIVING ROCK AT NAKSH-I RUSTAM, PERSIA   Photograph by Harold F. Weston

These carvings date from the third century A. D. and are cut on the face of the rock cliff near the great tombs of the Achaemenian kings.   That on the right portrays Varahran II and his courtiers.   The larger bas-relief on the left represents King Ardashir, founder of the Sassanian dynasty, mounted on horseback and receiving from the god Ormazd a ring that symbolizes the gift of sovereignty.   The pillar seen in the upper right corner of the photograph is hewn out of solid rock.

Photograph by Alfred Heinicke

WINNOWING GRAIN IN SOUTHERN PERSIA

Although a part of the table-land of Persia is well watered, it is often necessary to resort to irrigation. There are many districts which are accounted among the most fertile in the world, but the implements employed in agriculture are primitive in the extreme.

443

TANG-I ALLAHU AKBAR, THE PASS OF "GOD IS MOST GREAT"

Photograph by Harold F. Weston

After crossing the desert plateau, Shiraz seems a veritable island of emerald with its rows of dark cypresses and turquoise domes, bordered in the distance by purple hills.

THE INNER COURTYARD OF MEDRESSEH-I SHAH HUSEIN AT ISPAHAN,
PERSIA'S MOST FAMOUS COLLEGE

Medresseh-i Shah Husein, a college for the training of mullahs and dervishes, was built about the beginning of the 18th century by Shah Husein. Within its walls the students live in alcoves around a large court.
The scaffolding indicates that the exquisite turquoise tiling of the dome and the minarets is being repaired.

XVI

PERSIAN CARAVAN SKETCHES

The Land of the Lion and the Sun as Seen on a Summer Caravan Trip

By Harold F. Weston *

PERSIANS say, with a great feeling of envy, that the man who has seen the most of the world is the greatest liar. So when I am asked to tell "all about Persia," I·generally ask if I should not include Russia, too, having been there just six hours.

What counts most in enjoying, visualizing, or telling about the "romantic East," or any strange place, is your power of imagination. I can but bring together a few faded petals; the reader's imagination must arrange them and appropriately spray them with attar of roses.

Persia, for a surprising majority of people in America, is not much more definite than a hazy pink or green spot swimming around India — "Oh, you know, beyond Turkey."

Persia suggests Omar Khayyam, gardens and rugs, rugs remembered from colorful magazine advertisements or hasty glimpses into Fifth Avenue windows. Many dusty books lie on library shelves. All I propose to do is to offer a few sketches in rough outline, which may help to visualize Persia of today; to pin those green or pink spots onto the map by a few vivid incidents, border them with bleak mountain ranges, dot here and there with crumbling palaces and cypressed gardens, color with affable hosts in the form of rotund chieftains or fugitive brigands, enliven with mysterious veiled ladies and equally hidden but more numerous minute "critters"; then sweep it all over with dust, heat, decay, and almost unbroken desert.

Persia is almost as large as Germany, France, Italy, and the British Isles com-

bined. It is an arid plateau from 3,000 to 7,000 feet high, seamed by snow-capped mountains. The people are mostly Aryan, but—but you can read all this in any encyclopedia. So we will start in the good old-fashioned way.

IN THE BEGINNING

Once upon a time there was an armistice, and two young Americans, who since '16 had been with the "Y" in Mesopotamia, conceived the idea of crossing Persia by caravan. The British military authorities gave the casual information that the 2,000 brigands still in possession of the one main caravan route through central Persia might have something to say about it, though the British were sending assistance to the Persian Government to round them up. That was naturally the straw that led the camel to drink, and by May we had left Bagdad with permission to travel on the British military motor convoys through Kurdistan to the Caspian.

A Ḳurdish lad was obtained as a servant, some emergency rations, a sixty-pound tent, which was never used except to be reviled (by Persian muleteers, of course), and various other incidentals, such as medicines, camera films, and cash.

A Sir Somebody wrote about his trip through Persia, that he had uselessly carried two articles—a revolver and a large box of insect powder. In both cases he sighed, "Of what use is one against so many!" Yea, verily; but he who would venture into the land of "the lion and the sun," let him go well armed with a goodly supply of patience and faith, faith that all is the will of God, no matter what happens: water to drink in which countless pilgrims have performed perfunctory ablutions; mules that are to come tomorrow, you wait, but, "Inshallah" (if God wills it), there will be another to-

* Accuracy in reproducing the vivid tints and tones in Persian costumes, architecture, and skies in the preceding color plates has been obtained through the coöperation of Mr. Weston, who is an artist as well as an author. He not only furnished color charts for all of the illustrations, but eight of the photographs have been colored by him.

417

Drawn by James M. Darley

A MAP OF PERSIA

Consisting in the main of an arid plateau seamed with snow-capped mountains, Persia is almost as large as France, Italy, Germany, and the British Isles combined.

morrow—for God is indeed Great, "Allahu Akbar!"

THE GATEWAY TO KURDISTAN

Leaving Bagdad by the little railroad which ran almost to the Persian border, by the second day we arrived at Khanikin. There is only one passable route from Mesopotamia through Kurdistan into central or northern Persia; hence the importance of the towns on the road used by thousands of caravans and pilgrims.

Let me present a brief sketch of the setting as we were waiting to cross the frontier:

Tents pure white against the autumn-toned uplands; for, although it was May, the coarse grass, thistles, and wild flowers that carpet these desert hills in the spring had already been scorched by the sun. Behind, the dark purple ranges that border the Persian plateau were still spotted with silver streaks of snow.

Black lines were slowly moving—sup-

Photograph from H. D. Baker

THE ARRIVAL OF THE FIRST TRAIN IN TABRIZ, CONNECTING PERSIA BY RAIL WITH
THE OUTSIDE WORLD

The Persian portion of the line runs from Tabriz to Julfa, on the Russian frontier. It was
built by a Russian company (see also illustration on page 364).

Photograph from Fritz Morris

WHEN URUMIAH ENTERTAINS PERSIAN ROYALTY

This city, near the northwestern frontier, attained notoriety during the World War as
the center of the Syrian or Nestorian massacres. The photograph shows the eager interest
of the populace in the arrival of the Shah on a tour of inspection.

THE FRUIT MARKET IN URUMIAH

The reputed birthplace of Zoroaster, this city, 12 miles west of Lake Urumiah, is situated in the midst of gardens and orchards. It is noted as a center of missionary activity.

VEILED ARMENIAN PEASANT WOMEN OF JULFA, ON THE RUSSIAN FRONTIER

Photograph by Harold F. Weston

In many Persian towns the Armenian women, whose forefathers have lived for three centuries in a Mohammedan country, have adopted the custom of veiling their faces with kerchiefs to protect themselves from the insulting gaze of Mohammedan men. Silver trinkets and coins—dowry presents—are hung in chains from their ears and around their necks.

A CARAVAN OF PILGRIMS EN ROUTE FOR MECCA

Photograph by Lt.-Col. Alfred Heinicke

Note the pen-like hampers in which the travelers are seated (see also illustration on page 363). Pious Persians also make pilgrimages to the holy cities of Kerbela and Nedjef.

WASH-DAY ON THE SHORES OF THE CASPIAN SEA

Photograph from A. N. Mirzaoff

Photograph from A. N. Mirzaoff

COLLECTING SILKWORMS IN THE PROVINCE OF ENZALI, ON THE CASPIAN SEA

Persian silks, like Persian shawls and Persian rugs, are world-famous. The chief centers
of production are Khorasan, Kashan, and Yezd.

ply carts and cavalry of the British forces marching up to cross Persia to the Caspian. Squads of khaki-clad figures on the parade ground near the camp, balanced on the other side by the dark-brown forms of camels, which were being loaded with bales of fodder to the accompaniment of an intermittent series of pathetic, enraged, impassioned roars and raucous gurgles from the protesting beasts.

Half a mile below lay the little mud-built town of Khanikin, half Arab, half Persian, brilliant in the sun against a dark fringe of date palms. Along the dusty road between the high-walled gardens there came out of the town a straggling group of donkeys and blue-clad men, returning Persian pilgrims from the sacred cities of Kerbela and Nedjef or caravans of merchandise for the bazaars of Hamadan or Teheran, all with tinkling bells, jangling bells, and clouds of dust.

At last our convoy of Ford cars was ready to leave, and, bumping and chugging, we wound along the white line of the new macadamized road toward the Persian hills.

The journey to Hamadan, some 300 miles, was by stages of twenty miles a day, accomplished in the early morning, before the heat of the day. The cars were driven by unskilled Indian mechanics, which fact added zest to the scenery of successive mountains and rolling valleys. On one day, out of thirteen cars (blessed Fords), one turned turtle, one burned up, one broke its steering gear on a steep pass, and one ran over a Kurd!

The following outline of a combined two days' journey is quite typical of the scenery: First, along the wide Kangavar Valley, past a small village with fine poplars and deliciously scented sweet brier. Over a three-arched brick bridge, which, though built some hundreds of years ago and of little more than a foot's thickness at the top of the arch, was so well constructed that loaded two-ton motor lorries could cross with safety (see Color Plate XII). Then up through the narrow defile of a pass, leaving a magnificent view of a snow range behind us, onto an undulating plain, where brown and white oxen were pulling crude wooden plows. Skirting another insignificant village with a picturesque ruined "château" perched on the top of a steep crag. Down to the side of a swift-flowing stream, with witch-elm, wild almond, and clusters of fruit trees—apricots, peaches, and cherries—where we camped.

With the dawn, out again onto the barren plateau, up and down a second pass to a deserted valley with shimmering salt deposits. Around a promontory of the range we were encircling, and, from the height of a bluff, there lay the village of Huseinabad below us. A characteristic heat or dust haze turned the clouds shell-pink, the clouds that browsed on the towering snow form of Mt. Elwend, which shouldered out the northern sky.

THE KURDISH HORSEMEN

The Kurds are racially quite distinct from the Persians and have rarely been submissive to the central government. They are in reality semi-barbaric, nomadic tribes that live on their flocks and by hunting in these wild mountain valleys. They have their own national costume, which is perhaps the most picturesque in all Persia.

Almost always armed to the teeth, these tribesmen look particularly romantic when dashing down a boulder-strewn hillside on their sure-footed ponies: the gleam of a rifle slung over a shoulder;

flowing purple turban loosely bound around a huge black felt hat; broad, colorful scarf about the waist, half hiding two or even three bandoleers and above which projects hilts of a knife and a locally made revolver or perhaps a German automatic Mauser; baggy trousers, gaily tasseled and embroidered saddle-cloths, and a certain air of bravado withal that vividly recalls an Oriental, a more brilliant Velasquez, or those gallantly attired heroes so naïvely shown in old Persian miniatures.

A KURDISH WEDDING CELEBRATION

The Kurdish women are generally somber in dress, but do not hide the beauty of their faces under veils as strictly as the Persian women. We were, however, lucky in seeing a gathering all decked out in their 'Sunday best. The occasion was a wedding.

It was evening. I was seated on a grave-stone, painting the dilapidated town of Kasr-i-Shirin, sprawled out over the brow of the opposite hill, ending in the ruins of a third-century castle. I could look into a courtyard over the enclosing walls and see a noisy wedding crowd.

"Hi, ya, ya, ya, ya," the women cried, emphasizing the first and last syllables, to the accompaniment of a big drum. There was an orchestra, too, consisting of four weird instruments — a guitar-violin, a piccolo-flute, a six-foot brass trombo-horn, and kettledrums — which were being played apparently at random and intermittently. Now and then one or more of the players would stop for refreshments, and then resume hastily and with much added gusto, catching up, I suppose, the part of the unwritten score that he had missed!

The men and women had formed in separate lines, and with locked arms were swaying backward and forward in a sort of folk-dance.

Finally a group of men guests left the wedding, trotting down the hill, still keeping in step and singing in unison that monotonous refrain of the Kurdish wedding march. They were going to a pile of merchandise under some willows by the banks of the river. Soon they would call their camels from where they were

WHERE THE RUSSIAN ROAD CUTS THROUGH THE ELBURZ RANGE DOWN TO THE CASPIAN SEA

The southern side of the mountains is absolutely barren, reminding one in color and formation somewhat of the Grand Canyon of Arizona. In contrast, on the northern side of the crest a thick jungle is watered daily by rain-clouds from the Caspian.

ONE OF THE BEAUTIFUL CITY GATES OF TEHERAN IN WINTER

Photograph from Faye Fisher

The snow stays on the mountain peaks all summer, and the quantity of the fall in winter determines the amount of water the city will have during the hot season. A mild winter is frequently followed by famine.

WHEN THE PERSIAN MOHAMMEDAN SAYS HIS PRAYERS (SEE ALSO FOLLOWING PAGE)

The Koran commands: "Wherefore glorify God when the evening overtaketh you, and when ye rise in the morning, and unto Him be praise in Heaven and earth; and in the evening, and when ye rest at noon." The two "evenings" are interpreted to mean at sunset and after sunset.

is commanded, "Wherefore observe fast when the evening ... is established, and when ye rise up in the evening, and when ye rest at noon." The two "evenings" are likely ...

Photographs by Lt.-Col. Alfred Heinicke

FOUR DISTINCT PRAYER POSTURES ARE TAKEN BY THE MOHAMMEDAN WORSHIPER (SEE ALSO PRECEDING PAGE)

Besides his daily prayers, made while facing Mecca, the orthodox Mohammedan is enjoined to perform four other acts of worship—recite his creed, fast in the month of Ramadham (the ninth month of the lunar year, "wherein the Koran was revealed"), give alms, and make a pilgrimage (*hajj*) to Mecca.

429

Photograph by Harold F. Weston

FALLEN AND NEGLECTED, THE GUARDIAN LION OF HAMADAN STILL RECEIVES
LIBATIONS FROM CHILDLESS WOMEN

This battered stone image, the only sculptural relic left at Hamadan, was spoken of by
Masudi, over a thousand years ago, as "very ancient." Numerous legends and traditions are
attached to this lion—pilgrims visit it and childless women pour oil upon its brow to remove
the curse of sterility.

grazing on the near-by hills; their caravan was to move on with the setting sun.

AMONG THE RUINS OF THE GOLDEN PALACES OF DARIUS AND XERXES

Delay in transport at Hamadan caused a never-to-be-forgotten week in one of the American Mission homes and provided ample time for climbing Mt. Elwend (about 12,000 feet) and exploring the historical sites. Queen Esther's supposed tomb is thoroughly uninteresting, and so is the Musallah, the renowned Median Acropolis, unless redecked by your imagination with the departed splendors of antiquity.

The crumbling mound of the Musallah was the site of the palaces of the kings of the Medes and Persians, of the summer capital of Darius, of the golden temple of Xerxes, of the seven-walled citadel overlooking Ekbatana (each wall tiled with a different color, the inner two being silver and gold).

Here the gold and the treasures from sacked Nineveh were brought by the victorious Medes; here Cyrus carried the untold riches of Crœsus; here Darius found the decree of Cyrus ordering the rebuilding of the Temple at Jerusalem; here Alexander the Great reveled on his return from the conquest of India.

Once mighty pillars and lofty halls stood here, and now there is nothing left but a mound of rubble, bits of hewn rock, and small pieces of pottery, a few pits where treasure-seekers have dug, and, on the summit of the mound, the machine-gun trenches used alternately early in the World War by the Russians, Turks, and Russians.

From the pomp and splendor of the court of the Achæmenian kings at the height of their glory to the sentinel's implacable tread or the rat-a-tat of the machine-guns of foreign armies fighting on neutral soil—such has been the descent of Persia.

PICTURES OF PERSIANS AND SHY MAIDENS

When I sketch, if near a town, I invariably draw a crowd of loquacious ob-

Photograph by F. E. Murray

A MODERN ROCK CARVING IN ANCIENT RHAGES

Fath Ali Shah, one of the early rulers of the present Persian dynasty, endeavored, perhaps in emulation of the great Achæmenian kings, to perpetuate the memory of his reign by this carving in the rock which forms a portion of the north wall of the ancient Median city of Rhages, whose ruins lie five miles south of modern Teheran, Fath Ali's capital. Just below, and not at all abashed by the overshadowing presence of the great Ghajar on his Peacock Throne, Persian women do their weekly laundry, rug merchants freshen their stock in trade, and tanned bathers enjoy the cool waters of the famous Chashmah-i-Ali (see also text, page 389).

servers. I was trying to catch the last rays of sunlight on the Musallah, with the Elwend Range in the background, when the owner of a lime-kiln in the foreground thought I had done enough and wanted to buy the picture.

The Persians, far more than the Arabs or Kurds, show a great interest in an "achs" (a painting or photograph). Not only small boys, as in Mesopotamia, but older men stop to look and linger and hold lengthy discussions among themselves as to what the poor creature is trying to do. As some one remarked, the bazaars of Persia are full of people very busy doing nothing, and in a country where the password is "Fardah" (tomorrow), everybody has time to watch what others are about.

The village people are often afraid of a camera, for the only thing they are used to having pointed at them is a gun. They also generally fear the "evil eye," which is somehow connected with the one eye of a camera. But Persians are, as a rule,

most vain, and those who have seen a camera before are so anxious to pose that often you have to pretend to take their picture just to satisfy them. Then ensue lengthy explanations why you cannot open the box and give them a copy at once, or else they expect to see the finished product by copying your example and looking in the finder.

Women, on the contrary, are most difficult to photograph. This is partially because, though veiled, they have the idea that a camera has almost X-ray powers—will show their very souls and "sans habits." One day, wishing to snap some peasant women working in a field, I tried to get my camera out speedily. Too late: the bevy of shy females had disappeared under a stack of hay.

A THREE-HUNDRED-MILE DIGRESSION—
TO THE CASPIAN SEA AND RUSSIA

Instead of going direct to Teheran, we decided to keep on northward to the Caspian. There is much silly romance about

Photograph by Harold F. Weston

THE PERSIAN GOVERNMENT'S "LIMITED MAIL EXPRESS" FROM TEHERAN TO ISPAHAN

This vehicle is a combination sleeper, day coach, and dining-car. Three hundred miles were traversed in it along a rough caravan track, bumping day and night, with opium-reeking drivers and relays of horses—hitched four abreast—every ten or twelve miles (see text, page 437). It was escorted by road-guards controlled by the notorious brigand, Mashallas Khan (since hanged), whom the Persian Government is said to have subsidized to protect the mails and official personages from robberies.

Photograph from Faye Fisher

A TYPICAL PERSIAN CARAVANSERAI, WHERE POST-CARRIAGES AND CARAVANS STOP

In the middle distance note the Persian mail stage drawn by four horses. The average road-house along a Persian caravan route has a large, dingy, smoke-filled room where the traveler can purchase tea, unleavened, pebble-baked bread, buttermilk, cucumbers, and melons (see text, p. 441).

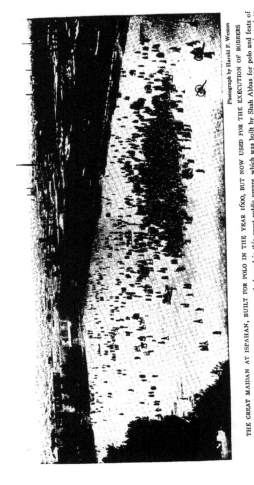

THE GREAT MAIDAN AT ISPAHAN, BUILT FOR POLO IN THE YEAR 1600, BUT NOW USED FOR THE EXECUTION OF ROBBERS

Two hundred and fifty brigand leaders were recently hanged in this great public square, which was built by Shah Abbas for polo and feats of horsemanship. Particularly notorious criminals were sometimes tied to the mouths of the cannon seen in the foreground and blown to pieces (see illustration on opposite page). The crowd shown in the photograph has gathered around the dead body of Jaffar Khouli (see illustration, page 436). The spires in the distance are not factory chimneys, but minarets of mosques built four centuries ago.

Photograph from Roland Woods

READY TO BE SHOT FROM THE MOUTH OF A CANNON—A PERSIAN ROBBER AWAITING HIS EXECUTION

The gun is loaded with powder and scraps of metal and the condemned man is roped to it, the muzzle touching his back. The white cloth shoes worn in Persia are distinctly shown here; also the tall, hard, felt hats.

Photograph by Harold F. Weston

JAFFAR KHOULI EXPOSED ON THE GALLOWS TO SHOW THE PERSIANS THAT THIS NOTORIOUS BANDIT HAS ACTUALLY BEEN KILLED

Jaffar swore he would never be taken alive. The main band of 2,000 brigands had been broken up or captured. The leaders escaped, but, after ten days' pursuit in the mountains, were surrounded. Jaffar, badly wounded in the fighting, died before he could be hanged. Across the Maidan can be seen the turquoise dome of the mosque of Sheikh Lutfullah (see page 447).

a name such as the Caspian that allures one. You look at a map and find an irregularly shaped "lost sea" in the unknown wilderness, the beyond, the uncharted main, where your imagination boards pirate junks and sweeps on cyclonic cruises uncurbed. "My," you think, "what a wonderful place that must be!" You get there at last and are amazed to find that it is just flat water, like a sea anywhere. But often it is the getting there that is the real joy, and so it was with the Caspian.

To reach the Caspian we had to cross the Elburz Mountains. In sixty miles the road drops 5,000 feet. Through a twisting gorge that a river has cut, you plunge from barren uplands down to impenetrable forest jungle. On the south side of the Elburz there is hardly a sign of unirrigated vegetation, for all the copious rains from the Caspian fall on the northern side of the crest. One of the blights of Persia is that the rainfall is excessively meager, in the central and southern regions being reported about 5 to 6 inches annually. But in this belt of malaria-infested rice-fields along the Caspian, in contrast, it is considered a dry year if more than half a dozen days pass without rain.

The trip to the Caspian was considered dangerous, as many British had been ambushed in the jungle-lined pass through the Elburz. We got through safely, but were startled to learn later that the driver of the car following was shot through the head and his companion badly wounded. So that, when returning (after crossing the Caspian on that precipitous invasion and retreat from Russia), we were sent in a convoy with an armed Indian escort.

Turning to the map (page 418), you will see that we have followed the British military road from Bagdad to the Caspian.

We now returned to Kazvin, where I shall let the two color plates (VI and XI) describe for me the tiled gateway by which we reëntered the city and one of the mosques, which seemed to me the most beautiful of northern Persia.

A day's run, still by the inevitable Ford, brought us to the capital, Teheran.

PORTENTS OF PERSIA'S REBIRTH

We threaded the intrigues of Teheran and its spacious avenues, lined with chinar trees and embassies, for a week.

Visits to the Shah's gorgeous palaces, tea with ex-potentates, now plotting against the weak government; discussions with the enlightened Persian official in charge of the suppression of the opium trade; a trip through the institution for training Persians to make permanent rug dyes and replace the cheap German aniline materials; dinner with the editor of the leading radical newspaper, whose revolutionary father was not long ago assassinated; inspection of the big American Mission school—all these I skip, though here one could see new blood coursing through Persia's atrophied veins, and hope became conviction in the eventual rebirth of the nation.

These I pass, as they look to the future, while I am describing the present. So I hasten to the more interesting parts of our trip, from Teheran to Ispahan by mail stage, and from that great city to the Persian Gulf by mule caravan.

What of those roving brigands we were to encounter? It was now a month since we left Bagdad. The authorities at Teheran announced that the robber band was being surrounded; that only if we hurried would we get there in time to participate in the last of the fighting. We were, let it be whispered, approaching the fray from behind the robbers, and, though some Americans may imagine themselves capable of handling ten enemies each, we were doubtful of the results if two opposed two thousand bloodthirsty fugitive brigands. Alas, history never will know! But more of this anon.

THE PERSIAN GOVERNMENT'S LIMITED MAIL EXPRESS

Anxious to cover the 300 miles to Ispahan as rapidly as possible, we decided to take passage by the Persian Government mail stage, the one regular link, aside from the telegraph line, between the capital and the great cities of central and southern Persia. It is analogous, one might well say, to a Washington-Chicago Limited, provided but two trains ran a week, consisting of one combined sleeper, Pullman, day coach, dining and mail car, the size thereof equaling that of a single hay wagon (see illustration, page 432). The following is a faithful portraiture and strictly not a caricature; in truth, I write feelingly of this memory:

An old, uncovered, springless hay wagon, with big, creaking, five-foot wheels in back and smaller ones in front; four horses, all abreast, that were changed at each road-house, located at intervals of ten or twelve miles; eleven Persians besides us (one more could have been embraced had there been a communal lap) rocking about on top, hanging on for dear life, as we swayed down a ditch or jolted over a rock; baggage and mail bags beneath us—billet-doux even can become callous when pounded upon continuously; alternately half frozen at night or blistering in the sun, as the hours slowly bumped by, for one drove at a trot along the ill-fashioned road across the uneven prairie, in the night carrying no other light than the consciousness of the stars.

Every stone in creation seemed to be strewn in the way during the first night ride; sleep was out of the question. At last it began to get light, and two peaks, largely covered with snow, loomed out of the east, gray sentinels in the cold of the early dawn.

As the sun rose a stop was made to change horses at a lone roadside inn on a desolate slope sweeping back to the

THE CHAHAL SITUN, OR HALL OF FORTY COLUMNS, AT ISPAHAN

Photograph by Harold F. Weston

A garden pavilion and throne room, where Shah Abbas in 1600 held his sumptuous court. The roof is brilliantly decorated with inlaid mirrors and painted wooden designs. The hall, or throne room, contains large mural scenes depicting the Shah and his courtiers. The author's one disappointment was that he saw only 20 pillars. "Oh," said his Persian friend, "you have not counted all. Look also in the pool!"

438

THE POOL, IN FRONT OF THE CHAHAL SITUN (SEE ALSO PRECEDING PAGE)   Photograph by Harold F. Weston

A fitting scene for a gorgeous reception by a medieval Shah in the days of Persia's glory! The carved stone column in the foreground probably dates from the 17th century, but in form and execution it is reminiscent of the great Sassanian carvings.

439

A GATEWAY OF THE RUINED PALACE OF SHAH ABBAS AND THE PAVILION OF THE ALI KAPU AT ISPAHAN

From the porch pavilion overlooking the Maidan, Shah Abbas and his court watched the tournaments in the great square below (see page 434). This building still contains mural paintings, crumbling but beautiful. The women in the foreground are completely veiled, according to the Mohammedan custom.

Photograph by Harold F. Weston

440

mountains twenty miles away. A customary "connaught," or channel dug underground, brought water from a distant mountain stream. Our good Mohammedan companions performed their ritualistic ablutions and morning prayers (see illustrations, pages 428 and 429).

By the next relay it was hot, with choking dust blowing from across the desert, and progress was slow, owing to tracts of deep sand. In desperation we drank some filthy tea, slightly flavored by the saline character of the only available water. Such was our initiation into the delights of traveling by *fourgon*, Persian Government hay-wagon mail service.

A PERSIAN WAYSIDE INN

Toward noon we arrived at a village of low, mud-built houses, clustering around a miniature mosque, with a sparkling blue-tiled dome. A thermometer would have registered well over a hundred in the shade; so with silent relief cramped legs crawled down from the top of the wagon the moment its creaking and lurching had ceased.

Inside the road-house (*menzil*) we found one large smoky room. A wide platform seat, covered with coarse, ragged rugs and lounging occupants, skirted the edge of the room. The "guests" were effectively indistinguishable from beggars, and our entry had roused most of these habitués from their noonday siesta—or was it a stupor caused by that drug which is the curse of Persia, for there was a smell of opium in the stagnant air.

The innkeeper, identified by a griminess surpassing that of the others and by the fraternal manner of his welcome, had started blowing up the charcoal in the invariable Russian samóvar. The smoke curled unconcernedly up to the flat, blackened roof of poplar logs covered with matted branches and earth; it wantonly dissipated.

A pilgrim—one could tell it from his blue hat, shaped like an auk's egg—was chatting in low guttural tones to a group by the doorway, probably telling the latest gossip (*gufti-gu*, the Persians call it) from the bazaars of Bagdad. Several were smoking a *kalian*, water-pipe (see Color Plate VII). Each inhaled deeply

a draught or two and the overworked mouthpiece, on the end of a coiling tube, was passed on, while the contented inhaler spat lustily on the earthen floor. Others were sipping tea from diminutive glasses with a loud guzzling noise.

A filthy beggar-like chap, who, to judge by the badge on his felt hat, was a Persian gendarme, was drinking from the mouth of a teapot used as the dipper from the kerosene canister in which the daily water supply was kept. I have heard that a Persian's idea of a teapot is that it is a vessel the spout of which is especially adapted to drink from.

Our lunch of tea, unleavened, pebble-baked bread (see page 455), a thick buttermilk replete with traces of its goatherd origin, raw cucumbers, and a melon, had been placed on the platform beside us. The hot sunlight was streaming through the one doorway and the few green trees outside looked particularly attractive. I said a prayer with each mouthful and sighed, "Well; this is Persia."

SMALL TRAVELERS BECAME CLOSELY ATTACHED

To return to a more absorbing aspect of our perch on the hay-cart, we had been warned that one of the annoying features of riding by mail stage was the Persian attitude toward cleanliness. In a country where it is "considered effeminate to be clean, any man who is obtrusively so is despised," and where "special resentment is harbored against any one who indulges in more than one shave per week," one is not apt to be particular about appearances. But the intimacy of our companionship did not let matters rest there. I remember a British Tommy's pathetic complaint during the "hunting season" down in Bagdad: "It ain't their blinkin' bitin' wot gets me fed up; it's their bally walkin' about!"

Needless to say, when, after three and a half days of this disquieting method of conveyance, we decided to wait for the next post-cart at Kashan and had finally said *Khouda hafiz-i-shuma* (the Persian farewell) to our fellow-travelers, we most involuntarily took a great many of them with us, so closely had they become attached.

One of the magical charms of Persia

Photograph by Harold F. Weston

CARAVAN MULES WAITING THE SETTING OF THE SUN WHILE STANDING IN THE BED OF THE ZENDA RUD AT ISPAHAN

Across the river lies the Armenian town of Julfa (population 3,000), a suburb of Ispahan, once a sort of Persian Versailles, the royal pleasure grounds of the Safavid kings, given by Shah Abbas in 1603 to Armenian refugees whom he transported there to save from massacre by the Turks. Note the huge camel bales and bells in the right foreground. This photograph was taken from the Pul-i-Ali Verdi Khan, the longest bridge in Persia, of 34 arches and 388 yards in length.

SURMEK, A MEDIEVAL STRONGHOLD, SACKED BY THE AFGHANS

Photograph by Harold F. Weston

This mound is called Kasr-i-Bahram, or the castle of Shah Bahram Gor, a famous Sassanian king of the fifth century. The crumbling mud-brick ramparts are probably far more recent. It was formerly used by the shahs as a hunting lodge, but has been in ruins since the Afghans overran Persia, in 1789.

443

A VIEW OF BIRJAND, IN THE PERSIAN PROVINCE OF KHORASAN, FAMOUS FOR ITS CARPETS

The town stands on the great Iranian desert plateau and suffers frequently from sand-storms, some of which are of such violence that occasionally houses exposed to their blast are completely buried. In the photograph a Cossack and a Persian are shown standing on a sand-hill blown up by the wind to such a height that it reaches to the level of the roof of the adjoining house. Birjand is one of the towns in Persia famous for its carpets.

is that it continually reveals glorious un-expected contrasts: the cool green of a garden breaking the barren iridescent plain; the sight of a majestic snow peak when you are plodding through dust and sand at 110° in the shade; brilliantly chiseled bas-reliefs on an abandoned mountain side; the shimmering, opales-cent dome of a mosque soaring above a drab city of crumbling mud-built houses.

THE MYSTERY OF A PERSIAN DAWN AND THE SINGING CARAVAN BELLS

I had not suspected that the lumbering post-wagon would be the means of first revealing to us the subtle wonders of nights of caravan, moving on the desert Persian plateau as on a silent, limitless sea, under the stars—oh, stars of Persia! nowhere else are there such stars!

The dawn disclosed a huge caravan coming up the long, undulating slope of the plateau out of the night. The vari-toned bells of the camels dinging and donging, first sounded like distant bugle calls or lurking snatches of some for-gotten orchestral rhapsody brought to us by the breeze. The high notes blended in a constant ripple of lucid tones, while the plodding "thung, thung" of the low, rich-toned bells of the leaders could be heard, fainter but still distinct, even after the last of the caravan had disappeared over the brow of the hill.

Slowly they went by, some 500 camels in all, with Afghan and Baluchi drivers loping along by their beasts or bobbing sleepily high up on a perilous nest among bales of merchandise. One thought of Vansittart's:

"Ding! dong!
Fugitive throng.
Out of the dark
Into the night,
Silent and lonely,
Gone!—the bells only
Tell us a caravan once was in sight."

Suddenly the sun, a pale gold disk, broke the rim of the horizon and outlined the sharp conical pearl-gray peak of Mount Demavend just to the north of it and fully 120 miles away (see also text, pages 393-400). Then I first fully real-ized the grandeur, the godliness, of its nineteen thousand feet of height. Half an hour later this vision was lost.

The sun rode high above the nearer barren ranges and the horizon was wrapt in the usual all-enveloping dust and heat haze rising from the desert *lut*, the deso-late salt swamps beyond.

KASHAN, FAMOUS FOR HEAT, SCORPIONS, RUGS, AND OTHER THINGS

At Kashan, the reputed home of the Wise Men of the East who set out for Bethlehem, we waited two days for the next mail stage. The heat, for one accus-tomed to the 120° in the shade of "Mes-pot," was not remarkable; the eight-inch scorpions, I suppose, slept under the many layers of dust; the rugs were in the hands of profiteers, judging by the prices; Mohammedan "Salvationists" made the narrow, arched passageways of the ba-zaars reverberate with their wails and discoursed to a sleepy rabble—these and a reputation for cowardice are the claims of Kashan for renown.

Let Kashan and the remaining hundred miles by mail stage sleep with the scor-pions; for the reader will begin to think those much-harassed brigands were only a smoke screen, anyway, to lure him to disillusion—"the man who has seen the most of the world is—." So we will jump down to the refreshing home of a mission doctor in Ispahan, plunge into a bath, and go out to look for those illusive robbers.

THE FATE OF THE CAPTURED BRIGANDS

While walking out to photograph the turquoise-domed Shah Abbas Mosque, the morning after our arrival, we were startled by the sound of a bugle. A crowd congested the narrow street. Soon a company of white fur-capped Persian police swung into view. In their center marched a tall, gaunt, black-bearded man with hands bound behind his back. It was one of the captured brigands being taken to the great central square to be hanged.

Seven men were hanged the day before, we were told, and nearly two hundred more were to be disposed of that way. They were the leaders of the band that for ten years had terrorized the roads and villages around Ispahan. We had passed villages roofless and deserted that they had plundered. Hundreds of

Photograph by Harold F. Weston

THE TOMB OF CYRUS THE GREAT AT PASARGADÆ (SEE PAGE 465)

Note the size of the stone blocks and the bush growing between the first and second tiers on the eastern (right) side. The road-guard at the right carried a magnificiently carved, silver-sheathed sword.

innocent peasants are said to have been killed. Countless wealth had been taken from caravans. Some 40,000 tomans (about $80,000) in specie, recently seized, had been recovered and the Persian authorities were trying to find out, by the aid of promises and tortures, the hiding place in the mountains where the bulk of the loot had been stored.

NAGAR ALICHE, A FOURTEEN-YEAR-OLD DESPERADO

Nagar Aliche, the fourteen-year-old son of one of the robber chieftains, was among the handful who knew where this treasure lay hidden. His deeds were more the talk of the bazaar than those of any of the older brigands. Popular rumor accredited him with over two hundred human killings by his own hand or rifle. When the Governor of Ispahan threatened to have him blown from the mouth of a cannon unless he told where the booty could be recovered, he is reported to have replied arrogantly:

"I kill others every way. Watch them die fast, slow. Myself not yet killed. Like best to be blown from cannon. See quick what comes after."

There are other stories of this lad which confirm reports of his fearlessness. When a thousand or so British-trained and officered Persian soldiers had arrived from the south to help the local Bakhtiari, and the main gang of bandits had been rounded up, captured, or exterminated, a tiny band of leaders escaped through treachery to their own men. For ten days they were chased among the mountains.

Finally all but the boy, Aliche, and two followers were killed or captured. Four days later the three were cornered behind a garden wall. The two men were anxious to give in. "What's the use?" they said. The youngster turned, gave them a scornful glance, shot them dead, killed three more of the attacking Bakhtiari with his last three cartridges, and then gave himself up.

When Aliche, led shackled before the Governor, was asked how many men, women, and children he had killed, he haughtily replied:

"If I had imagined you were such a fool as to want to know, I would have stopped to count."

WHERE BANDITS SWING SLOWLY IN THE BREEZE

We drove to the Maidan-i-Shah, where Shah Abbas and his courtiers used to compete in polo and feats of horsemanship. In the center of the square a great crowd was assembled around a lone gallows. There the dead body of Jaffar Khouli, one of the most notorious of the brigand chieftains, was hanging, turning slowly in the breeze. He had died early that morning from wounds received in the fighting—he had boasted he would never be taken alive. But unless he was exposed thus the Persians would probably not believe that he had been caught or killed (see illustration, page 436).

Imagine a street in size and congestion like the central aisle of a department store during a Christmas rush, only—. The blazing electric lamps are shafts of sunlight that pierce the small apertures in the successive brick domes overhead and filter down through the hanging dust, and under foot the earthen street is soft with dust or soggy, with occasional holes which drip down to sewer pits.

You must not be surprised if you feel something hairy brushing your cheek. You start to jab an indignant elbow into a red-bearded worthy (they dye their beards with henna as soon as they begin to get gray) and find it is only the mangy hide of a passing camel, for, I must hasten to add, the bazaar smells can even stifle at their very birthplace the pregnantly masterful aroma of those stealthy-footed beasts.

BAZAAR SNAP-SHOTS, SMELLS, AND ONE-EYED MAIDENS

Diminutive asses with choking loads of hay flatten you against a wall or drive you into the opening of one of the tiny counter-like stores that line the way; mules with reeking piles of raw hides fresh from the tanneries (Ye gods, how much attar of roses would it take to drown a Persian tannery?) obstinately refuse to give you the right of way; the head of a horse, champing on the cruel Persian bit, unexpectedly projects over your shoulder, as you leap to safety with

Photograph by Harold F. Weston

FLUTED SHAFTS MARK THE AISLES OF XERXES' AUDIENCE HALL, AT PERSEPOLIS, WHERE LEVEES WERE HELD WITHIN ITS ONCE TAPESTRIED WALLS (SEE TEXT, PAGE 465)

Of the 72 columns only 11 remain. The terrace which leads up to the hall is covered with an elaborately carved frieze and cuneiform inscriptions. The low ruins in the center are of Darius' palace, some 50 yards to the rear.

THE ROCK CARVINGS OF SHAPUR, THREE MILES SOUTHWEST OF KAZERUN

Photograph from Roland Woods

These bas-reliefs record the victories of three Sassanian kings, one of whom (Shapur I) captured the Roman Emperor Valerian and kept him a prisoner to the day of his death, then stuffed his skin with straw and presented it as a trophy to a Persian temple. The two men seated to the right are tea merchants. Note the samovar in which the tea is made.

A VIEW OF SHIRAZ FROM ABOVE THE KORAN GATE (SEE ALSO COLOR PLATE XV)

Photograph by Lt.-Col. Alfred Heinicke

The capital of the Province of Fars is situated in a fertile plain at an elevation of a mile above sealevel. A mud wall four miles in circumference surrounds the town, which has a population of 60,000.

450

curses for its unconcerned rider; then you are pushed against the oozing side of a goatskin bag which the migrating human soda fountain of Persia carries (see illustration, page 369), the seller of "drinking" water for half a penny a cupful and freshly drawn from the—but I spare you.

Color, movement, shouts, brayings, smells—all through a jazz of dust—such is the heart of a Persian metropolis!

The labyrinth of bazaar streets in any large city covers several square miles. The various trades are concentrated in wards. In the cloth bazaar I remember seeing, beside some of those charming "Persian prints" (cloth decorated with pen-drawn designs of grotesque men and beasts, or pressed by hand-cut wood blocks with quaint flower figures), great piles of cheap Manchester or Birmingham cloth in blatant European patterns.

In the section of confectioners' shops a large variety of odorous fly-covered sweetmeats was spread out on sloping shelves to catch the eye of the crowd, and incidentally dust. In the middle of the side street, over a four-foot copper platter, seven men, stripped to the waist, pulled out a huge mess of thick taffy to a chorus of shouts and laughter.

Around the corner was a Persian "Quick Lunch": mud-plastered seat along the wall of a smoky recess; samovar and many liqueur glasses from which to sip tea; open fire, over which on a metal spit Persian "hot dogs," called *khobobs* (minced meats), were being roasted to the tremulous delight of only Persian nostrils (see illustration, p. 370).

These bazaar details are but snapshots, blurred, distorted. In the murky whirl of the bazaar posed pictures would be out of place. Of verbal snap-shots I could fill endless pages, for this is the market-place, school, movie, ball field, and home for most city Persians.

And those one-eyed maidens? In the gold and silver bazaar were gold filigree ear-rings and jeweled brooches which buckle a woman's face cloth at the back of her head, and silver armlets which are amulets, or prayer holders. There were also many engraved miniature silver pots for *khol,* or lampblack, with which Persian women underline their eyes and as-

sist their eyebrows to meet over their nose "like the horns of a gazelle," and rouge, even rouge, without which no well-bred lady of the harem would be content.

Most of the customers here were figures completely shrouded in baggy, formless blue or black gowns, which propelled themselves with unseen feet, guided by unseen eyes. As a Persian poet sang of his fair one, "She has a face like a full moon, but she waddles like a goose!"

Oh, full moons of Persia who have attained that zenith of praise for a Mohammedan beauty, let the silken clouds continue to veil your pride, lest we find those languorous eyes, the "Cheshumkhoomar" of the Persian poets, rule but the nights of our imaginations! Oh, tender, painted faces, lurking under white lace-worked veils, veils that you lift surreptitiously to get a better look at us as we pass, disclosing one furtive eye, one lonesome eye escaped from the darkness of an unknown harem! Oh, one-eyed maidens of Persia, half moons of mystery, beware, lest too much be revealed and our vision of delight fly from us, lest beauty be stolen with the veil! For such is present-day Persia: a land of hidden treasures for our dreams, but of appalling disillusions when we are confronted with realities, when our expectations are brought into the light.

RETAINING AND REGAINING A THRONE

By the end of a week at Ispahan we had secured a muleteer, four mules, and a donkey with which to negotiate the next 300 roadless miles to Shiraz.

Naturally, the day we were to start and sat waiting on our baggage our muleteer, true to his race rather than his many-times sworn promise, did not turn up. So we philosophically reflected on what Persia would become without the hope of "tomorrow," and had another day to wander through the miles of bazaars, gaze at the vanishing mural paintings on the ruined walls of the great palace of Shah Abbas, and search vainly for the much-sung "Roses of Ispahan."

Riding on a mule Persian style is about as exciting for the novice as a first trip on six-foot stilts. Each mule will carry

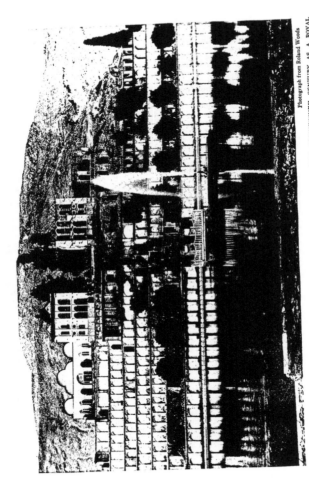

THE "BAGH-I-TAKHT," TWO MILES NORTH OF SHIRAZ, BUILT AT THE BEGINNING OF THE SEVENTEENTH CENTURY AS A ROYAL CARAVANSERAI BY SHAH ABBAS THE GREAT

This building of plastered brick, now used by the South Persian Rifles as British instructors' quarters, is supposed to be one of a series of 999 erected by Shah Abbas. It has suffered considerably in recent years at the hands of raiders. In 1918-19 the wooded park which surrounded it was cleared by the Bushire field force to make an aeroplane landing ground. Below the two main buildings are many baths, store-rooms, and dungeons.

up to 300 pounds; so a great bulk of fodder sacks and baggage is first piled on, with bedding thrown on top to mold these into a less irregular and hence more slippery saddle. These are strapped to the beast's pack with a single cord. You then get enthroned by means of stepping on a bent Persian shoulder, with one cord to the animal's head to rule your destiny.

If the cord holding the load does not slip, if the faithful animal never exceeds a dignified walk or does not stumble, or if in the drowsiness of the long little hours before dawn you do not relax vigilance and irretrievably slip to the ground, all is well, unless, unless—as my docile "Maude" did one day—upon coming to a nice, murky stream, your mount decides it would be refreshing to take a bath.

The main problem, once you are deposed, is to remount unassisted. My companion, Donald B. Watt, in his first day of ignorance, tried it from a rear attack, got half-way seated on the ledge of the mule projecting beyond the baggage, when she, disapproving, went tearing down the hill, with tremendous kicks at each leap, until all my friend had to restrain his unruly subject from absolute freedom was the bitter end of the tail.

So we often preferred to walk part of the twenty-odd miles which was the usual length of a day's journey. Furthermore, it was often penetratingly chilly at night, even though it was too hot for the animals to travel regularly in the daytime.

PROTECTION AND HOSPITALITY FROM LOCAL CHIEFTAINS

We had letters to the principal chieftains along the route and were provided by them with road-guards to protect us from the attacks of stray bands of robbers.

The trip from Ispahan to Shiraz took about three weeks, and every day brought new experiences. Space will allow me to relate only a few typical experiences, while the reader's imagination is given free scope to deck the rest of the way with more vivid incidents yet untold.

It was on the third night of travel, our mules being loaded and ready to leave by sunset. We walked ahead of our caravan, telling our road-guards to follow with the baggage.

We were crossing an uninteresting plain with parallel mountain ranges some five miles away on either side. The route was only distinguishable by dim white streaks, paths trodden by years of caravans. A few hours later the half moon sank. We were alone. Something seemed to have delayed our guards and caravan. We walked slowly on until about 2 a. m., when, all traces of a caravan track giving out, we realized we were lost.

We tried to find our way back and wandered about aimlessly among the strangely quiet hills for a time, until— dull at first, then sharper and sharper— came the pound of horses' feet across the distant plateau. We thought we heard our road-guards out searching for us, and I was just about to fire my revolver to attract their attention, when I realized that shots would probably be the answer. I happened to have a pocket flashlight. I signaled with this. The pound of the hoofs stopped as we stumbled across a dried water-course toward them.

SAVED BY A FLASH

Suddenly there was a shout in Persian to halt. It was a chill moment. By the wavering light of my electric torch we could see the supposed guards standing with rifles lowered at us. We feared they were robber Bakhtiari, for they were the wildest-looking gang of ruffians I have ever seen.

We tried to explain who we were and the chieftains to whom we were going; that we were lost, but that our guards would surely be back any minute looking for us.

Some of them apparently wanted to strip us and leave without further ceremony. Others seemed to have acquired a curious reverence for my flashlight, which I kept turning on and off, to their terror. They thought it would explode and blow them out into the distant gardens of eternity.

The fears of this group fortunately triumphed and we were set in the right direction, as they hastened off into the silence and safety of the night.

Toward dawn we met our road-guards frantically galloping across the plain, with our poor servant breathlessly running after them. There followed much

A FRUIT SHOP IN THE SHIRAZ BAZAAR

Photograph by Lt.-Col. Alfred Heinicke

Shiraz is noted for its wine, made from the famous grapes of the Khullar Valley. Its melons, too, have more than a local reputation.

Photograph by Lt.-Col. Alfred Heinicke

SELLING THE PANCAKE-LIKE FRESH BREAD IN THE STREETS OF SHIRAZ

In most of the Persian cities bread is prepared in the large ovens of public bakeries, the dough being spread on huge mounds of red-hot pebbles. Instead of being baked in loaves, it is made in thin, crisp sheets (see text, page 383).

Photograph by Lt.-Col. Alfred Heinicke

A BAKER SHOP IN A SHIRAZ STREET

The Persian Government takes drastic steps to punish the baker who overcharges his customers, one method of punishment being to bake the malefactor in his own oven.

gesticulation and attempted explanation, mixing in, as we always did, Arabic, Hindustani, or even English words where Persian failed. We borrowed their horses and rode on toward our destination, Kumishah.

THE BLOOD SACRIFICE: OUR HONOR WAS SAVED

We came to cultivated fields, where, even before the sun was up, here and there one could see puffs rising that floated off and vanished like the smoke from a tug down the harbor on a frosty morning. It was the dust from the grain tossed high in the light breeze by the winnowers (see Color Plate XIV). We drew nearer.

All at once my horse bounded, as one of our guards lowered his gun and took several shots at the peasants.

"What are you doing?" my companion shouted, horrified, forbidding the guard to reload.

"Must kill one those men. Insulted you," said the road-guard in Persian, and he made a determined effort to fire again.

"What do you mean?" my friend asked, seizing the guard's arm.

"Come to them early this day when you lost. Ask where are *firangi-sahibs* (foreigners). They say, 'Don't know. We thresh grain. What for *firangi-sahibs* come bother us in Persia?' Insulted you. We fire them. Dark, no kill. Come back. Kill one now."

He was dissuaded with great difficulty.

The worst part of this is that undoubtedly the charge of insult was false. The road-guards knew that the khan, their overlord, would hear that we got lost. This was entirely our own fault, but he might think we said so to shield our guides. They might be bastinadoed or more severely punished, for the khan has practically life-and-death power over all in his region.

YEZDIKHAST, THE MOST EXTRAORDINARY TOWN IN PERSIA

These guards, therefore, were going to kill an innocent peasant, a serf who was bought with the village, in order to say to the khan that we had been avenged,

Photograph by Harold F. Weston

BEING RECEIVED AT THE PORTAL OF A PERSIAN VILLAGE MANSION

The serfs or poor peasants of Persia generally live in hovels of stone and mud, in villages inclosed by high walls. Windows are crude holes. Doors are blankets hung across the openings. The smoke wanders out gradually. These people are extraordinarily ignorant and primitive, but are good-natured and hospitable.

our annoyance had been atoned for, our honor saved.

Three days later, after an exceptionally weary night of caravan, our eyes, accustomed to the unbroken sequence of rolling plateau, were abruptly presented with a view of Yezdıkhast. It is only by realizing this contrast that the picture of Yezdikhast can give one something of the thrill that we felt on coming upon it across the waterless, treeless, almost trackless, uplands of Persia (see Color Plate X).

Yezdikhast, which means in old Persian "God wills it," is the most strikingly situated town in all Persia. It has been

compared to a petrified ship left stranded on one side of a deep river valley.

Approaching it from the plain, one sees only the tops of a few houses and the cracked dome of a single mosque; but on reaching the edge of the ravine, a quarter of a mile broad and fully 200 feet deep, formerly a river-bed and now covered with rich grain fields, one is unexpectedly confronted with the most remarkable picture of a city of the dead—a sheer rock cliff topped by half-ruined mud-and-stone-built houses piled four stories high on its narrow crest, projecting beams of broken wooden balconies that thrust their arms against the sky like decaying gib-

A KASHKAI WOMAN AT DEHIBID MAKING A RUG

Photograph by Harold F. Weston

This type of rug loom is portable and is generally used by the nomadic tribes. The woolen yarn is dyed locally, but in recent years German aniline dyes, blatant in color and liable to run or fade, were widely used. The small rugs only can be made. On it were dumped upon Persia and The Persian Government has made strenuous efforts to stop the sale of these dyes.

458

Photograph by Harold F. Weston

A KASHAN RUG PRODUCT OF TODAY

A young Persian nobleman out for a constitutional in his garden of cypress and roses! In modernity of conception and humor of execution this product of the rug loom is far removed from the Persian masterpieces of old. Many prominent grandees, from the Shah downward, have their portraits done this way (the design is more apparent if the illustration is held in a vertical position). When the author was in Kashan he was asked for a photograph of the President of the United States to immortalize thus on a Persian rug.

459

Photograph by Harold F. Weston

AN OLD WOMAN OF YEZDIKHAST SPINNING YARN (SEE COLOR PLATE X)

She holds the primitive spindle steady with her bare feet, turns the wheel with her right
hand, and guides the yarn with her left.

Photograph from Faye Fisher

A PERSIAN MOTHER AND HER CHILDREN

In the hammock-bed is a boy. The little girl wears the familiar head covering like her
mother. As in Japan, the shoes are removed upon entering the room and chairs are seldom
used. When a foreign visitor arrives, several high cushions are piled together, so the
stranger will not have to sit on the floor.

Photograph by Lt.-Col. Alfred Heinicke

A TEA PARTY IN PERSIA

Everybody carries something—one the charcoal, one a teapot, one the sugar, another the samovar. Having selected the spot for the social hour, the women squat down and prepare the tea and water-pipe.

bets, a single drawbridge that spans the deep breach between the town and the former river bank and affords the only possible means of entrance.

All forlorn, it stands baking in the hot sun. One expected to see vultures soaring above it and could not refrain from thinking of the time when a tyrannical shah years ago flung the best of its young men to death in the valley below.

AN AFFABLE HOST AND A TYPICAL FEAST

The Khan of Yezdikhast had come out to meet us, and hospitably escorted us to the cool courtyard of his house. in the valley at the further end of the town.

That evening we sat with him and a few notables of the village on rugs spread out on the flagstones of the porticoed platform overlooking a little enclosed garden.

A sumptuous feast had been prepared. There was, of course, pilau, the *pièce de résistance* of any Persian dinner. It consisted of rice cooked in grease and meat cooked in pomegranate juice with nuts

and fruits of various kinds. We had also great slabs of Persian bread. other dishes of rice, buttermilk and "sherbet" (sweetened tepid water), which we sipped from a communal wooden spoon.

We ate with the fingers of our right hands, as is the Persian custom, and all would have been well but for the etiquette which required our host to pluck off those fatty portions which he deemed choice morsels, roll them in a huge ball, and deftly force them between our unwilling lips.

The pleasure of dining with this khan was greatly diminished for me by discovering, shortly after our repast was over, the usage of a little domed building placed over a small stream, the only apparent source of water supply. "For washing the bodies of the dead," I was told. This would not have worried a Persian, however, who firmly believes that "all running water is pure."

One of our road-guards shot a gray quail with feathers as varicolored as those on the necks of the male doves in Bag-

A NOMAD WOMAN PHYSICIAN BLEEDING A PATIENT BY "CUPPING."

When leeches are not to be had for blood-letting purposes the cup method is employed. Messengers have been sent as far as 180 miles on foot to get the blood-sucking worm for a wealthy patient (see text, page 465).

fact I gathered from him was that he had seven sons and (many) daughters—he did not bother to count—and two wives, one at Abadeh and one at Surmek. I found most Mohammedan worthies agreed that though allowed four legal wives, one is generally enough, and if two are owned they are placed in separate localities, as the saying runs. "Better two tigresses in a single den than two brawling wenches."

We paused, a week later, during the heat of one day at a small encampment of Arab nomads. These tribes live in black tents made from camel's hair. With these they move from the borders of Mesopotamia, where they pass the winter, up onto the high plateau of central Persia. Here they graze their flocks and reap one meager crop of grain (see illustration, page 391).

They are as primitive as the old Semitic wandering tribes before the days of Babylon. They are, however, peaceable and hospitable.

dad. We often stirred up small herds of gazelle, which our guards generally went wildly chasing after and once killed one; but this was the only feathered creature that was shot. When the road-guard presented it to me, he started a long discussion about America. My monosyllabic answers seemed intelligible.

A LOQUACIOUS ROAD-GUARD

Persians, by the way, often call America *Yangi Dunya* (New World), which has a curious phonetic resemblance to Yankee Doodle. The most interesting

AN ARABIAN-NIGHTS HOST

We were taken into the tent of the petty chieftain who ruled this obscure tribe.

Every one has retained from youth a few vivid pictures from hearing the stories of the Arabian Nights. Our host, except for his clothes, which were far too drab, looked indeed as if he had just escaped from one of those jars in which Ali Baba's thieves were hidden.

Of enormous proportions, about six

Photograph by Lt.-Col. Alfred Heinicke

YOUNG PERSIA

feet three, with huge arms, he had great drooping mustache and a large nose with similar proclivities, eyes that bulged so that nearly half an inch of white showed around the small brown irises, shaggy black hair spreading out some six inches over either ear, head shaven on top, giving the effect of semi-baldness, and a tiny flat-topped black cap perched on the back of his head, which artistically gave scale to his romantic features.

He had come back from killing and skinning a little brown lamb that had been given to us by the Khan of Dehbid (see Color Plate IV). Standing there in the sun, peering into the tent with a broad grin, a blood-dripping, curved knife in one hand, in the other the lamb-skin glistening red as his hands—who would not swear that this was the vision of a nightmare from boyhood dreams of the Arabian Nights?

We had been told that the native Persian villagers have the belief that every European who appears is a *hakim sahib* (doctor), so great is the reputation of the few mission doctors; consequently we carried quite a stock of simple remedies in order to humor them. After our arrival at Gabarabad, a walled village of unusual filth (Color Plate VII), a great crowd of women carrying anemic babies, of cancerous-looking individuals with festering sores, and of many of all ages partially blinded by that terrible eye disease of the East, gathered about us. They said nearly a third of the village folk had died recently after three days of being *kheili garm* (very warm). It seemed to be malignant malaria, so we gave out as liberal portions of quinine as possible.

It is pathetic what suffering people have to endure who have no ideas of

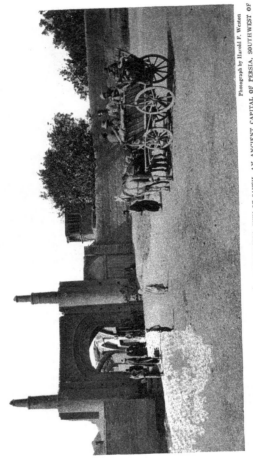

Photograph by Harold F. Weston

THE PERSIAN MAIL STAGE (SEE ALSO PAGE 432) ENTERING THE CITY OF SAVEH, AN ANCIENT CAPITAL OF PERSIA, SOUTHWEST OF
THE MODERN CAPITAL OF TEHERAN

To the right of the gateway can be seen a picturesque wind tower or shaft built on the roof called "bad-girs" (wind collectors), constructed
to carry the prevailing summer breeze down to the water-sprinkled rooms in the basement of the buildings, where the Persians recline during the
heat of the day.

464

cleanliness and are hundreds of miles from the nearest doctor. For eight days a peasant from Yezdikhast attached himself to our caravan. I wondered at first whether he was starting out on some pilgrimage, but I learned later that he was walking some 180 miles and back to get some leeches for a rich and prominent citizen of his town who was desperately ill, but who most probably did not need leeches at all. But I digress from the incident at Gabarabad.

An old man came pleading to us, calling loudly on Allah the Merciful for a miracle. He was almost blind with cataracts on both eyes. I tried to explain that there was nothing we could do for him, but he followed us and sat outside our hut, howling pitifully, calling strenuously for the mercy of Allah. Our caravan had to move on that night. We needed sleep. So, finally, thoroughly annoyed, as we could not persuade him to leave and wishing to give the old man at least one peaceful night, I poured out a large dose from a bottle and gave it to him.

"You will see in the morning," I said. The label on the bottle was "castor oil."

"THE PATHS OF GLORY LEAD BUT TO THE GRAVE"

Three famous historical sites lie on this caravan route: Pasargadæ, where the only building left intact is the tomb of Cyrus (page 446) ; Naksh-i Rustam, where the tremendous tombs of the Achæmenian kings and Sassanian carvings are cut in the face of a great cliff (see Color Plate XIII), and Persepolis. All of these are so well known, have been so adequately described by every archæologist and famous writer who has visited Persia since the days of Marco Polo, that I hesitate to attempt even a brief sketch of how the most important of all Persian ruins, Persepolis, looks today (see illustration, page 448).

As the traveler crosses the plain of Mervdasht, the slender columns of Persepolis grow steadily taller and more distinct. The ruins lie on a great platform built out from the promontory of a mountain range. The stately palaces of Darius, Xerxes, and Artaxerxes, once wonders of the world, can still be clearly distinguished, and many of the great stone architraves and portals, sumptuously covered with bas-relief, are standing today, lonely but vividly impressive ruins, as left after they were destroyed in that tremendous bonfire set by the torch of Alexander the Great in drunken celebration of victory (Color Plate I).

Far more interesting to me than the ruins of Babylon or Assur, Persepolis is preëminently satisfactory in giving still a graphic idea of the vast scale of the buildings, in possessing astonishing bits of bas-relief as clear as the day they were hewn from the stone, and, above all, in leaving enough intact so that the imagination can without difficulty span the great gaps from high-flung column top to column top, raise the fallen fluted pillars, and resurrect the former glory of those world-famed palace halls.

Would that the glory and ambition that once was Persia's had not so completely disappeared! In the character of the Persian peoples today, except for a handful of enlightened radicals, there is little trace of their ancient heritage. To sense the grandeur of the days of the Achæmenian kings, the leadership and power of the Persian Empire that was, is not, and may never again be, one must turn to the silent and neglected ruins of Persepolis.

After threading caravan roads under barren mountains, across the desert plateau, sweltering days in filthy caravanserais, nights under the stars when the monotony was broken only by the mystical sound of caravan bells passing in the darkness, the first sight of the emerald island of Shiraz against the blue-violet hills is so impressive that one unconsciously exclaims, *Allahu Akbar* (God is Most Great). Thus we first saw Shiraz, with its rows of dark cypress and the turquoise domes of mosques, coming through the Koran Gate into the city from the north (see Color Plate XV), and our eyes continued to behold far greater beauty in this squalid town than it deserved, for it came to mean a week's rest in the British consulate garden (see pages 450, 454, and 455).

I must leave to the imagination our visit to the tomb of Hafiz, the best-loved poet in Persia, and a week full of enter-

HUSBAND AND WIFE ON THEIR WAY FROM PERSIA TO MECCA

Photographs by Lt.-Col. Alfred Heinicke

THE WAY A RICH PERSIAN TRAVELS: HE IS IN FRONT, CANE IN HAND, SPEAKING TO
HIS SERVANTS, HANGERS-ON, AND GUN-BEARERS

tainment by the uncle of the present Shah of Persia, by a British general, and other officials, for (shall I confess?) we were thought to be secret agents for the American Government traveling through Persia in those unsettled times, and the more we said to the contrary the more it was believed. I cannot leave Shiraz and Persia, however, without describing at least one of the famous gardens.

"THE ENVY OF HEAVEN"

Before breakfast we cantered out to the garden of the British Resident. He had asked us to come out for a swim in the adjacent garden of a Persian grandee of the neighborhood.

It was the most attractive garden I had seen in Persia. The main avenue was well over a hundred yards long, with superb cypresses on both sides, most of them thirty or forty feet high. There were also double side alleys with chinar, pine, and fruit-trees.

At the lower end of the central grass-covered lane was a pillared garden-house, open, as the Persian name for it (*Chahar Fasl*) implies, to the four winds, the four seasons. At the other end there was a series of terraces with silent fountains, stagnant pools of water, and forgotten beds of flowers.

This terrace led up to a huge tank, recently repaired and sparkling with clear blue-green water, which acted as a doorstep and mirror to a house of a particularly attractive style of Persian architecture.

We were about to undress when our friend the Resident advised us to take to the bushes and don improvised bathing suits, for the ladies of the harem were wont to watch the proceedings from the darkened recesses of the latticed windows! With visions of rows of unseen flashing black eyes, we plunged into the protecting shade of towering pine and cypress.

The name of this garden is, I think, *Resht-i-Behesht* (Envy of Heaven). Persian gardens, so praised by Persian poets and glowingly described by travelers, can hardly be expected to live up to their names. "The Garden of the Thousand Nightingales" is typical of Persian exaggeration.

One garden is called "The Garden of the Forty Colts" because, so the legend runs, it was formerly so vast that a mare which had been lost was not found until she had reared a brood of forty colts.

Again the charm lies in that deceiving power of Persia—contrast. After the desert and dust of weeks of slow caravan, the coolness and refreshing greenness of these little walled gardens, intensified by the *dolce far niente* of days of rest, makes one almost believe these are veritable "Gardens of Eight, Paradises."

To reach the Persian Gulf we had still a week of caravaning. The British, by aëroplane attacks on their strongholds, had disposed of most of the robber bands in this neighborhood and had erected small garrison forts along the more unsettled lower section of the route. It was formerly dangerous crossing the precipitous passes by which one leaps down 6,000 feet over jagged, serrated ranges to the sea.

The hoofs of beasts of burden from ages past have worn steps in the steep face of the rock; it is impossible to pass a fellow-traveler except at special places, and a slip means oblivion. The British have, however, started building a road and have blasted a remarkable path winding up the sheer cliffs.

Our one misadventure was that half way to the gulf I had to stop at a tiny British fort for a week of malarial fever.

The last three days of caravan I did laboriously, with a "sick convoy" of Indian soldiers bound for the gulf port of Bushire to await transport by hospital ship to India.

THE LAST NIGHT ON THE CARAVAN ROAD

My last night on the Persian Caravan Road will never be forgotten. It was at the caravanserai, then used as a British garrison fort, at the top of the pass at Kamarij. My cot had been placed on the roof. It was a hot night. I had fever and did not sleep.

From the courtyard below came the sound of the *tablas* and *dholkis* (drums) of the Indians. They sang, about thirty of them, an endlessly repeated chorus to an endless verse, taken up by various leaders at various pitches. When they

Photograph by Lt.-Col. Alfred Heinicke

A STALL IN THE COPPERSMITH'S BAZAAR OF A PERSIAN CITY

had finally ceased and the moon had set, a dog, five dogs, ten dogs barked furiously in a near-by camp of nomads.

A night caravan passed, with much tinkling of bells and the usual gruff calls of the muleteers. Later, the stillness of the night was abruptly broken. The sentry at the corner tower had challenged and incidentally scared the life out of a Persian who passed too close with, judging by the sounds, three or four donkeys.

A breathless pause, a volley of unintelligible shouts from the Indian, and this lone quivering Persian stole off into the night. But then, his fear overcome, to show his truly Persian bravery, he burst forth into the characteristic long moaning warble of a Persian melody.

There was something very sad and yet fascinating about that wailing refrain sung to the grayness of a desert gravel plain at night, with ghostlike mountain ranges, sharp irregular peaks, still catching the faint light of the moon, and with the stars—myriads of stars—overhead.

Contrast again and mystery. Silence and then barren night mothering at her bosom those weird notes of that intriguing Persian rhythm. It will always remain typical of the Persian Road for me, one of its greatest charms, that lonely Caravan Song fading into the night.

A Map of Asia in six colors (size 28 x 36 inches) will be issued as a supplement with the
MAY GEOGRAPHIC

Notice of change of address of your GEOGRAPHIC MAGAZINE *should be received in the office of the National Geographic Society by the first of the month to affect the following month's issue. For instance, if you desire the address changed for your June number, the Society should be notified of your new address not later than May first.*

Vol. XXXIX, No. 5   WASHINGTON   May, 1921 46⁹

WESTERN SIBERIA AND THE ALTAI MOUNTAINS

With Some Speculations on the Future of Siberia

By Viscount James Bryce

AUTHOR OF "IMPRESSIONS OF PALESTINE," "THE NATION'S CAPITAL," AND "TWO POSSIBLE SOLUTIONS FOR THE EASTERN PROBLEM," IN THE NATIONAL GEOGRAPHIC MAGAZINE

SINCE JULY, 1914, no American, Englishman, or Frenchman, except those officers who were sent out on military missions, has had a chance of traveling along the great Transcontinental Railway which connects western Europe with China and Japan; so before I come to speak of the Altai Mountains, it is worth while to say something of this wonderful highway of commerce, along which I passed in 1913, on my return to England from Japan.

From Calais, on that arm of the Atlantic which we call the English Channel, to Vladivostok, on that arm of the Pacific we call the Sea of Japan, it is more than 7,000 miles, while from New York to San Francisco it is only about 3,000 miles.

An interesting comparison may be made between these two transcontinental roads, on opposite sides of the world, linking the Atlantic with the Pacific.

Each when it leaves the Atlantic coast runs for more than 1,500 miles through civilized and thickly peopled regions, mostly agricultural, though studded with cities. Each when it approaches the center of the continent climbs a mountain range and passes over vast tracts of wild and thinly inhabited country, sometimes through deserts, sometimes through forests. Each crosses great rivers; each coasts along the shore of a large and beautiful inland sea. Each emerges finally from the solitudes of its middle course into a rich and prosperous land and finds its end at a famous harbor—the American Transcontinental at San Francisco, the Asiatic Transcontinental at the equally spacious and well sheltered, if less beautiful, port of Vladivostok.

Along both roads there is a great variety of scenery, much of it striking, but the Asiatic line has an interest that is all its own in the variety of the peoples also through which it passes. One language only rules from the Hudson to the Golden Gate, whereas between Calais and Vladivostok many tongues are spoken and many races of men—Hollanders and Germans, Poles and Lithuanians and Russians, Bashkirs and Buriats, Manchus and Chinese—have their homes.

FROM CALAIS TO MOSCOW

The best way to enjoy the Asiatic Transcontinental journey is to begin at the west end and travel east, whereas the American Transcontinental should be taken from the east toward the west, and for the same reason, viz., that it is more interesting to start from civilization and pass by degrees into wilder regions, more solitary and more picturesque, which keep curiosity constantly alive, than it is to reverse the process.

So, although it was my own fortune to have to travel from the east to the west, I will venture to conduct the reader the

KIRGHIZ CHILDREN AT PLAY

There are more than three million members of the Kirghiz family—a Mongolian people inhabiting an area of nearly 3,000,000 square miles, extending from northwest China to the lower Volga River. There are two main branches of the family—Kara, or Black Kirghiz, so called from the color of their tents, and the Kazaks, or riders, from which term the word Cossack is derived.

Photograph courtesy Department of Commerce

A GROUP OF SIBERIAN CHILDREN IN FRONT OF A COÖPERATIVE STORE ESTABLISHED BY
THE SIBERIAN GOVERNMENT FOR THE DISTRIBUTION OF PROVISIONS TO THE NEEDY

other way, *i. e.*, from western Europe to far eastern Asia. (See the National Geographic Society's Map of Asia, issued as a supplement with this number of THE GEOGRAPHIC.)

Of the comparatively familiar 1,600 miles or thereabouts from Calais to Moscow, nothing need be said, except that so far as the aspects of nature are concerned they are comparatively monotonous, for the surface is an almost unbroken level, only one group of low mountains in Westphalia rising out of the sandy plains of western and central Germany.

From Moscow onward the land, though generally flat, has its undulations; but to the eye of a naturalist it continues to be somewhat uniform, for there are very few deep railway cuttings to indicate the rocks that lie beneath the surface, and as the country traversed is nearly all either cultivated or forest-clad, few wild plants are seen, and these, the latitude being the same, are of the usual Central European types.

The first striking view is reached at the town of Samara, where the broad

Volga, greatest of European rivers, is crossed by a long and lofty bridge, more than five hundred miles above the point where it enters the Caspian Sea. Here for the first time one feels a change in the air, for here begins the dryness of the Asiatic steppes.

CONSUMPTIVES CAME TO DRINK MARES' MILK

Thirty-seven years ago, when I sailed down the Volga, the railway ended at this point. Thither, in that day, consumptive patients used to come from northern and middle Russia to drink mares' milk and gain strength in the invigorating breezes that came from the southeast over arid plains. It was then the summer sanatorium of Russia, as the south coast of Crimea was the winter resort of those rich enough to travel so far.

A hundred miles beyond the Volga blue heights appear on the eastern horizon, and we quickly enter the foothills of the Ural range, their gently rounded slopes descending into charming **valleys**, pasture alternating with open woods which

SETTLERS FROM EUROPEAN RUSSIA ARRIVING AT A RAILWAY STATION IN SIBERIA

The conquest of Siberia by Russia began in the reign of Ivan the Terrible, who gave to two merchants the right to build forts on the rivers Tobol and Irtish. These merchants, Jacob and Gregory Stroganov, hired 800 Cossacks, under the leadership of the Volga River pirate, Yermak, to protect their recently acquired territory. This band penetrated far into the interior and in 1581 captured Sibir, capital of the Siberian Tatar Empire.

distantly suggest those of the "Parks" of Colorado—woods not thick, because the climate is dry, but scattered in picturesque clumps over hill and dale.

As the line pierces deeper into the mountains, the glens are narrower and are filled with a denser forest, out of which bare summits rise to heights of three or four thousand feet. It is a lonely land, with few and small villages, but it is rich in gold and silver, copper,

coal, and platinum—from here comes nearly all of the world-supply of that metal—with an extraordinary variety of rare and valuable stones.

The train takes about seven hours to traverse this picturesque region, stopping here and there at a busy mining town, and passing an obelisk which, at the summit level, marks the frontier of Europe and Asia. Thereafter it emerges suddenly (for the Asiatic slope is shorter and steeper than the European) on the boundless plains of Siberia, here bare and almost waterless as are those of Arizona, but drearier, for there are here no rocks or hollows to diversify the surface, no glimpses of distant peaks to break the level line of the horizon. It is the dullest part of the whole journey from ocean to ocean.

IRTISH, THE WESTERNMOST OF SIBERIA'S FOUR GREAT RIVERS

But presently one comes, at the thriving town of Omsk, which was in 1918 the headquarters of Admiral Kolchak in his campaign against the Bolsheviks, to the first of the four great Siberian rivers, the Irtish, which, having risen far away to the south in the hills of western Mongolia, is here on its northern path to join the Obi and send its waters into the Arctic Sea.

To the Obi itself, an even fuller stream, we come in eight hours more, and see a flotilla of steamers moored to its bank. But of it more anon, for up it one voyages to the Altai. From this point onward the country is rougher and thinly inhabited, for much of the land is the sort of forest swamp which the people call *Taiga*.

On each side of the railway track the woods have been cut back to leave an open space of fifty to one hundred yards wide, so that sparks or coals from the locomotive will not start a conflagration. This open, wide grassy belt is in summer covered with a luxuriant growth of tall flowers on each side of the line, giving the effect of what gardeners call a "herbaceous border," with the railroad track for the gravel walk between the two flower beds.

Behind stand the pines, with their tall, straight, reddish trunks, contorted boughs, and dark-green foliage, beautiful as are those of the Scottish Highlands.

THE YENISEI, GRANDEST OF SIBERIAN RIVERS

After many hours' journey through this delightful parterre, the traveler sees beneath him in a valley, three hundred feet deep, the grandest of all the Siberian rivers, the Yenisei, with the city of Krasnoyarsk lying on the slope between the station and the stream.

This is the finest view of a river from a railroad I can remember to have seen anywhere. The Mississippi at St. Louis and the St. Lawrence at Montreal are as wide, and may have as great a volume; but their banks are comparatively low. Here the *coup d'œil* of the bold heights and the mighty stream filling the long hollow that winds away to the north between rocks and thick woods, is magnificent.

The stream is seen to advantage from both sides, for the track stoops down more than a hundred feet to cross the valley by a lofty bridge, and rises again as much on the eastern slope, making a wide semicircle.

Thirty hours more bring us to the fourth river at Irkutsk, that capital of eastern Siberia for which the contending Bolshevik and anti-Bolshevik armies fought so long in 1917 and 1918. It is the Angara, bearing down a tremendous torrent of clear green water from Lake Baikal, which the train reaches before long.

BAIKAL, ONE OF THE WORLD'S GREAT INLAND SEAS

Lake Baikal is one of the great inland seas of the world, nearly as long as Lake Superior, though not so wide, for in clear weather the eye can reach from the one shore to the other. It fills a bow-shaped depression four hundred miles long, between high mountains dipping steeply into its waters; and on its coasts there are only wood-cutters and fishermen, with a few hunters.

Till long past the middle of last century, some while before the Transcontinental railroad was built, there was no way from the west into the lands of the Amur River and Manchuria except by a ferry across the lake of some twenty or more miles in the summer, or by sledging over its icy floor in winter, and the travelers of those days loved to describe the midnight drive under a brilliant moon.

Now the line runs for many miles along its southern shore on a shelf cut out of the steep mountain side, high above the waves, with frequent tunnels through projecting cliffs.

It was supposed, when fighting began there in and after 1917, that any retreat-

A SIBERIAN PEASANT

His is a land forty times as large as Great Britain and Ireland, but with only one-fifth the population.

ing force would be likely to cover its rear by destroying the track in some part of the shelf, behind or above which no passage could be found. I have not, however, been able to learn whether this has happened.

The line runs high along the curving shores for forty or fifty miles, affording a succession of splendid prospects. Beneath are woods, mostly of birch and aspen, richly yellow in autumn, and wherever they have been cleared the space is filled by a profuse growth of the tall willow herb (*Epilobium angustifolium*), called in North America the fireweed, whose deep pink blossoms make a waving sea of color, stretching mile after mile till all tints melt into the blue of distance.

Solemn and lonely in its mountain setting, the Baikal yields in grandeur to only one other fresh-water sea, Lake Titicaca, on the plateau of Bolivia, above which tower the peaks of the Cordillera Real, the finest line of snows in all the ranges of the Andes.

WHERE THE TRANS-SIBERIAN DIVIDES

Presently the railway, leaving the lake, turns south up the valley of the Selenga River, and thence climbs the slopes, and threads for many miles the ravines of the great mass of rugged and almost uninhabited highlands which figure on our maps as the Yablonoi Mountains. Beyond these come wide plains, and beyond these plains another mountain range, till at Harbin the line divides, one branch turning southwest to Peking, the other southeast to Vladivostok.

Henceforward there are no more Russians to be seen, nor the Buddhist or spirit-worshiping tribes over whom Russia rules, for we are now in Manchuria, where the population is mainly Chinese.

From overcrowded China the industrious Celestials, no longer wearing pig-tails (for the Republic abolished that custom), swarm out in all directions; and had not the Russians in the middle of the last century established their power in the country south of the Baikal and all down along the Amur River to the sea, these regions would have soon been peopled by Chinese emigrants.

The last part of the way from Harbin to the Sea of Japan is, perhaps, the most beautiful, for the soil is fertile, the pastures excellent, the landscapes charming, and the wealth of flowers surpasses that of western Siberia. Even after seven or eight days of unbroken travel from Moscow, the summer tourist comes reluctantly to the end of such a journey.

All that stopped in 1914. When will any tourist find the journey possible once more?

So soon as peace and order have been restored, under whatever government may rule, that government will begin to repair and equip the railroad; but to do this from end to end, through a country impoverished by years of war and blockade, will be no short or easy task.

So much for the Transcontinental Railway, the one great factor in the social and economic life of Siberia which those who wish to understand the country must keep always in mind.

Now let me speak of western Siberia in particular, and of the excursion into the Altai Mountains which I have to describe.

Photograph from Horace Brodzky

TWO RUSSIAN GYPSIES OF IRKUTSK

PREPARING FOR A TRIP TO THE ALTAI MOUNTAINS

In 1913 Siberia was just as open to travelers as was European Russia, but everywhere in the Tsar's dominions whosoever sought to diverge from a regular railway or steamboat route found that he could not get along without facilities granted by the government.

Before starting for the mountains it was therefore necessary to obtain letters of recommendation to local authorities, and the official permission to call for horses at post stations. To get these indispensables I went to Tomsk, the administrative capital of western Siberia, to present to the Provincial Governor the credentials I had brought along with me.

Tomsk lies fifty miles north of the Transcontinental Railroad, to which it is joined by a branch line. Why, considering the importance of the city, was not the main line made to run through it, there being no engineering difficulties to prevent this?

Every traveler asks this question, and receives—so, at least, I was told—the same answer. The Tomsk people did not pay a sufficiently high "gratification" to those officials with whom it rested to prescribe the course of the railway.

I was reminded of a like question and a like answer when, three years before,

FLOWER GIRLS AND BOYS AT A WAY STATION ON THE TRANS-SIBERIAN RAILWAY

being on the west coast of South America, I inquired why large sums were being expended on the construction of harbor works at Antofagasta when, only a few miles away to the north, there was a better sheltered bay at Mejillones. "Because," was the reply, "there was nobody at Mejillones to put up the money that was needed to outbid the people who wanted the harbor to be at Antofagasta."

WHERE COURTESY BECAME A TRIAL

We arrived at Tomsk at 1 a. m. and on stepping out of the cars were received by a bevy of uniformed officials, headed by the chief of the police, a grave personage, decorated with seven medals and six orders (crosses and ribbons). In Russia under the old régime orders and medals were distributed according to length of service and the satisfaction given to the superiors in the department, and the medals determined and indicated the salary paid, a useful method in a

bureaucracy both of securing perfect subservience and of impressing the mind of the undecorated private citizen.

We were driven three miles through woods to the city—in Siberia, as in India, stations are apt to be far from towns—and lodged in a passable hotel, where, however, though it was August, no window was open or could be opened, and baths were unattainable.

No one spoke anything but Russian, and as I had forgotten the little I had learned thirty-seven years before, the position was difficult. Our police chief's sense of duty and politeness compelled him to remain along with us, though it was now 2 a. m. and our grateful and frequently repeated bows did not seem to intimate to him that his further stay was needless.

Searching up and down through a Franco-Russian phrase book, I could find, as usually happens, no sentence that fitted the occasion, but many that seemed

THEY HAVE HAD FEW TRAVELERS TO BUY THEIR POSIES DURING THE LAST SIX YEARS
(SEE ALSO ILLUSTRATION ON OPPOSITE PAGE)

designed for occasions far less likely to occur, among which I recollect this: "Have you seen the crocodile?"—a question singularly inappropriate in an empire none of whose waters are warm enough for that animal.

At last, however, we found words the equivalents to "Many thanks," and "Farewell," and the highly decorated Tchinovnik (the Russian term for a member of the civil service) departed, returning next morning to bring with him a Danish gentleman, a mining engineer, who spoke English and proved very helpful, discovering for us an interpreter to accompany us on our journey. We were

surprised to find that in a city of sixty thousand people nobody, except one or two university professors, seemed able to speak either German or French.

CELEBRATING THE "NAME DAY" OF THE TSAREVITCH

When we awoke next morning all the bells were clanging, for it was the "Name Day" of the Tsarevitch, the delicate child destined one day, if his thin thread of life could be kept from breaking, to mount the imperial throne and rule over nigh two hundred millions of men.

All the functionaries of the city— military, civil, and educational, each

Rising in the Altai Mountains and flowing northward into the Gulf of Obi, an inlet of the Arctic Ocean, the Obi has a total length of 2,000 miles, 1,600 of which are navigable.

A FREIGHT-BOAT ON THE OBI, ONE OF THE FOUR GREAT RIVERS OF SIBERIA

Photograph by W. W. Cutler, Jr.

decked out with his orders and medals—flocked to the cathedral to attend the solemn service in honor of the day. The service was long, as those of the Orthodox Church always are, and only the sweet voices of the choirs relieved its tedium.

We knew that all over the Russian dominions, from the Baltic to the Pacific, every official and every priest and bishop was imploring the blessing of God upon the boy whose life was so precious.

As the worshipers bowed and knelt, as the voices sank and rose, what a wonderful thing, we thought, is this Russian Cæsarism, what a hold it has on the obedience, if not the affection, of its subjects, buttressed as it is by the Orthodox Church, with an omnipresent army of officials to execute its will! But within five years the innocent boy was, with his parents and his sisters, murdered in a cellar at Ekaterinburg, in the Urals, and not a Russian voice throughout what had been the Empire of the Tsars was raised in anger or in sorrow.

THE LEGEND OF TSAR ALEXANDER AS A HERMIT

Tomsk is a large, irregularly built town, straggling from a hill on which stand the cathedral, with its three bulbous domes, and the huge barrack-like university, where law and medicine were being taught to a thousand students, down to the river Tom, navigable for small steamers and carrying a considerable trade.

From the other side of the stream the place looks quite picturesque, brightened by the colors of the church domes and roofs, painted blue or light green, and the house roofs also often red or green; so the general aspect has, from without, a gaiety which the interior belies.

Of the inhabitants, all Russians, for the thinly scattered native tribes live far off to the north, about one-third are exiles or the descendants of exiles. Depressing as Europeans think life must be on a featureless plain, where snow lies more than half the year, they seem as cheerful as men are in Berlin or Rotterdam or London.

One strange tale is told, and universally believed, that the Tsar Alexander the First did not expire at Taganrog, on

Photograph courtesy Department of Commerce.

A VIEW OF OMSK, FORMERLY CAPITAL OF ALL WESTERN SIBERIA

Eighteen hundred miles east of Moscow, Omsk is the meeting-place of highways to middle Russia, Orenburg, and Turkestan. It is visited by violent sandstorms in summer and snow-storms in winter.

CROSSING THE RIVER OBI IN A ROWBOAT

Siberia has some of the largest rivers on the globe—the Obi, the Irtish, the Yenisei, and the Lena—but their value as freight carriers is greatly reduced by the fact that all have their outlet in the Arctic Ocean. For only a few weeks during midsummer can vessels from western Europe make their way through the Kara Sea, east of Novaya Zemlya, to the Gulf of Ob (Obi).

Photographs by W. W. Cutler, Jr.

A PASSENGER BOAT ON THE RIVER OBI

The cabins of these steamers are small and rough but clean. "From the deck one looks over a wide, smooth plain, the uniformity of the prospect in all directions broken only by the sweeps and curves of the mighty stream" (see text, page 485).

Photograph by D. A. Foster

AMERICAN ARMY OFFICERS IN A SIBERIAN WHEAT FIELD

Agriculture is the principal occupation of the natives of Siberia as well as of the emigrant Russians. It is estimated that half a million square miles of the empire are suitable for cultivation. The chief grain-producing regions are the Tobol and Ishim valleys, the marshy steppe between the Obi and Irtish rivers known as the Baraba, the territory around Tomsk, and the foothills of the Altai Mountains.

Photograph by Maynard Owen Williams

A GLIMPSE OF SIBERIA'S LUMBER INDUSTRY

South of the treeless northern tundras lies the almost illimitable forest zone of Siberia, abounding in birch, larch, and the conifers. With the improvement of transportation facilities, the lumber industry will prove an enormous national asset.

CHILDREN AT PLAY IN ALEXANDRIA PARK, IRKUTSK

Photograph by B. A. Foster

the Don, in 1825, as was supposed, but caused the body of a soldier who had died in the hospital at Taganrog to be represented and buried as his, while he himself secretly stole away in the garb of a pilgrim and made his way among a troop of emigrants to Tomsk, where he thereafter lived a religious life as a hermit till extreme old age.

I was told of men alive in Tomsk who, in their youth, had seen him, but no one could say whether the hermit encouraged the belief that he had been Tsar. If he did, he gained nothing from it, except freedom from molestation and additional veneration from the people. Slight as the evidence for the story seems to be, there was nothing in Alexander's character, pietistic and emotional, to make it impossible.

A MOUNTAIN LAND OF MYSTERY

Now, before I come to the journey, a few words on the Altai. It is the name given to the southwestern part of a great mountain mass which divides the lowlands of Siberia from the plateau of central Asia, sending forth on one side the great rivers that flow north to the Arctic Ocean, and on the other, the southern and drier side of the range, smaller streams that lose themselves in the lakes or marshes of Mongolia.

Most of this vast mountain land is unexplored, and only a small part has been surveyed for the purpose of locating the mineral wealth it is believed to contain. As a boy, I had sought to learn something about it

A VIEW OF IRKUTSK LOOKING TOWARD JERUSALEM HILL.

Photograph courtesy Department of Commerce

The Irkutsk Cathedral, originally built of wood in 1693, was rebuilt of stone 200 years ago.

Photograph by B. A. Foster

THE VOZNESENSKI MONASTERY, FIVE MILES FROM IRKUTSK

This ecclesiastical establishment was founded in 1672 and contains the bones of St. Innocent. Most Russian churches are rectangular in form, with five brightly colored domes, the largest of which is in the middle, and each is surmounted by a Greek cross. In the lofty detached tower or campanile are the bells, which are not swung but fixed, only the clappers being movable.

from books of travel, and been able to discover scarce any that had aught to tell; and when I began to read the history of the East, curiosity was reawakened by finding that from the very beginning of history all these regions north and east of the Black Sea and the Caspian had remained unvisited and unknown from the days of Homer down to those of Marco Polo.

Unknown and mysterious, but also terrible, for out of the mists that shrouded them there came from time to time hosts of fierce horsemen, who broke like sudden thunder-storms on the civilized peoples of the eastern Mediterranean and of Europe.

As the Cimmerians and Scythians had descended on Media and Asia Minor and Syria long before the Christian era, so in the fifth century Attila led his Hunnish hordes across Germany into Italy and Gaul, followed by Avars and Bulgarians and Magyars, and in the thirteenth century there came the tremendous invasion of the Mongols under Genghis Khan.

Of this region of mystery and the great mountains that rise in its midst, it was possible to receive some impressions by diverging from the line of homeward journey along the Transcontinental Railway, so we seized the opportunity. Everybody told us that we should have plenty of discomforts or even hardships to encounter, but, being seasoned travelers, we were not deterred, and even, perhaps, put upon our mettle to see whether we could not still "rough it" as in former days.

To reach the glaciers and climb the great peaks would be impossible, for we had no tent or other equipments for high mountaineering, but we could at least have a glimpse to make the mountains live as realities in memory.

A MUSHROOM TOWN IN SIBERIA

Our point of departure was the town of Novo Nikolaevsk, a mushroom growth of the years since the opening of Transcontinental line, for it stands at the meeting point of two great lines of trade— that of the Obi, which brings down the minerals and the grain and the butter from the south, and that of the railway which carries these products eastward to Irkutsk and beyond to the Pacific, westward to Russia and Germany. It reminded me of the new cities in the newest parts of America, with its big warehouses rising fast along half-finished roadways, while the untouched prairie, dotted here and there with scrub birches, lay just outside the houses.

In another ten years, had peace continued, Novo Nikolaevsk would have become the most populous place in all Siberia. By now it may have gone to pieces. Steamers lay thick along the river bank; and in one of these we embarked. The cabins were small and rough, but clean; the food, scanty and unappetizing, was sufficient to support life; and though the days were hot with a strong August sun, the nights were cool, the dry air of the steppe deliciously fresh and invigorating.

From the deck one looked over a wide, smooth plain, the vast dome of heaven resting on a level horizon, the uniformity of the prospect in all directions broken only by the sweeps and curves of the mighty stream.

THE GRANDEUR OF A GREAT RIVER

Nothing in nature is grander than a great river. It embodies the irresistible strength of the forces of nature and their changeful activity, ever the same and yet ever different, here with a glassy surface, there swirling with deep eddies, making and unmaking islets, here eating away the bank, there piling up sand to enlarge it. It is older than man, and will outlive him: it is a part of his life, serves him in many ways, but it heeds not his coming or going.

These great Siberian rivers specially impress the imagination, because their sources lie in unexplored snowy solitudes, and from their middle course in habitable lands they descend into a frozen wilderness—*terra domibus negata*—to find their ending in an ice-bound sea.

We had just come from a long voyage up and down another famous river, the Yangtze, singularly unlike its Siberian sisters in this, that it is the central avenue of commerce through a highly cultivated country, passing on its way many cities swarming with people, and bearing on its bosom not only steamships, but fleets of sailing craft such as can be seen nowhere on Rhine or Danube or Mississippi, or even on the Nile, where once they carried all the traffic of the country.

Here, on the Obi, not a sail was to be seen and hardly even a rowboat. The steamer calls rarely, and then it is to discharge or take in freight, for passengers are few.

FEW VILLAGES ARE SEEN ON THE BANKS OF THE OBI

Like the Mississippi and Volga in their middle courses, the Obi has scooped out for itself a wide flat or depression about seventy feet below the general level of the steppe and swings itself hither and thither across this flat, so that when it is close under the high bank of the steppe on one side it is far from the high bank on the other side. The banks are of alluvial soil, and usually bare, but the low shores and the islands are covered with a growth of willows and poplars.

The few villages on the banks, usually where a small side stream comes down, are clusters of rough wooden huts, irregular and dirty, with the blue cupola of a whitewashed church rising in the midst.

The peasants, stalwart fellows in colored flannel shirts, crowd down to the landing place when the boat puts in; the women, not handsome, but with pleasant kindly faces, wear gaudy blue or red or yellow skirts, with handkerchiefs, mostly white, tied round their heads. All are Russians; it is only in the town of Barnaul, a commercial center to which all the minerals are brought, that one sees now and then an aboriginal nomad from the steppes to the south, over which hills, outliers of the Altai, begin to show themselves.

Photograph from Horace Brodzky

THE FERRY AT IRKUTSK, ON THE RIVER ANGARA, CHIEF OUTLET OF LAKE BAIKAL

In winter traffic is carried over the ice; in summer there is a pontoon-bridge (see illustration on opposite page). When the ice is breaking up and before the pontoon-bridge is swung into position, this ferry-boat is used.

A day and a night from Barnaul brought us late in the evening to Biisk, a place of some importance, to which all the butter coming from the vast pastures which lie all round is brought, and to which timber from the vast mountain forests beyond is floated down the river Biya, which, joined a few miles lower down by the river Katun, issuing from the Altai, forms the Obi. It lies at the edge of the steppe, here rising nearly two hundred feet above the stream, and is a brisk, thriving place. with a good many people of the middle class, traders and government officials.

Through one of the latter (whose tardy action I ought, perhaps, had I better known the "manner of the god of the land," to have accelerated in the proper

Russian fashion) I managed to engage a tarantass, the only kind of vehicle for travel that is suited by its structure for the country it has to traverse. It seats four persons (two behind and one beside the driver), has four low wheels, and short poles supporting the low frame, which play, however inadequately, the part of springs in reducing the jolts and shocks of the rough cart tracks, full of stones and holes, which are here called roads. Beside the two horses, a third, running outside, is usually harnessed.

Our party included an interpreter and a police sergeant, told off to accompany us, not for protection, since the region is perfectly safe, but rather to insure our getting horses at the post stations on the way into Mongolia. We set off on Au-

Photograph by Maynard Owen Williams

THE PONTOON-BRIDGE AT IRKUTSK BEING PUT INTO PLACE IN SPRING (SEE ALSO ILLUSTRATION ON OPPOSITE PAGE)

The history of Irkutsk goes back to 1652, when winter quarters were established here by the Russians for the collection of fur taxes from the Buriats (see illustration, page 494).

gust 18, crossing the broad stream of the Biya in a large ferry-boat.

FLOORS OF POST STATIONS BETTER THAN THEIR BEDS

Each post station, which is bound to provide horses for travelers presenting a "Crown Podoroshna," has one or two small rooms reserved for the use of officials and called the Zemstvo Quartier; and passably furnished. There are usually two beds, but into these we never ventured, preferring to sleep on the light mattresses which, according to custom, we carried with us and laid on the floor. After a long day's jolting in the open air one can sleep on the hardest floor.

The people were always civil, and gave us what food they had, black bread, usually butter also, which was always good, and sometimes eggs, but vegetables were never, and meat scarcely ever obtainable. We had brought a tin of biscuits, with a little tea (needless, because it is the beverage of the country) and preserved meat and desiccated soup, the latter always to be recommended whenever hot water can be had.

We started every morning as soon as the horses could be got, and never reached the night's halting place till after dark, yet could seldom cover twenty-five miles a day, for the tarantass cannot, along such tracks, on an average, and allowing for the changing of horses, accomplish more than three or four miles an hour, and we might just as well, and with more pleasure, have journeyed on foot, but for the frequent swamps and occasional downpours of rain. Twenty-five miles is an easy day's walking in exhilarating mountain air, if one has no knapsack to carry.

A LAND OF MARVELOUS FLORAL BEAUTY

The first day's journey was over the rolling grassy steppe; the second brought us into soft valleys between the lower hills, valleys filled with flowers of many brilliant hues, such as one might find on the lower slopes of the Alps in July, for here the snow does not melt away till

Photograph by Maynard Owen Williams

EASTER MERRYMAKING IN THE GREAT SQUARE AT IRKUTSK

Irkutsk is the principal city of Siberia, 3,792 miles by rail east of Petrograd. It is the capital of the government of Irkutsk, which has an area 20,000 square miles greater than that of Texas. The building in the background is the Cathedral of the Virgin of Kazan.

May. They were mostly of west European genera, some of them British species—blue larkspur, columbines, and (if I remember right) the blue Jacob's ladder (*Polemonium*), purple and yellow aconites, campanulas, gentians, and the white grass of Parnassus (a plant widely scattered over the world), the tall pink willow herb, and, in great profusion, one of the most ornamental among British wild flowers, the purplish blue geranium (*Geranium pratense*). Such a wealth of color I have seldom seen.

On the third day we reached a charming hollow surrounded by cliffs, whose sheltered situation and pure air have occasionally drawn to it a few visitors threatened with tubercular disease. It would be an excellent spot for a sanatorium if the track were rendered passable for invalids and if there were an inn.

Not far off a Russian landscape painter had made a studio for himself in a hut. He was absent, and as it stood open, we saw the studies of Altaian scenery, which were decidedly clever, though rather hard in color. These were the only signs that met us during the journey to indicate that any one ever comes here from the plains except on official business or, for the slender trade in Mongolian wool.

It was a singularly beautiful valley, bold rocks rising out of the forest and the splendidly bright torrent of the Katun sweeping down through pastures gemmed with Alpine flowers.

Desiring only to convey a general impression of the region, I will not attempt to describe the course of our wanderings, nor the difficulties encountered on rocky tracks and along the crumbling edges of deep ravines, nor in plunging through swamps where stones hidden in the mud sometimes all but capsized the luckless vehicle into the water.

Worse still were the risks we ran of being overset in the mire of the track

Photograph by Maynard Owen Williams
FERRIS WHEELS IN IRKUTSK DURING EASTER WEEK
In pre-Bolshevik days Irkutsk was the home of the Siberian branch of the Imperial Russian Geographical Society.

where it led into and through the villages, for here all the space between the houses was a bottomless sea of black farmyard filth, immersion in which would have left the traveler's clothes "a thing to dream of, not to tell."

ALTAI'S LOFTIEST PEAKS ARE AS HIGH AS THE MATTERHORN

All these and many other drawbacks to an Altaian journey are outweighed by the views one gets from the heights, as well as by the wild charm of the woods and the sparkling torrents that foam down the glens. In particular there dwells in my memory one panoramic prospect obtained from the summit of a mountain above the Semenski Pass, a little over three thousand feet high. From it we looked out over an immense stretch of rugged ridges and bare peaks rising one behind another to where in the far southeast snowy summits shone in the sunlight.

The two Belukha peaks (14,900 feet), believed to be the loftiest points in the Altai (about the height of the Matterhorn), were hidden by nearer heights. They are the center of a mass of glaciers,

and round them the grandest crags and gorges are to be found.

Those gentler landscapes which we did see were always picturesque and sometimes charming, but not comparable in beauty to the finer parts of the Italian Alps or Pyrenees, or to the valleys of the Sierra Nevada of California, or to the majestic summits of the Caucasus.

Indeed, the Altaian scenery never reminded me of the Alps. It is more like that of the Canadian Rockies, or even perhaps the most secluded glens of the Scottish Highlands, though in the latter everything is, of course, on a much smaller scale.

What one found peculiarly impressive in the Altai was the sense it gave of an untouched primeval wilderness, remoteness and immensity. The very breezes, whistling or moaning through the trees, sound like

" . . . a wind that shrills
All night in a waste land where no man comes,
Or hath come, since the making of this world."

Civilization seems infinitely far away, for one is in regions where few signs

Photograph by Graham Romeyn Taylor

PEASANT CHILDREN IN A WESTERN SIBERIAN VILLAGE

Photograph from Horace Brodzky

VOLUNTEERS FOR THE WHITE GUARD CAVALRY AT IRKUTSK

"The communist doctrines of the Bolsheviks find no support in Siberia except among the few town workers" (see text, page 499).

Photograph by C. S. Stilwell

ITS HANDSOME RAILWAY STATION IS ONE REASON WHY IRKUTSK IS CALLED THE
PARIS OF SIBERIA

"Nothing forbids the hope that the natural action of economic forces will, perhaps within a few years, install some sort of settled government in Siberia, able to enforce order and to permit men to resume their daily work in a normal way" (see text, page 507).

Photograph by Hugh A. Moran

THE IRKUTSK MARKET PLACE

Before the World War 100,000 industrious Russian peasants immigrated to Siberia annually. This is destined to become one of the great food-producing countries of the world.

Photograph by Hugh A. Moran

PEASANTS IN THE REGION OF FORMER PENAL
COLONIES IN THE LAKE BAIKAL
DISTRICT, SIBERIA

meet the eye to show that man has cared
to dwell, or will ever care to dwell, in
this wilderness, save to fell the woods
and hunt the wild creatures that shelter
therein. It is a land not to be thought of
in terms of time and space, for in it
nothing has ever happened to measure
time by, and in space it exists only as the
gathering place of the waters that feed
the great rivers, as yet receiving nothing
from without and as yet producing hardly
anything to send elsewhere.

THE KALMUKS DWELL IN CONICAL HUTS
MADE OF BARK

From a station on the route into Mongolia we were forced to turn back, for
the tarantass, which had been frequently
repaired, was pronounced unfit
to carry us any farther along a
track described as worse than
that we had traveled; so it was
evidently impossible to reach the
central snows of Belukha.

Taking a more westerly track
on the return journey, we passed
between bare, bold mountains
over several high table-lands, in
some of which we met nomad
Kirghiz, with their herds; in
others Kalmuks, dwelling in
round conical huts of bark, not
unlike the Indian wigwams. The
former were Mussulmans of
Turkic stock, the latter Buddhist
Mongols, but in both there remains much of the old Shamanist
spirit worship, which prevailed
over all northern and central Asia
before the spread from Arabia
and from India of the two great
religions aforesaid.

The people are wild and unkempt, many of them wearing
sheepskins or bearskins, but they
are peaceable in aspect and with
good, simple faces, not wanting
in intelligence. Round the Kalmuk huts birch poles are fixed,
from which flutter strips of white
linen, apparently meant to ward
off evil spirits.

Both races live off their sheep,
cattle, and horses, drinking the
milk of all three, but loving best
the *koumiss,* made of mare's
milk, which, when fermented, becomes
intoxicating. Sometimes they cultivate a
little patch of ground.

KALMUKS AND KIRGHIZ NEVER WALK

A Kalmuk or a Kirghiz never walks;
like an Icelander, he jumps on his wiry
little horse to go a hundred yards.

On these high plains we saw swarms
of little burrowing creatures called tar-
baghans, resembling the marmot of the
Alps, scurrying to their holes as our vehicle approached, and at a spot where the
ground was covered with the Alpine edel-
weiss (*Gnaphalium leontopodium*) for
fully a square mile, we saw a long train
of camels stalking over the pasture, a
strange juxtaposition of the plant that in

Europe grows beside the glaciers with the denizen of the Arabian desert.

Once we came suddenly on a huge eagle, bigger than the sea eagle of North America or the golden eagle of Europe, sitting on a low rock surrounded by a parliament of crows. He rose very slowly at our approach and sailed deliberately away while the parliament dispersed. He may possibly have been a lammergeier, but did not seem to me quite the same as that splendid bird, which I once saw circling over my head on the top of a peak in the Engadine.

Of hawks and falcons there were plenty, as there are of wolves, bears, and lynxes; but the tiger, though he can stand cold—for he is sometimes seen on the shores of Lake Baikal, and puts on a thick coat of fur in northern Korea—does not in this region come farther north than the marshes of Lake Balkash, some hundreds of miles to the southwest.

A few days more through picturesque rocky valleys brought us down to the foothills, and thence over the steppe to the town of Biisk, whence we had started. There, after a farewell view of the mountains from the high bank above the river Biya, we embarked on a steamer even smaller than that which had carried us up.

Having now the current of the river to speed our downward course, we came in three days, making a long halt at Barnaul, a dreary place in the dreary hours of rain we had to spend there, back to Novo Nikolaevsk, where we were hospitably

Photograph by D. A. Foster

ALONG THE BAIKAL MOUNTAIN SECTION OF THE TRANS-SIBERIAN RAILWAY

"The line runs for many miles on a shelf cut out of the steep mountain side, with frequent tunnels through projecting cliffs" (see text, page 473).

entertained by the representatives of the great American firm which supplies agricultural machinery to half the Russian world.

THE ALTAI MOUNTAINS AS A POSSIBLE PLAYGROUND OF ASIA

Here we rejoined the railway; here we boarded the train which was to carry us to Omsk and through the Urals to Moscow and Petersburg (Petrograd), and Königsburg, ancient capital of the Teutonic Knights; and so on to Berlin, where, ten months before the fatal days of July, 1914, we were told (and, as I

A BURIAT MAN AND FOUR WOMEN ASSOCIATES

Photographs by D. A. Foster

A GROUP OF BURIATS NEAR CHITA, EASTERN SIBERIA

The Buriats are a broad-shouldered, stalwart people of nomadic tendencies. They are noted for their devotion to their horses, and when a Buriat chief dies his steed is tied to a stake at his grave, there to die by starvation; but the thrifty heirs usually confine the horse by slender cords in order that it may break away. Note the members of the American Expeditionary Force to Siberia in the background.

believe, honestly told) by a high official of the Foreign Office, that the diplomatic relations between Germany and England had been steadily improving.

Whether the Altai will ever become the mountain playground of Asia, as Leslie Stephen called the Alps the playground of Europe, may be doubted, for the Altaian landscapes, varied and charming as they are, and sternly grand as is the high glacier region, have not the more exquisite charm and the more inexhaustible variety of the Swiss and Italian Alps. But they and the lofty ridges that continue the great line of elevation as far as the river Amur are the only Asiatic ranges in which mountaineering can be enjoyed as we enjoy it in Europe or as Americans enjoy it in the Sierra Nevada and Rocky Mountains.

The Himalayas are incomparably grander, but there the summers are wet and intensely hot, and both the heights and the valleys are on a scale too vast for average human powers. To cross one single gorge like that of the Teesta below Darjeeling, descending 7,000 feet, and mounting another 7,000 to the opposite edge, is work enough for a long day under an Indian sun.

Nevertheless, less interesting as are the Altai, there will some day be much delightful exploration; and some fine climbing will be done in the thousand miles of lofty Siberian mountains, east of the 85th meridian of east longitude. Doubtless, American and Canadian, as well as British climbers, will be found to do it.

Hunters, also, will come, and for a time at least they will find a fair number of wild creatures to destroy—deer, though scarce any elk, as well as wolves and bears and lynxes, and the ibex that haunts the high crags, and in some spots on the Mongolian side of the range, the rare mountain sheep (*Ovis ammon*) with the great curved horns, a creature which, it is to be hoped, they will not be allowed to extirpate.

Though disappointed, owing to the difficulty of making the preparations requisite, at not having been able to make an effective reconnaissance of the approaches to the great peaks, we returned to civilization—half famished, indeed, but sound in health—with the satisfaction of having seen new and most interesting aspects of nature and having caught glimpses of the life of ancient nomad races.

From Siberia we carried back the recollection of a land of large, free, breezy, sunlit spaces, beautiful in summer, with a glorious abundance of flowers, and the impression of a people more cheerful and prosperous than we had expected to find in a country hitherto associated with the cruelties of a tyrannical government and the sorrows of lifelong exile.

THE ECONOMIC FUTURE OF SIBERIA

Something must now be said of the economic future of Siberia, a subject that will become of high significance to the world, for the country contains the one hitherto imperfectly developed region in the temperate zones that has the greatest possibilities of future development for the production of food.

Omitting the districts in eastern Siberia, comparatively small districts, that are fit for agriculture, and omitting, also, the larger and more fertile regions along the river Amur, which Russia acquired seventy years ago, there are between the Urals and the river Yenisei thousands of square miles available either for pasture or for cultivation.

Into this region there had been flowing, mostly from central Russia, a steady stream, averaging 100,000 per annum, of industrious peasants, to whom the government gave farms. Though there were some large estates, Siberia has been, broadly speaking, a land of occupying farmers, very ignorant and living very rudely, but intelligent and laborious.

The cultivated area was being steadily extended, and beyond it, especially along the Obi and the middle course of the Irtish, the rich pastures were supporting an increasing number of cattle, so that an immense trade in butter had sprung up. Most of it was bought by Danish merchants and dispatched in refrigerating cars to western Europe, to be there sold as Danish butter.

Thus, in 1913, the country was thriving, with every prospect of a rapid growth in wealth and population. There were few manufacturing industries, but the minerals hidden in the long mountain

© Underwood & Underwood

A STREET SCENE IN VERKHNE-UDINSK, 300 MILES EAST OF IRKUTSK

This prettily situated town is the headquarters of the Western Trans-Baikal mining administration. It is the junction point of the two great highways of northern Asia. Southward stretches the road to Kyakhta, to Urga (the capital of Mongolia), and to Peking.

range that divides Siberia from Mongolia are believed to be of immense value. That they had not been better ascertained and exploited, and that all the resources of the country had not been more swiftly developed, was attributed to the incompetence and, above all, to the corruption of the imperial administration—a deeply rooted evil, which neither well-meaning emperors nor an energetic minister, if one now and then appeared, had been able to cure.

Had Siberia been in the hands of Americans or Canadians from 1870 to 1910, its revenues and population would have been double what they were in the latter year, for the internal river communications would have been improved and the railway tracks into European Russia would have been duplicated or triplicated.

In 1913 men were discussing one expedient for increasing the trade of the country which a more enterprising government would have done its best to

A TUNGUSIAN GIRL OF EASTERN SIBERIA

The Tunguses are at home throughout central and eastern Siberia, from the Yenisei River to the Pacific. Broad, flat features, with small nose, thin lips, and dark, oblique eyes, are characteristic of this Mongol-Tatar people. They are a cheerful, brave, self-reliant, modest, and hospitable race of hunters, making their home chiefly in the midst of dense forests.

favor. The greatest want of Siberia is cheaper transportation for its heavy products to European markets, especially to those of Germany, France, and England, and that which most reduces the value of its great rivers as freight carriers is the fact that their mouths in the Arctic Ocean are difficult of approach, even in summer, because vessels may get caught in the Kara Sea east of Novaya Zemlya, and even if they reach the Gulf of Ob

(Obi), may be unable to return with the cargoes they have loaded.

The employment of wireless telegraphy may be expected to immensely reduce this risk, for by means of it approaching or departing ships could be kept informed from various points regarding those parts of the sea which may be open—and there are always such parts in July and August—and their course could be directed accordingly.

Photograph by C. S. Stilwell

A TRANS-SIBERIAN TRAIN-DE LUXE IN WAR TIME: NOTE THE MEANS OF INGRESS TO
THE BOX-CARS

The total length of the Trans-Siberian Railway with its branches is more than 5,400
miles, including the more than 1,000 miles in Chinese territory. It was begun in 1891 and
virtually completed by 1902, the total cost being estimated at more than $435,000,000. It is
the one great factor in the social and economic life of the country which those who wish to
understand Siberia must keep always in mind.

Photograph by D. A. Foster

THE CHINESE EASTERN RAILWAY STATION AT NIKOLSK USSURIISKII

Nikolsk Ussuriiskii is a town of 50,000 inhabitants, sixty miles northwest of Vladivostok.   In
peace times it was a popular resort for sportsmen seeking musk-deer and wild boar.

Photograph from Horace Brodzky

A CROWD IN FRONT OF AN OPEN-AIR THEATER AT A CHINESE FAIR IN KYAKHTA, SIBERIA

The crowd contains Chinese, Mongolians, Russians, Tatars. Buriats, and a few Europeans. There is no scenery at these shows. The "announcer" tells the spectators what scene they must imagine. The costumes of the actors, however, are very beautiful, elaborate, and costly.

As far back as the days of the Tsar Ivan the Terrible and the English Queen Elizabeth, a bold English captain tried this route with success, and there is reason to believe that nowadays vessels fitted with wireless apparatus could make pretty sure of safe voyages to and fro.

NOT A FERTILE FIELD FOR BOLSHEVIK DOCTRINES

The economic progress I saw in 1913 was arrested by the war which broke out in Europe just a year later; and in 1917 there was fighting in Siberia itself between the Bolsheviks, who had then seized power in European Russia, and their opponents, who were organized for a time under Admiral Kolchak.* The Bolsheviks prevailed, not because the Si-

* See "Glimpses of Siberia, Russia's Wild East," in the NATIONAL GEOGRAPHIC MAGAZINE for December, 1920.

berian peasants adopted communist doctrines, for those doctrines find no support in Siberia except among the few town workers, but because the men who surrounded the unfortunate and entirely well-meaning Kolchak made themselves detested; so that his forces, at first successful, ultimately melted away before the Bolshevik advance with very little resistance.

What is happening in Siberia as I write these lines, in January, 1921, few people in western Europe know, and I am not one of them, but evidently the economic conditions must have gone sadly back since 1913. When will progress be resumed? When will the setting up of a stable and tolerably enlightened government make progress possible?

No sensible man will venture to prophesy about Russia; but one thing at least may be said: In the long run, eco-

Photograph from Horace Brodzky

THE "HOT-DOG MAN", AT A MONGOLIAN FAIR, AT KYAKHTA, ON THE SIBERIAN-CHINESE BORDER

Note the oddly notched rim of the cart-wheel in the left background—ancestor of our modern non-skid tires. Kyakhta was for many years the great emporium for the caravan trade between Russia and China. The completion of the Suez Canal and the construction of the Trans-Siberian Railway robbed it of its importance. Over the border, on the Chinese side, is the town of Maimatshin.

MONGOLIANS AT MEAL-TIME ON THE STEPPES OF SIBERIA

Photograph from Horace Brodzky

The man on the right is a Russian. The chief item on this menu is boiled meat. A Mongolian never eats bread and seldom vegetables. He drinks the soup from the boiled meat and then lifts out a chunk of flesh, holds one end in his mouth and cuts off a piece with his big knife. A Mongolian eats only once a day, at noon or in the evening, when he consumes from five to ten pounds of mutton or beef. In the winter he carries the meat under his saddle, which keeps it from freezing.

© Underwood & Underwood

SACRED BREAD: SIBERIA

The greatest feast in the Russian year is Easter. Religious ceremonies are held in the churches lasting several hours after midnight and terminating with the curious ceremony of blessing the Easter cakes. Instead of the customary cross, the bread shown in the illustration is capped with the Mohammedan symbols of moon, sun, and flame.

A SHAMAN FEAST: SIBERIA

Shamanism is the primitive religion of certain tribes in Siberia, but in most cases practiced under an outward show of newly introduced religions, either Christian, Buddhist, or Mohammedan.

© Underwood & Underwood

Photograph by Hugh A. Moran

THE LAST EXPRESS TRAIN TO GO THROUGH FROM VLADIVOSTOK TO PETROGRAD IN FEBRUARY, 1918: BAIKAL, SIBERIA

nomic factors are sure to prevail. They assert themselves, because revolutionary disorders never last very long, since it is the general interest of the vast majority in every people to see a stable administration established; and when some strong man, or group of men, possessing the gift for rule have established it, the self-interest of the rulers prompts them to occupy the energies and promote the well-being of their subjects by extending facilities for trade and industry.

SIBERIA'S HISTORY IS ALMOST A BLANK

A few concluding words may be said, as to the future generally. The history of Siberia was almost a blank, and had little interest for the world at large, till 1917. The men of Great Novgorod had occasionally sent trading or raiding bands across the Urals in the eleventh and twelfth centuries, and the first invasion was under a robber chief, named Yermak, who led his followers into the country in 1580. But thereafter the process of conquest and colonization went on unnoticed by Europe, with no serious resistance from the aboriginal inhabitants, who were weak and loosely scattered savage tribes.

Thus there were really no events for historians to record; the process went on gradually and unobserved. The racial character of the Russian immigrants has (except in the Far East) been scarcely affected by any infusion of aboriginal blood, and so far as the Siberian Russians differ from the Russians of Europe, they are nowise inferior.

Serfdom never existed in Siberia. The immigrants were mostly more enterprising than their brethren who stayed at home. The exiles, banished for political offenses, real or alleged, often came from the intellectual élite of Russia, while the descendants of criminal convicts did not permanently stain the population, although some of those who escaped used to range the country as robbers.

Taken as a whole, the Siberians, if not fit to work democratic institutions, are quite as capable of local self-government as are the peasantry of European Russia and just as unlikely to become Communists of the Marxian or any other stripe.

WILL SIBERIA REMAIN A PART OF RUSSIA?

So far as I could learn, the only class in which political discontent or any signs of an interest in politics existed had been the students in the university, who occasionally "demonstrated" or "struck work" when some particularly offensive piece of tyranny proceeded from the university authorities acting at the instance of the ecclesiastic authorities. There have been ferments among the students everywhere in Russia for the last half century; and now and then professors have been dismissed or exiled.

THE SPIRIT OF SIBERIA

Yellow sunset and boundless expanse of fine country, in the midst of which a raging forest fire is sending up a wisp of smoke more than twenty miles away. The photograph was taken from a swiftly moving train.

Whether Siberia will remain politically a part of Russia, it is impossible to predict. An able English observer, who traveled there forty years ago (the late Mr. Ashton Dilke), told me he thought Siberia would break away, peaceably or otherwise; but nothing I could learn in the country confirmed that forecast.

The Transcontinental Railroad has become a bond of union, and the Ural Mountains, though they would form a good natural boundary if the peoples living on each side differed in race, speech, and religion, do not, the facts being what they are, constitute a barrier worth regarding.

It is much to be wished that they were such a dividing line, for the Russian Empire before 1914 was an unwieldly mass, too big for any one set of men to govern, even had such men been more capable than any Russian ministry has ever been.

Yet in 1913 the Russian Government, moved by that insane impulse which induces states to extend territories already too large, was trying to establish political control over Mongolia as far as the frontiers of China.

The portentous expansion of Russian dominion and the growth of Russian population had become a danger to the world. It was a danger much reduced by the stupidity and corruption of the government, but if a malign fate had set a genius like Frederick the Great or Napoleon on the throne of the Tsars, things might have gone ill for Europe.

PERHAPS A UNITED STATES OF SIBERIA

For its own sake, as well as for the world's sake, it is much to be desired that Siberia as well as Transcaucasia should be disjoined from Russia; and if the inhabitants of Siberia were capable of working a system of federal govern-

MONGOLIAN OFFICIALS SEATED BEFORE LAMAS WHO ARE CONDUCTING A RELIGIOUS CEREMONY

Ordinarily the queue is concealed under the hat, but it must be let down while religious ceremonies are in progress.

Photograph by Eugene Lee Stewart

ment, such a system, consisting of five or six federated states between the Urals and the Pacific, would be better than one huge unitary empire or republic.

What sort of political future may this or the next generation expect to see?

Neither Russia nor Siberia is likely to enjoy free popular constitutional government within any period which conjecture can now assign. But neither is it likely that the economically ruinous despotism which now rules both countries will long endure, or that the incompetent despotism of the Tsars will return.

There may be a time of strife, for the habit of obedience has been broken, and there is now no legally constituted authority for the citizen to obey. But anarchy never lasts long.

Nothing forbids the hope that the natural action of economic forces will, perhaps within a few years, install some sort of settled government, able to enforce order and to permit men to resume their daily work in a normal way.

So soon as Siberia obtains such a government, her economic resources and the industry of her people will enable material progress to start afresh, and she will some day become what western America became fifty years ago and Argentina became thirty years ago—one of the great food-producing countries of the world.

THE PEOPLE OF THE WILDERNESS

The Mongols, Once the Terror of All Christendom, Now a Primitive, Harmless Nomad Race

By Adam Warwick

THE cancellation of Mongolian autonomy by China in November, 1919, and the subsequent trouble at Urga, their capital, have once more drawn attention to the "People of the Wilderness," as the Chinese, with thinly veiled contempt for all who dwell beyond the borders of their own civilization, call their neighbors, the Mongols.

Mongolia is a land with a great past. Seven hundred years ago Genghiz Khan set out from its barren steppes to conquer the world, and swept all before him from the Yellow Sea to the Adriatic. Dazzled though we may be by the magnitude of modern warfare, we stand aghast at the unexampled record of his hundred thousand horsemen, who made a three days' victorious march across a hostile land from the Carpathians to Budapest, with minor expeditions deep into Bohemia, Germany, and Serbia.

ALL CHRISTENDOM FEARED THE TATAR CHIEF

In those stirring times the world so feared the Great Captain that a special prayer, "Save us from the fury of the Tatars," was introduced into the Christian litany.

It was no idle dread. But for the death of his successor, which imposed three years of mourning and inactivity on the troops, the Mongol forces could not have been stopped by any earthly power until they had reached the limit of the continent of Europe.

Forty years after the disappearance of the Mighty Conqueror (1227), a grandson, Kublai, crowned his triumphs by becoming, not only the master, but the enlightened, magnificent monarch of the whole of China, Indo-China, Burma, Korea, Borneo, and Sumatra. Unfortunately for his dynasty, the settled life of ease and luxury in Peking sapped the vigor of his followers in a single century.

One more great leader was to appear among them in the person of Timur the Lame (Tamerlane), born to subdue Iran and Turan, defeat the growing power of the Turks, and fire Moscow, thus blazing the way for his last descendant, the kindly knight-errant and poet, Sultan Baber, to found the Empire of the Great Mogul.

Photograph by John B. Rogers

THE TOMB OF TIMUR THE LAME (TAMERLANE) AT SAMARKAND, RUSSIAN TURKESTAN

The body of this renowned oriental conqueror, who died in 1405, was embalmed with musk and rose water, wrapped in linen, laid in an ebony coffin, and sent to Samarkand, where it was placed in a sarcophagus of jasper. The tomb consists of a chapel crowned with a dome. Both time and earthquakes have left their mark on the structure, but the interior is still beautiful with turquoise arabesques and inscriptions in gold.

Photograph by Graham Romeyn Taylor

THE DOOR TO THE TOMB OF TIMUR THE LAME

The Mongol leader has been variously painted by his biographers. To some he seems merely "a deformed and impious person of low breed and detestable principles"; to others, a sort of demi-demon, like Richard III. The magnificent but now crumbling summer palace, mosques, and colleges in his capital city of Samarkand bear witness to the fact, however, that Timur and his immediate successors were splendid patrons of architecture.

A MONGOL PRINCESS IN FULL COURT DRESS

A MONGOL PRINCE, DESCENDANT OF GENGHIZ KHAN

Photographs by Adam Warwick

They are the scions of a race of conquerors who for a short time struck terror to the hearts of Christendom. Only a custom which required the hordes to observe a mourning period of three years following the death of their leader saved Europe from a Mongol inundation during the first half of the thirteenth century.

Photographs by Eugene Lee Stewart

MONGOLIAN BOYS OF THE BARGOO TRIBE

The cut and pattern of their robes about the neck are distinctive of their tribe.

SONS OF MONGOLIANS OF THE OFFICIAL CLASS

The youngster at the left is not at all frightened by the camera, but sand blew in his eye when he was trying to pose his best.

Photograph by Eugene Lee Stewart

A MONGOLIAN OF THE OFFICIAL CLASS

He is wearing a handsome brocaded yellow silk robe and an ornate necklace of wooden beads. Woven into the robe is an elaborate dragon pattern.

But the fall of the Mongols was scarcely less rapid than their rise. In China they were able to hold sway only eighty-eight years. Elsewhere their empire crumbled quickly, leaving only isolated remnants under their dominion, until today the unhappy heir of Sultan Baber sits forlorn and impotent beside the Ganges in a badly cut frock coat, with the crown of the King of Delhi on his head.

MANCHUS SUCCEEDED MONGOLS IN CHINA

When events shattered their dream of world power, the Mongols retired once more within the confines of Mongolia proper, where they have lived for centuries in peaceful isolation. The population, thinned by war to 2,600,000 souls, is spread over a vast territory embracing 1,367,000 square miles —an area more than one and a half times as great as the United States east of the Mississippi.

The Manchus, who took over the rule of the Mongols in China, exercised only a nominal control over them. Indeed, Manchu official feet never trod many parts of interior Mongolia, and the local Mongols were allowed practically to govern themselves, preserving their original tribal organizations, headed by native princes.

When the Manchus disappeared, the People of the Wilderness, who had never recognized any Chinese rights over them, declared their autonomy and lived contentedly under their self-appointed ruler, the "Hutukhtu," or Bogda Khan (the third Living God in the Lamaist hierarchy, whose temple palace is at Urga), until the fatal day when the Chinese Republic canceled their right of self-determination.

The stage on which this drama of Far Eastern politics took place was too remote to awaken the interest of the powers. Location, climate, and, above all, lack of communications—for Mongolia cannot yet boast a single railway, although there are plans for a line from Kalgan to Kyakhta and only recently a motor-car service has been started across the steppes to Urga (550 miles)—shut off the country from the rest of the world.

We still know little of Mongolia's resources. Gold mines certainly exist there (one of which, the "Mongolor," is beginning to be developed by American capital), as well as silver, copper, and coal mines. The rivers abound in fish, the forests in valuable timber and fur-bearing animals, while the great tablelands have farming potentialities equal to Texas and Nebraska.

But the primitive Mongols derive little benefit from these riches. Like

Photograph by Eugene Lee Stewart

A CHINESE MAN IN FANTASTIC GARB, MOUNTED ON STILTS

He is participating in a Chinese New Year's parade in Hailar, a town in Manchuria near the Mongolian border (see also illustration on page 520).

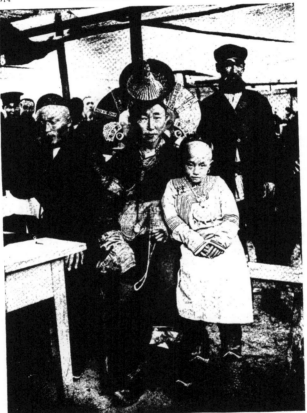

Photograph from Horace Brodzky

A MONGOLIAN WOMAN OF THE WEALTHIER CLASS WITH HER DAUGHTER

Note the string of beads in her hand, with which she toys; also note the talisman the girl is wearing. The mother has on her summer hat of black velvet. It is not easy to get a picture of a Mongolian child, for there is a superstition that Europeans use the eyes of children to make the lenses of their cameras.

Photograph by Eugene Lee Stewart

A CAPTAIN IN THE MONGOLIAN ARMY

He is wearing his state robes and wooden beads and is ready for attendance upon an
outdoor religious ceremony. The chief difference between the garments of the men and the
women is that the former gird themselves with a belt. The usual word for woman in Mongolia is "beltless."

MONGOLIAN SCHOOLMASTERS

The man on the right is Mr. Gobel. He has had four years of college at Peking, speaks Chinese, Mongolian, and Russian fluently and quite a bit of English. He is principal of the Mongolian school at Hailar and has seventy boys learning Chinese and Mongolian and studying geography, history, reading, and writing from Chinese text-books. His assistant (at the left) studied at Tsitsihar, in Manchuria.

ipal affairs do not concern them, for they build no cities, leading a migratory, care-free existence. They need no roads: the illimitable steppe is a natural highway where nations can pass without crowding. They require little water: in their climate men cannot wash. They want no electricity: at sundown, after a long day in the saddle, they are ready to lie down and sleep. Finally, the increasing cost of living does not trouble them, since it costs them nothing to live except the care required to guard their herds from wolves.

These herds provide them with clothing, with food, with transport, with fuel even. All over the grass-lands the flocks graze freely, watched by men unchanged since the days of the Great Genghiz.

Living an easy, open life—a life of true liberty, remote from courts of justice and police— the Children of the Wilderness willingly abide by his ancient code of laws, simple, logical, humane, and admirably suited to their nomadic habits. Their lumbering ox-carts were designed in his day; their sheep and horses are the original native breeds; the ancestors of their camels

the lilies of the field, "they toil not, neither do they spin." They are a shining example of how men by reducing their needs reduce their anxieties.

Financial crises cannot affect them, for money as a medium of exchange is little used on the plains, where brick tea has more value than minted dollars. Munic-

carried the silken tents of the Conqueror.

"SHIPS OF THE BURNING SANDS" GAMBOL IN THE SNOW IN MONGOLIA

Mongolian camels are superb beasts, very different from the ugly, flea-bitten, one-humped Arabian variety. In all the glory of their winter coats—for, strange to say this species thrives in the cold and even delights to gambol in the snow—they are pictures of stately dignity, though in summer, when the long hair falls off in patches, they become repulsive-looking.

Winter or summer, however, these camels retain the objectionable character that Kipling has immortalized. Their breath is so poisonous that it is said no camel-driver lives long. Their kick will overturn a motor car. Their bite, followed by a twist of the lower teeth, generally induces blood-poisoning.

Particularly vicious males are marked with a piece of red cloth tied to the forelock, warning the stranger to beware. The Mongols know from experience that even a horseman is not safe from the determined onslaught of a furious camel, who can outgallop a pony and has a nasty, effective trick of throwing beast and rider and then rolling on them.

Photograph by Eugene Lee Stewart

LAMAS POSING FOR THEIR PICTURE

The tall man is a Tibetan and the smaller a Mongolian. The Tibetan Lama is arrayed in gorgeous yellow silk robes and hat and the Mongolian Lama in red silk with a jacket of yellow silk and red hat. Before posing for his picture the latter repaired to his tent and procured the handsome necklace which he is holding.

THE MONGOL HALF-SOLES HIS CAMEL

Though the largest camel will bear only a comparatively small load, lying down and squealing if an extra pound be added, he is the only freight-carrier that can cross the desert, and even he, after some days traveling in the sand, wears his feet to the quick. When this happens the Mongols throw the limping animal on his side, put his foot on a low stool, and cover the tender part with a patch of leather attached by thin thongs drawn

through the adjacent callosities of the sole, much as a cobbler mends a shoe.

The camel may be useful, but the horse is much more popular among the Mongols. The native breed, indigenous to the country, is seldom over thirteen hands high and rarely beautiful. But for endurance, cleverness, and originality, the little Mongol pony has few rivals. In the depth of winter his owner neither feeds him nor provides him with shelter.

An extra growth of hair and thickness of hoof (for he is never shod) protect

Photograph by Luther Anderson

COYNESS IS NO PREROGATIVE OF OCCIDENTAL WOMANKIND

One would never suspect from her jovial demeanor and her gala costume that this Mongolian woman leads a rather lonely life in her felt-covered yurt (seen in the background) while her lord and master is tending his herds on the steppe. Her neighbors are few and far apart, and her chief opportunity for social intercourse is found during the fairs (see text, page 541).

him against the bitter cold. As for his food, this intelligent little beast learns to scrape away the snow with his forefeet and find the sparse remains of the summer grass underneath.

WILD PONIES OF MONGOLIA ARE CAUGHT WITH POLES

Though mares, easily distinguished by trailing manes and tails, are kept at home for breeding purposes, Mongol ponies are exported in great numbers to China. They roam the plains freely until wanted, and are then captured in a curious way.

Two or three Mongols start out together on fleet mounts specially trained for their work. The riders carry long birch poles, like fishing-rods, with a rope noose at one end. When the chosen animal is overtaken, the noose is slipped over its neck with a dexterous twist (see illustrations, pages 528, 529, and 530).

One man then dismounts and, squatting upon his haunches, seizes the terrified animal by the tail. Like the proverbial dog with the tin can, he starts off at a run, dragging the man behind him. The latter slides along over the grass in his soft felt boots till the beast stops exhausted. Then he is easily thrown and a saddle fixed on his back.

Again the animal struggles, squealing like a pig meanwhile, but a strong rider manages to mount him, and after a few minutes the pony is considered tamed and fit to be ridden—by Mongols, at least.

FIELD-MICE OFTEN CAUSE DOWNFALL OF EXPERT HORSEMEN

Expert horsemasters from childhood, both men and women are equally at home in the saddle. In fact, the pony is man's inseparable companion on the steppes, and the Mongol, who will never walk if he can help it, develops an unsteady, rolling gait when ill-luck forces him afoot.

The plains, stretching for miles and miles, would be ideal riding country but for one defect. Falls with fatal results are sometimes occasioned by hollow ground. Field-mice and marmots excavate galleries a few inches below the surface of the earth, and a pony passing over these must go through. Such dangerous spots are usually distinguished by a different color and appearance, but sometimes even a practiced eye may be at fault, especially in early spring, when large tracts of the plain are accidentally fired by passing caravans.

James Gilmour tells an amusing story of an accident due to a few moments' inattention—quite enough to bring down the best horse and rider.

"My guide was before me," he says in his "Travels," "and we were going at a rapid pace when all of a sudden I saw his horse with its head turned toward me, its four feet in the air and its rider undermost.

"My guide was a large man and was considerably crushed, though it is strange he was not more hurt by so bad a fall. Both his saddle girths were burst, but, true to his Mongol instinct, he held fast to the bridle. After a little while he recovered and set about repairing the damage.

"As we had no spare strings or straps with us and were far away from human habitations, I wondered how he would do this. But without hesitation he took a handful of hairs from the tail of his pony, twisted and plaited them together, and in a few minutes the straps were as strong as ever.

"Luckily for the careless Mongol horseman on a journey, who is apt to forget anything that can be left behind the tail of his horse, which is never docked,' makes up for all deficiencies."

PONIES ARE GRADUALLY STARVED BEFORE A RACE

As is only natural among such confirmed horse lovers, races are a popular amusement in Mongolia. But races on the steppes are conducted quite differently from ours. Even the preliminary training of the entries seems topsy-turvy to us.

Each competing pony is tied to a rope picketed on the grass plain. This rope is shortened every day by a certain number of inches, thus reducing the circle in which the animal can graze. Yet, strange as it may seem, this method of gradual starvation—tested by centuries—appears to increase rather than diminish its endurance.

On the day of the race fetlocks are clipped and manes and tails tightly plaited with varicolored ribbons, so as to offer as little wind-resistance as possible. The jockeys are children, and it is not unusual to see lads of nine or ten start on the exhausting stampede which a Mongol race really is.

No saddles are allowed, but each rider is given a heavy whip and a handkerchief. With the latter he leans over and wipes the dust from the eyes and nostrils

Photograph by Eugene Lee Stewart

A LAMA PAYING HIS CAB FARE IN A MONGOLIAN BORDER TOWN

His name is Yoongadoong. He belongs to the Hkalka tribe, and is attired in new flaming red robes, cap, and black boots trimmed in green.

Photograph by Eugene Lee Stewart

MONGOLIAN WOMEN OF HAILAR IN HOLIDAY ATTIRE WATCHING A CHINESE NEW YEAR PARADE

In the right foreground are a Chinese man, a Chinese girl, and a Mongolian man.

of his mount, as otherwise the dust of the steppe might injure wind or eyesight.

RACING PONIES LEAN ON EACH OTHER FOR SUPPORT

The straight course is so long that often there is a good deal of difference in time between the arrival of the ponies.

Enthusiastic owners or spectators, including bishops and archbishops of the Lamaist faith (for the church does not frown on racing on the plains), gallop out to meet the contestants and assist in whipping them in. But sometimes two favorites arrive at the finish literally leaning against each other shoulder to shoulder. Thus they support each other on the run, though both are so exhausted

that if they were suddenly separated they would drop in their tracks.

Wonderful tales are told of the distances covered by famous Mongol racers at a stretch. Fifty, even a hundred, miles at full gallop are claimed. This is doubtless exaggeration, but fifteen- and twenty-mile races at great speed are well authenticated.

After the "meeting," the crowd usually adjourns to some neighboring monastery, where a festival in honor of the day is held.

A RELIGION OF TERROR

In Mongolia monasteries are the great centers of amusement, interest, culture, and—wickedness (see illustration, page

Photograph from Horace Brodzky

A MONGOLIAN SOLDIER SMOKING HIS PIPE

The Mongolian never drinks water or anything cold. He uses brick tea, made of tea dust from Chinese factories. A "brick" is as hard as granite. When a Mongolian wants a drink, he wraps the brick in an old cloth and places it in hot cow dung until it gets soft enough to enable him to break off a piece. This piece he grinds in a mortar and boils up in a horrible mess with milk or lumps of fat. He uses salt instead of sugar in his tea. He has a method of boiling his milk that makes it look like custard.

544). The priests exercise complete sway over the people by their unlovely religion of terror, the Tibetan form of one of the later sects of Buddhism known as the Tantric—a revival of the morbid Indian cult of Siva.

This repulsive creed, with its hideous demonology, is so well suited, however, to a land where cruel and tremendous atmospheric phenomena make man appear a helpless atom struggling against the mighty natural forces of existence, that it prospers.

Like frightened children, the People of the Wilderness desire to see their terrors embodied in idols which may be placated, and the cunning monks are quick to take advantage of their fears. Thus monasteries arise and grow rich.

These establishments are the finest buildings in Mongolia. Resplendent from afar in colors and gold, their lofty square buildings, often standing on some rise of ground, lose their effect somewhat on closer view, owing to the dirt in which these communities of monks are content to live.

"THE ABODES OF THE LIVING BUDDHAS"

The most famous and best kept monasteries are the abodes of the Living Buddhas. The current belief is that these men are gods incarnate, and when they die, or, as the Mongols put it, "change the vehicle," are reborn into this world with the power to remember their former existence and prove their identity by using phrases characteristic of the last Buddha, selecting things that belonged to him from among many which were not his, etc.

Great parade is made of the testing of a new candidate. Of course, the chief Lamas arrange everything and coach "the

Photograph by Eugene Lee Stewart

MONGOLIAN MEN WAITING FOR THEIR DINNER TO COOK

The average evening meal of the Mongolian "tenthold" consists of mutton boiled with a kind of millet. When done, the morsels are fished out of the pot with fire-tongs and served in a rude basin or on a board (see also illustration, page 501).

Photograph by Eugene Lee Stewart

A MONGOLIAN OF THE BURIAT TRIBE (AT THE RIGHT) BARGAINING WITH A CHINESE

The men of the Buriat tribe usually shave their heads and wear a pigtail. In summer they wear silk and cotton gowns; in winter, sheepskins and furs (see also illustrations on page 494).

Photograph by Adam Warwick

MONGOLIAN WOMEN IN GALA ATTIRE

On ordinary occasions the chief article of apparel for both men and women is a wide, roomy coat of plain material, but on holidays the women wear elaborately embroidered gowns. Note the headgear with curtains of red coral, turquoise, and pearl.

successor," but the common people, perhaps even the majority of the monks, believe the hoax most implicitly.

The crowd that makes its way to the monastery after the races gives abundant proof of the part religion plays in Mongol every-day life. Many a man will be saying his prayers or counting his beads as he rides along. Follow him to the temple itself and you will find him, as soon as he dismounts, joining a company with dust-marked foreheads to make the rounds of the sacred places, visiting at every shrine, bowing before every idol, prostrating himself on sloping wooden platforms.

Fanatical devotees may be met performing the "falling worship"—that is to say, throwing themselves flat on their faces and marking the place of their next prostration by their foreheads—a very exhausting form of piety, which soon wears out hands and clothes unless (as generally happens) wooden sandals are fitted to the hands and sheepskin pads to the knees.

Even little children may be seen turning prayer-wheels filled with written prayers, the idea being that any devout believer who turns the wheel acquires as much merit by so doing as if he had repeated all the prayers thus set in motion.

Photograph from Luther Anderson

A MONGOLIAN LADY SITS FOR HER PORTRAIT BEFORE A CHINESE PHOTOGRAPHER

Note the primness of the pose, the beauty spot on each cheek, and the cup-and-saucer and clock decorations at the lady's right.

One of the greatest festivals of the Lama church is the Devil Dance, which takes place each spring and represents the chasing out of the Spirit of Evil. The dance is simply a series of posturings of men and boys in rich costumes and fearsome animal masks accompanied by an impressive chant.

But a far more interesting survival of the primitive nature cult is known as the Midsummer Festival. It attracts crowds of pilgrims. Caravans begin to arrive many days in advance, when the sur-roundings of the temple present a busy scene of activity, with Chinese traders crying their wares and itinerant restau-rants preparing food.

The richer and more prosperous visi-tors arrive in camel-carts, with an im-pressive train of outriders, and camp in their own tents. Some families come in bullock-wagons, which, with a few mats fixed over them, make admirable shelters for their stay.

But by far the greatest number appear on horseback, solitary or in companies,

Photograph from Horace Brodzky

TWO MONGOLIAN BEAUTIES, THE WIVES OF WELL-TO-DO CATTLEMEN

They wear their wealth upon their hair, including a profusion of silver ornaments.

men and women, respectable characters and notorious thieves, Lamas and laymen, dressed—some well, some poorly, but nearly all gaudily—in yellow, blue, red, white, or green.

On the day of the ceremony the monastery is astir before dawn. The monks of various grades assemble soon after cock-crow, gorgeous in purple hieratic gowns, red waistcoats, scarlet or golden togas.

The Living Buddha appears in his fringed orange felt helmet, the abbots in their fat lacquer hats, the lesser Lamas in silk or gold brocade skull caps, the lay

officials in the old Manchu hats topped with colored buttons to denote their rank.

The whole company rides out of the monastery gate on ponies well groomed for the occasion and crosses the steppe to the obo, or sacred mount.

Such elevations, crowned by piles of stones with a flagstaff and fluttering prayer banners in the center, are landmarks all over Mongolia (see illustration, page 543). They represent the ancient totems to the nature spirits which have been adopted by Lamaism from the "Black Faith" (Shamanism) and dedi-

A MONGOLIAN PEASANT WOMAN RETURNING FROM A MIDSUMMER FESTIVAL

Note the typical Mongolian coiffure. The precise nature and shape varies according to the tribe.

cated to some saint of its own creed. Every Mongol stops to worship at them, leaving some kind of an ex-voto to attest his piety—a bit of rag or fur from his clothing or a handful of hairs from his horse's mane or his own head.

Having ascended the hill, the priests gather round the stone cairn, which has been previously decorated with leaves and branches. A tent is set up near by for the Living Buddha, the high Lamas, and the civil officials. Lesser dignitaries squat upon the ground in a circle.

Then the weird service begins, accompanied by all the strange paraphernalia of the Lama cult—huge bronze trumpets six feet long, flutes made from sea shells, and libation cups from human skulls.

A FEAST OF COLOR IN A RADIANT LANDSCAPE

The ceremony must be completed by sunrise, when the participants return to the monastery for the popular festival, some riding rapidly, some riding cautiously, but all converging on one center, and many of them ending up with a furious gallop to show their fine ponies at their best.

The crowd is a feast of color in a radiant landscape. The background of distant blue hills merges into a translucent blue sky, where an eagle, sharply silhouetted in the clear air, circles above. Holy cranes watch solemnly beside the waters of a distant salt lake that reflects the azure of the heavens. Queer long-tailed mice prowl about. A spotted deer, whom it is a sacrilege to kill, stands fearlessly—an orange speck—beside the temple, and a herd of fleet "tserengs," or steppe antelope, knowing the hunter is less considerate of them, bounds away toward the horizon.

By this time a group of white tents has been erected in the meadow for the feast. The largest serves as a reception hall. Inside a big transversal bench has been prepared for the guests of honor, whose places are marked by double cushions covered with priceless old silk carpets

Photograph by Adam Warwick

A PAIR OF MONGOLIAN HORSE-CATCHERS

The riders carry long birch poles, like fishing-rods. Mongol ponies are exported in great numbers to China. The mares, which are easily distinguished by their flowing tails, are kept at home for breeding purposes.

Photograph by Luther Anderson

A MONGOL COWBOY

Expert horsemasters from childhood, both men and women are equally at home in the saddle. In fact, the Mongol so seldom walks that he develops an unsteady, rolling gait when ill luck forces him afoot.

Photograph by Adam Warwick

TAMING AN UNBROKEN PONY

The man on the ground will, in a moment, seize the pony by the tail. The animal will start off at a run, dragging his captor behind him. The tamer slides over the grass in his soft felt boots until the pony is exhausted (see text, page 518).

Photograph by Ethan C. Le Munyon

A MONGOLIAN HORSEMAN

The pole with the slip-noose is used in place of the lariat and is almost as effective, as the rider will ride into a herd and "cut out" the horse he wants, drop the noose over his head, and soon subdue him.

LASSOING A MONGOL PONY

The native breed of horse is seldom more than thirteen hands high, but for endurance and cleverness it has few rivals. It is never a source of care to its owner, who provides neither food nor shelter. A long coat of hair in winter suffices for the latter, and when the snow covers the pasture land the little beast paws its way to the sparse remains of summer grass underneath.

Photograph by Adam Warwick

530

level of horse is seldom over a thin thirteen hands high, but for endurance and cheapness it has few ... of little cost power its way to the sparse remains of summer grass underneath.

A HARDY, SELF-RELIANT, HOSPITABLE MONGOL OF THE STEPPES

Photograph by Luther Anderson

He cares not for wealth or power. His flocks and herds provide him a livelihood, and his temporary abiding place (usually for ten days at a time) is wherever he chooses to pitch his tent.

A MONGOLIAN LAMA POSING FOR HIS PICTURE

Photograph by Eugene Lee Stewart

The Lamas, and in fact nearly all Mongolian men, are glad to have their pictures taken, but it is a decidedly different story attempting to obtain photographs of the women.

531

Photograph by Eugene Lee Stewart

THE SON OF A HIGH MONGOLIAN OFFICIAL, WEARING HEAVILY PADDED WINTER
CLOTHING

The Mongolians might be said to be born on horseback. Women ride as well as men.
In the races, the children jockeys ride without saddles, but carry handkerchiefs to wipe the
dust from their ponies' eyes and nostrils.

Photograph by Eugene Lee Stewart

A MONGOLIAN OF THE OFFICIAL CLASS AND HIS NEPHEW, MOUNTED ON A
MONGOLIAN PONY

from the treasury of the monastery. Two choirs of singers in bright robes kneel on either side of the broad entrance and chant a welcome.

Soon the feast begins. A cup made of the precious "zabia" wood, which will make water boil and has the power to detect poison, is placed before each distinguished visitor, with smaller cups for the "airak" and "koumiss"—liquors made from fermented milk.

The principal meat dish is mutton. Sheep are served whole on large platters, the four legs arranged around the rump, the skull on top.

As a kneeling attendant passes each dish to a guest, the Lama host makes a cross on the skull, which is then taken away. A second serving Lama, acting as butler for the occasion, then cuts up the

meat. The rumps and tails are given to those whom the monks especially delight to honor.

This curious custom originates from the fact that, a sheep having but one tail, the presentation of this delicacy to a person necessitates the slaying of a sheep especially for him, and it must also be a good one, for none but a fat sheep has a tail fit to be seen.

"TABLE MANNERS" IN MONGOLIA

To the foreigner a Mongol feast is a doubtful pleasure. He dislikes the idea that the unfortunate sheep have been slaughtered in the barbarous native way (said to preserve the flavor), by ripping open their bellies, after which operation the butcher puts his hand into the viscera and snaps the aorta. It is difficult also

A MONGOLIAN CAMEL-DRIVER AND HIS CHARGE

It is a tradition among the People of the Wilderness that the breath of the camel is so poisonous as to shorten the life of the driver.

Photographs by Eugene Lee Stewart

MONGOLIAN CAMELS FORAGING THROUGH THE SNOW AND OFF THE TOPS OF TREES

It will be noted that the Mongolian species is the Bactrian, with two humps, whereas the dromedary, or one-hump animal, is found in Arabia.

Photograph by Luther Anderson

A MONGOL OX CART

This is the same type of cart which was used by Genghiz Khan to transport his army supplies seven centuries ago.

Photograph by Eugene Lee Stewart

A CARAVAN STARTING FROM HAILAR TO URGA

These caravans number from ten to fifty camels. They carry blankets, clothing, tobacco, and other commodities to the native population of Urga and usually return empty.

THE FOREST ZONE OF OUTER MONGOLIA

JOURNEYING TOWARD MONGOLIA'S PASTURE LANDS

The most fertile portion of Mongolia lies to the north of the Gobi Desert, along the Siberian frontier.

Photograph by Adam Warwick

WEALTHY MONGOLIANS TRAVEL TO THE FAIRS IN CAMEL-CARTS WITH AN IMPRESSIVE TRAIN OF OUTRIDERS (SEE TEXT, PAGE 525)

Photograph by Eugene Lee Stewart

RUSSIANS SUPERVISING THE DEPARTURE OF A CARAVAN FROM HAILAR TO URGA

The two men with the fur caps are the owners of the caravan. As soon as it is under way, they will return to their homes in town, and at the end of a week or ten days will mount their ponies and overtake the caravan, in order to be present when it arrives at the Mongolian capital.

537

Photograph by Adam Warwick

A MONGOLIAN OBO ON THE DISTANT HORIZON

These elevations are landmarks in all parts of Mongolia. They represent the ancient totems to the nature spirits (see also illustrations on pages 542 and 543).

A COMPANY OF MONGOLIANS STARTING ACROSS THE STEPPE TO THE OBO, OR SACRED MOUNT (SEE TEXT, PAGE 526)

Photograph by Adam Warwick

A MONGOLIAN LAMA CAMP

Photograph by Eugene Lee Stewart

The priests are conducting religious services near by and will remain camped here for several days. A Mongolian school teacher and his nephew are seen in the foreground. The camels have been recently clipped.

Photograph by Adam Warwick

A WRESTLING BOUT AT A MONGOL FESTIVAL

At the conclusion of the match, which consists of three rounds, the winner will approach the grandstand tent in kangaroo-like leaps, as etiquette requires. He will kneel before the Lama, who will bestow upon him a robe of silk, an embossed silver cup, or an honorary scarf.

to eat comfortably, having to attack, with only the assistance of a knife, a great expanse of fat mutton spread on a brass dish nearly two feet in diameter.

Practice, however, makes Mongols expert, and in an incredibly short time each native has gobbled his share, seizing the piece of meat in his left hand and cutting it off close to his lips, while the knife flashes past so close to his face that, but for his short nose, accidents would certainly happen.

After all have gorged themselves and grown cheery with copious drafts of airak, hosts and guests mingle with the crowd in the meadow for the "fun of the fair." Many gather round a story-teller, who recites a legend drawn from the rich Mongolian folklore, some historical incident connected with the Great Khan or some quaint fairy tale inspired by the mystery of the steppes.

A MONGOLIAN WREST-
LING BOUT

The majority hasten to see the wrestlers (see illustration on this page). Dressed in a costume with stiff vest and short skirt, not unlike the garb of a Roman soldier, two champions face each other in the center of an open space. One is obviously a horseman, to judge by his

Photograph by Eugene Lee Stewart

MONGOLIANS BRINGING SIBERIAN SHEEP TO MARKET

The "shepherd's crook" is the familiar long pole with a noose at the end (see also illustrations on pages 528 and 529).

bowed legs. His length of arm and breadth of chest show him to be a redoubtable opponent. The adversary is a gigantic Lama belonging to the "tsang" (community of the Living Buddha) of the neighborhood. Three rounds constitute the match, and according to the rules neither wrestler may grip the other, but each must try to throw his opponent by laying hold of his belt.

The first bout is adjudged to the Lama amid great enthusiasm; the second goes to the horseman, and the third, which the crowd watches in a fever of excitement, is also won by him after a hard struggle.

Then the proud champion, much cheered, rises to his full height, expands his mighty chest, and approaches the entrance to the grand-stand tent in big jumps, as etiquette requires. Here he kneels before the Lama, who distributes the prizes, and receives a reward—a roll of silk, an embossed silver cup, a "Khadak," or honorary scarf, or a few bricks of tea, which he raises above his head in token of thanks. After that he retires

with more kangaroo leaps and another pair of athletes appears.

Apart from the amusements, there is also much visiting done at these fairs, which afford almost the only opportunities that neighbors, who live miles apart, have of becoming acquainted with one another. This applies especially to the womenfolk, whose lives of household drudgery are dull and lonely, while the men are away on the steppes rounding up the herds.

COSTUMES OF MEN AND WOMEN ARE MUCH ALIKE

The festivals also afford them a coveted opportunity to show off their finery. The dress of both sexes is much alike, as far as shape is concerned. The main difference is that the men gird themselves with a belt, while the women allow their long garments to hang loose from shoulder to heel; hence the common word for woman in Mongol is "beltless."

The outer robe of both sexes is a wide, roomy coat, which reaches to the ground,

Photograph by Eugene Lee Stewart

A SPECIAL TYPE OF OBO USED WHEN LAMAS HOLD SERVICES FOR MONGOLIAN SOLDIERS
(SEE ALSO ILLUSTRATION ON PAGE 543)

with sleeves so ample that the arms can be withdrawn from them and reintroduced without touching the buttons.

In this gown men and women are. for all practical purposes. inclosed in a little private tent from which only the head projects. It allows the wearer to dress and undress beneath it in perfect privacy, whether on horseback or surrounded by the crowded inmates of a full tent. Though on ordinary occasions wearing plain material, on high days and holidays the women don beautiful embroidered gowns often with quaint padded epaulettes.

MONGOLIAN WOMEN WEAR THEIR WEALTH ON THEIR HAIR

But the most remarkable features of Mongol costumes are the hair ornaments and head-dresses of the women (see illustrations. pages 525 and 526). Even a poor girl, once she marries, wears a profusion of silver ornaments on her head. The precise nature and shape of these varies with the tribes. One at least has a most ludicrous coiffure for its matrons, which projects so high that the cap. imperatively demanded by etiquette. is tied on above the ornaments quite clear of the head. Others adopt curtains of red corals or turquoise or strings of pearls reaching often to the waist.

When the wearers take their stand together in the picturesque veranda of some temple, the effect is most striking.

At the close of the festival, which may last two or three days, the crowds depart to their homes, sometimes hundreds of miles distant. A few must cross the Gobi Desert, that dreary stretch of sand and stones which taxes the endurance of man and beast, as mile after mile the weary camels tramp across a stretch of country where there are no tents, no wells, no inhabitants, through solitudes of sand and rocks.*

Alas for him who loses his way in a dust-storm here and wanders helplessly among the boulders, which in size, shape, color, and arrangement mock him with their resemblance to human habitations!

The more fortunate pleasure-seekers travel back across the steppes, where the noon mirages mock and beckon, where lakes glimmer and clouds on the far horizon give the illusion of mountain ranges,

* See "A Trip Across the Gobi Desert by Motor Car," by Ethan C. de Munyon, in the NATIONAL GEOGRAPHIC MAGAZINE for May, 1913.

rendering unreal a world of beauty and of dread.

THE MONGOL TENT IS MADE OF FELT

It is rare indeed that at nightfall the wanderers will find an inn, but the rule of the plain is that any traveler who demands hospitality at a tent must be lodged and fed, except in the rare cases where a tent is under prohibition because of sickness and it is therefore impossible to allow strangers to enter.

The lodging is nowhere luxurious, though the larger encampments—"ails," as they are called—have special guest tents with wooden doorways.

The average Mongol yurt is of the simplest construction (see illustration, page 550). Round a mud floor is built a trellis-work of laths about four feet high, from which a number of sticks radiate to a point at the top. These are covered with a single or double layer of felt, tied down firmly in winter by leather thongs, but raised in summer to admit the breeze.

The furnishing of the interior is equally simple. In the center is an iron fireplace, in which "argol" (dried cattle dung), the only fuel to be procured, is burned. The smoke finds its way through a hole in the roof, so much of it at least as does escape. Round the walls are a few chests and presses of the rudest Chinese manufacture and plain brass pots from Peking.

A few sheepskins and pieces of felt

Photograph by Eugene Lee Stewart

MONGOLIAN LAMAS CONDUCTING AN OUTDOOR RELIGIOUS CEREMONY

The obo, or shrine, is built up of stones, dirt, and tree branches liberally punctuated with paper banners bearing prayers to the spirits. The meat eaten during the feasts which accompany these religious celebrations is blessed by the Lamas. The flesh which cannot be eaten is placed on the obo to appease the appetites of the spirits.

represent bed, sofa, and chair. Sometimes the refinements of life are represented by a basket, a pan, or a broken bowl in which half a dozen carefully-tended heads of garlic are growing.

THE GUEST SLEEPS BESIDE THE LAMBS AND CALVES

If travelers are not proud and are willing to lie down beside the lambs and calves of the household, even the poorest Mongol gives a cordial welcome and the

Photographs by Adam Warwick

A TYPICAL MONGOLIAN MONASTERY

These establishments are not merely the centers of religion, but of amusement. The religion of the Mongols is one of terror—a revival of the morbid Indian cult of Siva.

Photograph by Eugene Lee Stewart

A CHIEF LAMA, IN YELLOW SILK ROBES AND BRASS HAT, AND A HIGH MONGOLIAN OFFICIAL RETURNING FROM RELIGIOUS SERVICES.

Photograph by Adam Warwick

THE DEVIL DANCE, ONE OF THE GREATEST FESTIVALS OF THE LAMA CHURCH

This ceremony takes place in the spring and represents the chasing out of the Spirit of Evil. The dance consists of a series of posturings by men and boys, who wear over their heads fierce-looking animal masks (see text, page 525).

Photograph by Luther Anderson

A BUDDHIST TEMPLE AT A TOWN ON THE CHINESE-MONGOLIAN FRONTIER. NOTE THE TYPE OF SADDLE ON THE CAMEL

MONGOLIAN SCHOOLBOYS AND THEIR MONGOLIAN TEACHERS IN CHINESE MILITARY UNIFORM

When the Mongols retired from China, the Manchus took over their rule and assumed a nominal control over Mongolia. But the Mongols were allowed to preserve their original tribal form of government. When the Manchus in turn disappeared from China, the Mongols declared their complete independence. Today, however, they are fearful of the inroads of the Chinese, whose thrifty agriculturists are already encroaching on the Mongolian steppes.

MONGOLIAN SCHOOLBOYS PLAYING A GAME RESEMBLING OUR "BLIND MAN'S BUFF"

The mud wall in the background surrounds the village of Hailar, on the Trans-Siberian Railway.

Photograph by Eugene Lee Stewart

547

Photograph by Eugene Lee Stewart

A MONGOLIAN CEMETERY

The Mongolians do not bury their dead, but set the coffin on the ground and cover it with earth, and in time the elements undo the work which has been performed to keep prying eyes away. The mound in the foreground has been recently made. Frequently the dead are merely cast out upon the steppe (see illustration below).

Photograph by Adam Warwick

HUMAN BONES WHITENING ON THE MONGOLIAN STEPPE AND A BOX CONTAINING A CONDEMNED PRISONER

While an occasional cemetery is found (see illustration above), the more customary method of disposing of the dead is to leave the bodies exposed on the plains to be preyed upon by wolves and dogs. Condemned criminals, fastened in wooden boxes, are left to die of hunger and finally to be eaten.

best he has. "Only observe etiquette, and every tent is yours," as the saying goes.

For even among the rough-and-ready People of the Wilderness there are a few essential rules of politeness. From whatever side a tent is approached, for instance, be sure to ride up to it from the front.

When within a short distance stop and shout "Nohoi" (dog). The Mongol dogs are very savage, and it would be dangerous to attempt to advance farther till the people of the village have come out to restrain them, which by law all Mongols are forced to do. Until they receive this protection, horsemen remain in the saddle, and those on foot keep the animals at bay as well as they can with sticks.

A Mongol dog has many wolfish qualities, and the disgusting Mongol habit of leaving corpses on the plains instead of burying them increases his savage instincts.

No white man can pass the human skulls and bones strewn over the steppes without a shudder, and he turns sick with disgust at the sight of the occasional wooden box in which a condemned criminal is left to die of hunger and finally be eaten by the wolves and dogs (see illustration, page 548).

Photograph by Eugene Lee Stewart

A MONGOLIAN COFFIN, PAINTED GREEN AND DECORATED IN WHITE AND GOLD

EXCHANGE OF SNUFF BOTTLES IS A TOKEN OF HOSPITALITY

Once a stranger enters a tent, however, the savage creatures will no longer attack him; so that to bring a stick inside is considered a lack of good manners. Having left his stick outside, then, the traveler on getting through the low doorway, says "Mendu" (greeting) to the people inside and proceeds to sit down on the left side of the fireplace cross-legged or, if he cannot do this, with his legs stretched toward the door.

The next thing is the interchange of snuff bottles. A Mongol visitor offers his first to the host and the people of the tent, and receives theirs in return; but, as foreigners do not carry snuff generally, the Mongol host offers his to the foreign visitor. Meanwhile the women have been warming tea.

Photograph by Adam Warwick

A MONGOL GUEST TENT WITH WOODEN DOORWAY AND A MONGOL FAMILY GROUP

Photograph from Horace Brodzky

THE INTERIOR OF A YURT, OR MONGOLIAN TENT-HOUSE

Note the god, an American lamp, and the glass-fronted cupboard with small cups, which contain flour, cakes, tobacco, and rice—offerings to the god. A yurt is never placed where the shadow of a tree can fall upon it, for this brings bad luck. It is made of trellis and covered with felt. The fire is laid in the middle of the one room, and the smoke allowed to find its way out through a hole in the roof. When it rains, this hole is covered up, with consequences easy to imagine.

About sunset the hostess, glancing up at the hole in the roof as if it were a clock, will inquire, "Shall I make dinner?" and her lord and master, nothing loath, will answer, "Make it." Accordingly she proceeds to the dog-proof cage outside the door and hews off a piece of frozen meat with an axe. This is boiled with a kind of millet, and, when pronounced done, the morsels are fished out with the fire-tongs and served in a rude basin or on a board.

TUCKED IN FOR THE NIGHT

Most Mongols retire immediately after this meal, and the servant's last duty is to pile up the fire and then tuck the sheepskin coats snugly around hosts and guests, while the master of the tent, in true Mongol fashion, indicates by the points of the compass the places where the tucking is deficient.

Next morning, on leaving, the traveler mounts his horse at the tent door with a bow and a smile, as the Mongols do not have any customs equivalent to our handshaking and good-bye. A few days later the village itself will have perhaps moved on.

Like some of our American Indians, whom they resemble, the People of the Wilderness cannot endure a settled life. All their belongings pack easily on the back of a camel; the few calfskin bags with provisions, the tent, the cooking pots, the grate, two water-buckets, and a few odd pieces of

Photograph by Eugene Lee Stewart

A MONGOLIAN BOY AND GIRL WITH LITTLE BROTHER ON BEHIND

The wealth of Mongols is reckoned, not in real estate, but in live stock and these children, as the heirs of an average *yurta*, or family, will possess fifty sheep, twenty-five horses, fifteen cows and oxen, and ten camels. Next to cattle-breeding the most important occupation of this people is the transport of goods. It is estimated that more than 1,200,000 camels and 300,000 ox-carts are employed in the country's internal caravan trade.

felt are all they need—except space to journey in.

What the Mongols most fear is the attempt of the Chinese to colonize their country, and they see with alarm how the tilled fields of these thrifty agriculturists are already encroaching on the steppe.

THE NEW MAP OF ASIA

MORE than one-half the human race lives in Asia, which has an area nearly six times as large as continental United States, approximating one-third of the earth's entire land surface.

Asia boasts the world's highest peak, Mt. Everest, 29,140 feet, and the ocean's deepest pit, off the coast of Mindanao, in the Philippines, 32,088 feet. Somewhere within its borders was probably the birthplace of man, and from those fastnesses within the shadow of its Himalayas began the migrations which resulted in the peopling of all the continents and all the islands of the seas. It is a land of teeming millions of men and of vast solitudes.

There are twelve rivers on the earth's surface which exceed 2,500 miles in length, and of these six rise in and flow through Asia.

The continent extends from Cape Chelyuskin, within twelve and a half degrees of the North Pole, to the Malay Peninsula, within one and a half degrees of the Equator; and from the Strait of Bab El Mandeb, separating Arabia from Africa, to the Bering Strait, separating Siberia from Alaska, is 6,700 miles—more than a fourth of the circumference of the globe.

A MEDIEVAL EMPIRE DISMEMBERED

To the casual observer, the New Map of Asia, published by the National Geographic Society and issued as a supplement with this number of THE GEOGRAPHIC,* may not present an appearance radically different from that of pre-war Asia; and yet the world conflict on the fields of Europe has wrought vast changes here, resulting in the dismemberment of a great empire, which had come down from medieval times, the creation of five new nations, the provisional creation of four others, and the possible evolution of half a score of semi-independent states from the wreck of what were once the proud provinces that gave allegiance to the Tsar under the collective name of "Russia in Asia."

*Additional copies of the New Map of Asia may be obtained from the headquarters of The National Geographic Society in Washington. Cloth edition, $1.50; paper, $1.00.

While, with the exception of the Turks, none of the ancient peoples of Asia participated in the World War to the same extent as European and American peoples, there were fewer neutral governments in the Orient than in Europe; for Siberia, as a part of Russia; India, Burma, and the suzerain states which cluster on the slopes of the Himalayas, as parts of the British Empire; Indo-China, as a part of the French Colonial Empire; Persia, as a battleground for contending armies; Arabia, China, Japan, and Siam in their own right—all were involved in the struggle.

EVERY ASIATIC NATION AFFECTED

Strictly speaking, Afghanistan and Mongolia alone of all Asia's vast dominions were untouched politically by the World War; and even these two nations were not wholly divorced from it, but were affected indirectly, as Mongolia from 1913 to 1919 was under the protection and guidance of Russia, and Britain's influence was paramount at the court of the Amir of Afghanistan.

As an ally of the Germans, Turkey by her defeat has lost not only most of her territory in Europe, but has been forced to surrender extensive and populous portions of her Asiatic empire, out of which have been set up the "independent states" of Syria, Mesopotamia, Palestine, Hedjaz, and Armenia and the autonomous province of Kurdistan.

France has assumed a guardianship (mandate) over Syria, and Britain exercises a similar office toward Palestine and Mesopotamia until such time as the three countries can be entrusted with their own affairs. Armenia, though created a separate state by the Treaty of Sèvres (the Turkish treaty), has not as yet had her boundaries definitely delimited.

Unless this treaty is radically revised, as is now contemplated, Greece will administer a large and prosperous district surrounding Smyrna, the most important port of Asia Minor, for five years, at the end of which time a plebiscite will be held to determine whether the inhabitants wish the area to be incorporated permanently as a part of Greece or resume its

Photograph by Plâté, Ltd.

ANTISKID AIDS FOR A POPULAR CLIMB IN CEYLON

The areca, or betel palm, shown in the illustration is a versatile provider. Ivory beads are imitated in betel-nut; cordovan shades are obtained in tanning and dyeing through the use of areca catechu. One out of every ten men in the world dyes his teeth and much of his environment brick red with the astringent juice derived from chewing chopped areca-nut, wrapped in a lime-smeared leaf of the betel vine.

ALL THE INSTRUMENTS TO EQUIP A KOREAN ORCHESTRA ARE FOR SALE IN THIS SEOUL MUSIC STORE

The stringed instrument which three of the shop-keepers are playing is the *komungo*. It is a sort of long, narrow bass-viol without a neck. The player plucks the strings with his right hand and fingers them with his left near the bridge. This instrument has been popular with Koreans for seventeen hundred years. There is no such thing as "time" in Korean music.

WHEN THE DESERT BURDENS ARE LAID ASIDE: A BOKHARA KHAN

Photograph by Maynard Owen Williams

Welcome as a friendly greeting after long separation is the sight of caravanserai lights at the end of the caravan trail. Samovars murmur drowsy contentment and even the disdainful camels gurgle with satisfaction. The gossip of a thousand trails weaves itself into one epic of inconsequence, and even sleep is not sweeter than the companionship of those who find themselves once more at the journey's end.

555

Photograph by C. H. Kragh

FEW CITIES OF CHINA ARE MORE PICTURESQUE THAN HANGCHOW

Fourteen centuries ago a small village of fishermen and salt-boilers occupied the site of this great city. It became the center of foreign trade in China in medieval times and was known as the "City of Heaven." Hither flocked merchants, travelers, and adventurers to enjoy the sights and indulge in its material pleasures.

Photograph from M. Branger and Sons

DISCUSSING THE NEWS OF THE DAY IN FRENCH INDO-CHINA

Until given the mandate over Syria, the chief interest of France in Asia lay in Indo-China, which includes the protectorates of Annam, Tonking, and Cambodia. the colony of Cochin-China, and part of. the Laos country. French Indo-China has an area more than a third larger than France in Europe.

Photograph by John Claude White

THE LAMAS OF SIKKIM WEAR APRONS MADE OF HUMAN BONES

These "Rugens," as the decorative ceremonial garments are called, are worn during monastic dances. Sikkim is an Indian state, 70 miles long and 50 miles wide, in the Himalayas, between Nepal and Bhutan. The people are agriculturists and there are practically no towns or villages, each family living in a house on its own land.

Photograph by Raja Deen Dayal and Sons

THE PRINCIPAL LANDMARK OF HYDERABAD, DECCAN, INDIA

The present ruler of India's largest native state has inaugurated an era of building which is doing much to beautify his capital on the Musi. But no building is more highly regarded by the Moslem inhabitants of Hyderabad than the Char Minar, with its four towering minarets. It was constructed in 1591 by Mohammed Kuli, the founder of the city.

Photograph by Vittorio Sella

IN THE LAND OF LALLA ROOKH

It is not strange that Kashmir, set among the serried peaks, the eternal glaciers, and placid lakes of the Himalayas, should be famed in song and story as an earthly paradise—a vale of peace and happiness. The "rose of Cashmere" has been incorporated into our language as a synonym for natural beauty.

THE PALKHOR CHOIDE MONASTERY, IN THE WALLED TIBETAN CITY OF GYANTSE

Gyantse lay in the line of the British advance into Tibet in 1904. Its jong, or fort, was considered impregnable by the Tibetans, but the attacking force under General MacDonald succeeded in hammering a breach in the walls and capturing the stronghold. This left the road open to the capital of the Dalai Lama, the Forbidden City of Lhasa.

Photograph by Eric Keast Burke

AN ARAB-MADE FERRIS WHEEL IN BAGDAD

On festival occasions such crude devices spring up in the outskirts of the City of Caliphs and prove a popular diversion for Arab children. The motive power is, of course, manual. Note the similarity of these vertical merry-go-rounds to the fête wheels of Eastertide in Irkutsk, Siberia (see illustrations, pages 488 and 489).

Photograph by John Claude White

STRANGE BEASTS OF STONE GUARD THE TEMPLE STEPS AT BHATGAON, NEPAL

The Kingdom of Nepal has the distinction of owning the world's highest peak, Mt. Everest. Bhatgaon, an ancient capital city, possesses some of the most interesting shrines in Asia. Most of these are built of fine red brick. The ornamentation is generally of molded brick of the same color, with dark wood for overhanging windows and doorways.

© Underwood & Underwood

IN KOREA, THE LAND OF "MORNING FRESHNESS"

Natives from the rural districts surrounding Seoul enter the capital by passing under the new Independence Arch. Note the queer umbrella hats and white suits of the men at the left and the familiar pill-box headgear worn by the man at the right. The ox-caravan is laden with fuel for the city.

Photograph by Maynard Owen Williams

MARBLE CENOTAPH TO A FAVORITE ELEPHANT OF A MAHARAJAH

Beside the road which leads from Jaipur to the ancient capital of Jaipur State at Amber, there is, among numerous monuments to deceased wives, this costly reminder of one of the dumb servants whose faithfulness is thus rendered memorable. There are a thousand living candidates for burial in this cemetery of the Maharajah's favorite wives.

Photograph by Dr. Edward Burton MacDowell

AT THE END OF THE ROAD TO MANDALAY

In Burma the chief objects of interest to the Western traveler are the graceful pagodas. The Burmese pagoda consists of a masonry terrace, above which rises a bell-shaped structure crowned by a "ti," or umbrella spire, formed of concentric rings of metal, from which are suspended innumerable tiny bells that sway and tinkle musically in the wind.

Photograph by Charles Martin

BENGUET IGOROT GIRLS OF THE PHILIPPINE ISLANDS

Many of the kinsmen of these gentle maids were formerly among the wildest of head-hunters in northern Luzon, but today the members of their tribe are in the main industrious and skillful farmers, planting grain and potatoes on ancient artificial terraces and successfully practicing irrigation.

Photograph from M. Branger and Sons

SUNSHINE AND SHADOW ON TEMPLE STAIRS IN ANNAM

Both the men and women of Annam wear wide trousers and long black tunics with
narrow sleeves. Their country is a narrow ribbon of land, 800 miles long, on the eastern
coast of the Indo-Chinese peninsula. France exercises a protectorate over the 6,000,000
natives.

former status as a part of Turkey in Asia.*

The Kingdom of Hedjaz, over which rules Husein Ibn Ali, Hereditary Keeper of the Holy Places (Mecca and Medina), extends along the northeast shore of the Red Sea, from the principate of Asir to the southwestern frontier of Palestine. It has an area about equal in extent to that of the State of Colorado and a population of some 750,000. Its capital and chief seaport is Jidda, a town of 30,000 inhabitants..

RUSSIA IN ASIA AN UNSETTLED PROBLEM

. Of that vast territory formerly known as Russia in Asia, embracing Siberia, Transcaucasia, Turkestan, and the Steppes provinces, aggregating an area of more than 6,290,000 square miles (more than three times as large as Russia in Europe), but sustaining a population of only twenty-seven millions — barely four persons to the square mile—little can as yet be said with respect to its political future, and even the reports concerning its present status are vague and contradictory.

Out of Transcaucasia three republics evolved — Azerbaijan, Georgia, and, in part, Armenia. Their existence was tempestuous and short-lived. Whether, when the tide of Bolshevism recedes, they will be able to establish their interdependence as a Transcaucasus governmental trilogy none can say. The boundaries indicated on the accompanying map are merely indicative of their approximate extent as originally conceived.

For a speculative discussion of the possible future of the "United States of Siberia and Russian Turkestan," the reader is referred to the admirable article by Viscount Bryce, under the title "Western Siberia and the Altai Mountains," in this number of THE GEOGRAPHIC (pages 469 to 507).

JAPAN ASSUMES KIAOCHOW LEASE

The transfer of German treaty rights in the Chinese province of Shantung (Kiaochow and Tsingtao) to Japan is a cause célèbre of recent diplomatic history. The town, harbor, and district of

* See also "The New Map of Europe," in the NATIONAL GEOGRAPHIC MAGAZINE for February, 1921.

Kiaochow, embracing an area of some 200 square miles, exclusive of the bay, which has an area of an additional 200 square miles, were seized by Germany in November, 1897, and in the following March were transferred by treaty to the latter under a 99-year lease. A month later the district was declared a protectorate of Germany and remained as such until November, 1914, when it was captured and occupied by Japanese forces.

Despite China's contention at the Paris Peace Conference that there should be a restitution of this leased territory to her, together with a cancellation of all Germany's economic privileges, by the Treaty of Versailles the "lease" was transferred to Japan on the ground of conquest. It was for this reason that China refused to become signatory to the treaty with Germany, and the whole question constitutes in Asia a "sore spot" similar to the score or more which now pock-mark the face of the political map of Europe.

Before the influx of the Japanese, the Kiaochow district had a population of about 200,000. Surrounding the district and bay is a so-called neutral zone of about 2,500 square miles, with 1,200,000 inhabitants.

AMERICA'S INTEREST IN YAP

In addition to her acquisition of the Kiaochow leased district, Japan's spoils from the World War include the formerly German-owned Marshall Archipelago, the Marianas or Ladrone Islands, the Pelew Group, and the Carolines, including the much-discussed Island of Yap, important from an American standpoint as a connecting link for cables between San Francisco, Hawaii, Guam, the Philippines, China, and the Dutch East Indies.

The Carolines consist of 500 coral islets, supporting some ten thousand inhabitants; the Pelews are about twenty-six in number, with 3,000 natives; and the fifteen Marianas have 2,600 inhabitants. The Marshalls consist of two chains of some twenty-five lagoon islands, supporting a population of 15,000.

When the American Pacific cable was laid, at the beginning of the twentieth century, from San Francisco to the Philippines by way of Hawaii, the Mid-

Photograph by Eugene Lee Stewart

A MONGOLIAN HORSE THIEF SERVING A
THREE MONTHS' SENTENCE

The wooden collar, which is bolted together
to make certain that it will not come off, is
not removed during the three months, and the
prisoner gets what rest he can by sitting down
and leaning back against a post or wall. On
the white strips of cloth is written in Mongolian an account of his crime. During the
daytime he is obliged to keep himself on display in front of the prison (under guard), so
that his fellow-men may observe the penalty
for stealing horses (see pages 507-551).

way Islands and Guam (7,846 miles), a
branch line was laid from Guam to Yap.
From the latter, lines now radiate to
Japan, to Shanghai, and to the Dutch
East Indies.

Yap is intrinsically interesting as well
as commercially important. The west-
ernmost of the Western Carolines, it is
situated some 500 miles southwest of
Guam and 800 miles east of Mindanao,
of the Philippine group. The arrival of
Australian troops in October, 1914, prevented the Germans from erecting a
wireless station here, which would have
been of great service in communicating
with the commerce raider *Emden*, then
abroad on the highways of the Pacific.

Although surrounded by an atoll, Yap
is of volcanic origin. Its only good harbor is Tomil Bay. Its pseudonym, "the
Island of Stone Money," is derived from
the fact that native wealth is reckoned in
pounds or tons of limestone discs,
brought from Babeltop, 300 miles distant.
A single "coin" four feet in diameter is
said to represent a value of 10,000 coconuts. The coconut is the unit of value,
for copra is the only article of export.

THE EAST IS NO LONGER "CHANGELESS"

The natives of Yap, some 7,000 in
number, are safely catalogued as Micro-
nesians, a term which embraces a variety
of peoples of Melanesian, Polynesian, and
Malaysian stocks. They have light coffee-
colored skins and wavy black hair, dark
eyes, and prominent cheek-bones, and are
neither as tall nor so strongly built as the
natives of Samoa, the Fiji Islands, or
Tahiti. They are a docile, kindly, indolent people.

Two portentous agencies are at work
in the Orient, and until they become
relatively quiescent, political boundaries
throughout the great continent will be
"subject to change without notice." One
of these is the disruptive force in peasant
Russia, now felt not only in Siberia, in
Russian Turkestan, and in the Transcaucasus, but also in Persia. The other
is Japan, whose natural desire, based on
need for territorial expansion, has resulted in a reaching out into China and
eastern Siberia and into the islands of the
Pacific north of the Equator.

Asia is no longer the "Changeless
East"; it is the Continent of Ceaseless
Change.

Notice of change of address of your GEOGRAPHIC MAGAZINE *should be received in the office of the National Geographic Society by the first of the month to
affect the following month's issue. For instance, if you desire the address changed
for your July number, the Society should be notified of your new address not later
than June first.*

TIFFANY & CO.

84 YEARS OF QUALITY

ILLUSTRATIONS WITH PRICES
OF JEWELRY AND SILVERWARE
SENT UPON REQUEST

FIFTH AVENUE & 37TH STREET
NEW YORK

"You don't believe in signs, do you, Cuthbert?"

CHANDLER SIX
Famous For Its Marvelous Motor

The Chandler Dispatch Is
a First Choice Car

THE automobile industry has produced no distinctive model more popular than the Chandler Dispatch. It is a preference with those who want a close-coupled four-passenger open car. Its comfort, its beauty of line and finish and its dependability are typical of Chandler character.

The Dispatch Is a Great
Open-Road Car

A car to get out and go places in. It is alive with Chandler power, than which there is no surer, no more flexible power.

The Dispatch attracts young folks and older ones alike. The comfort of its roadability and the charm of its beauty add to their satisfaction. It seats four persons in real luxury.

The cushions, deeply upholstered, and trimmed in genuine hand-buffed leather, are wide and tilted at proper angles. The driver's seat is so comfortable as to take away any strain there might be in driving. It is a car with which you will be delighted.

Chandler Following Grows
By Thousands

A hundred thousand owners know the real worth of the Chandler, and thousands more are joining the Chandler ranks each year.

The stability of the Chandler company, which manufactures the Chandler car, and the steadfastness of its adherence to the policy of manufacturing one model and only one, and making that one right, has assembled this big army of Chandler owners.

The Chandler leads all fine cars in low price. It sells for much less than others you might compare with it. Consider values carefully and you will choose a Chandler.

Cord Tires Standard Equipment

Seven-Passenger Touring Car $1930	Four-Passenger Dispatch Car $2010
Four-Passenger Roadster $1930	Two-Passenger Roadster $1930
Seven-Passenger Sedan $3030	Four-Passenger Coupe $2930	Limousine $3530

(All prices f. o. b. Cleveland, Ohio)

THE CHANDLER MOTOR CAR COMPANY, CLEVELAND, OHIO
Export Department: 1819 Broadway, New York   Cable Address: "CHANMOTOR"

DODGE BROTHERS BUSINESS CAR

Dodge Brothers' own wide business experience is reflected in the thorough and practical design and construction of this car

It is so efficient, so strikingly free from need of repair, and so economical to run, that it constitutes a real asset to any business requiring delivery

The haulage cost is unusually low

DODGE BROTHERS, DETROIT

"He Left His Chains Behind!"

Weed Tire Chains left in the garage never stop a skid

Weed Tire Chains

on your tires reflect your prudence and intelligence.

SOME drivers never *think* always to carry Weed Chains and never *think* to put them on the tires until they feel their cars skid—then it is usually too late to do anything, except pray.

Don't wait until you feel your car skid—don't wait until you feel that terror of utter helplessness.

Make up your mind now always to carry Weed Chains in your car, ALWAYS to put them on the tires at the first drop of rain.

AMERICAN CHAIN COMPANY, INC.
BRIDGEPORT CONNECTICUT

In Canada: Dominion Chain Company, Limited, Niagara Falls, Ontario
The Complete Chain Line—All Types, All Sizes, All Finishes—From Plumbers' Safety Chain to Ships' Anchor Chain
GENERAL SALES OFFICE: Grand Central Terminal, New York City
DISTRICT SALES OFFICES:
Boston   Chicago   Philadelphia   Pittsburgh   Portland, Ore.   San Francisco

Largest Chain Manufacturers in the World

Douglas Fir
Northern White Pine
Idaho White Pine
Western Soft Pine

Western Hemlock
Washington Red Cedar
Red Fir and Larch
Norway Pine

HOW EXPERT SELECTION OF LUMBER MAY SAVE YOU MONEY

EVERYONE admires the fine old wood structures that have come down from Colonial times.

"But," you hear people say, "You can't get lumber like that these days."

It's not the fault of the lumber. It's the way lumber is used.

Use the *right wood* in its *proper place*—and, granted that your construction is right, you will get as sound and durable a building as any built in Colonial days!

There is today available in most markets a greater variety of structural woods than ever—with the possible exception of hardwoods, which are now seldom used for building purposes.

This same thing is true of woods for industrial uses.

Many woods formerly sold only in local markets are seeking wider outlets of distribution. For instance, Douglas Fir, probably the greatest wood in the country for structural timbers, has only recently come into common use in the great markets on the Atlantic Seaboard.

There is available a great body of detailed and scientific knowledge about the qualities of these woods, their strengths, their proper treatment and application, and how they will act under given conditions of service.

Getting this knowledge and acting on it may easily double the service you get from lumber.

You cannot judge the service of lumber by its appearance. The "nice clear board" that looks so attractive may not be the right species for your purpose. A sound board of another species, even though knotted, may give you much greater value and service.

It all depends on the natural characteristics of the wood, and on the careful selection of the lumber for the service it is to perform.

Which is the most practical wood for a given purpose, and what grade will do the work most economically, can all be told by the scientific knowledge about woods which has accumulated through years of observation and experience.

The user of lumber is today in position to fill his requirements more efficiently and economically than ever before.

What we advocate is conservation and economy through the use of the right wood in its proper place.

To this end we will supply to lumber dealers and to the public, any desired information as to the qualities of the different species and the best wood for a given purpose.

This service will be as broad and impartial as we know how to make it. We are not partisans of any particular species of wood. We advise the best lumber for the purpose, whether we handle it or not.

From now on the Weyerhaeuser Forest Products trade-mark will be plainly stamped on our product.

When you buy lumber for any purpose, no matter how much or how little, you can look at the mark and know that you are getting a standard article of known merit.

Weyerhaeuser Forest Products are distributed through the established trade channels by the Weyerhaeuser Sales Company, Spokane, Washington, with branch offices and representatives throughout the country.

WEYERHAEUSER FOREST PRODUCTS
SAINT PAUL · MINNESOTA

Producers of Douglas Fir, Western Hemlock, Washington Red Cedar and Cedar Shingles on the Pacific Coast; Idaho White Pine, Western Soft Pine, Red Fir and Larch in the Inland Empire; Northern White Pine and Norway Pine in the Lake States.

"Mention The Geographic—It identifies you"

For the G

ONE of the most suitable
gifts you can make to the
high school or college gradu-
ate is an Eversharp Pencil.
No matter whether the recipient is a girl or boy he will
prize this present highly and
find use for it every day in
the year. Eversharp Pencils
are made in many attractive
designs both in silver and in
gold. They are priced as low
as $1.00 and as high as
$65. Made by Wahl methods
which means jeweler precision, these pencils give perfect
writing service and will last
a lifetime. Be sure you get
the genuine Eversharp—
the name is on the pencil.

THE WAHL CO., Chicago

EVERSHARP

Made by
The Wahl Company
Chicago

Around the World Tours

FROM MARCEAUX'S MASTERPIECE
BERNE—SWITZERLAND

ACROSS the Pacific, through the kaleidoscope of the Far East, into the mystic spell of India, on to the lands where man began and ancient civilization forever inspires. Arrangement of all details in the hands of an American business institution of the highest reputation whose influence, radiating from offices in many lands, assures the utmost in transportation and travel comfort, and makes smooth the pathway of the traveler.

Seven Wonderful Tours—four East, three West.
Parties limited to twelve, under experienced escort.

Departures from September to December

ILLUSTRATED BOOKLET ON REQUEST

AMERICAN EXPRESS COMPANY

65 Broadway, New York City

DUES

Annual membership in U. S., $3.50; annual membership abroad, $4.50; Canada, $4.50; life membership, $100. Please make remittances payable to National Geographic Society, and if at a distance remit by New York draft, postal or express order.

RECOMMENDATION FOR MEMBERSHIP

IN THE

NATIONAL GEOGRAPHIC SOCIETY

The Membership Fee Includes Subscription to the
National Geographic Magazine

PLEASE DETACH AND FILL IN BLANK BELOW AND SEND TO THE SECRETARY

- .192

To the Secretary, *National Geographic Society,*
      Sixteenth and M Streets Northwest, Washington, D. C.:

I nominate -

Business or Profession -

Address -

- -
for membership in the Society -

- -
6-21   *Name and Address of Nominating Member*

(It is suggested that you inform the Nominee of your recommendation and of the benefits of membership)

An old friend and a new one

—both made by the Lambert Pharmacal Company

YOU know Listerine, the safe antiseptic. You've known it and used it and had confidence in it for years. We believe you'll like Listerine Tooth Paste equally well. Listerine users everywhere, in fact, are rapidly and enthusiastically accepting it.

When you use Listerine Tooth Paste you will discover a delightfully fresh, clean feeling in your mouth. Experience the pleasure of knowing your teeth are *really* clean and that your tooth paste is doing *all* a paste can do to keep your mouth healthy.

LAMBERT PHARMACAL COMPANY, SAINT LOUIS, U. S. A.

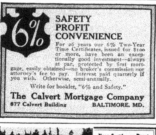

"Mention The Geographic—It identifies you"

The Victrola is to music
what gold is to commerce

—the one final standard of value. When, in
selecting an instrument for your home, you
choose the Victrola and Victor records, the
whole world confirms your judgment.

Victor records and Victor instruments are
one. Each is made to operate with the other
and no other combination can be made to
produce as satisfactory results.

Victrolas $25 to $1500. New Victor records demonstrated at all dealers in Victor
products on the 1st of each month.

"HIS MASTER'S VOICE"

This trademark and the trademarked
word "Victrola" identify all our products.
Look under the lid! Look on the label!
VICTOR TALKING MACHINE CO.
Camden, N. J.

Victor Talking Machine Co.
Camden, N. J.

VOL. XXXIX, No. 6   WASHINGTON   JUNE, 1921

THE
NATIONAL
GEOGRAPHIC
MAGAZINE

COPYRIGHT, 1921, BY NATIONAL GEOGRAPHIC SOCIETY, WASHINGTON, D. C.

ACROSS THE EQUATOR WITH THE AMERICAN NAVY

By Herbert Corey

AUTHOR OF "ON THE MONASTIR ROAD," "COOTIES AND COURAGE," "SHOPPING ABROAD FOR OUR ARMIES IN FRANCE," AND "A UNIQUE REPUBLIC, WHERE SMUGGLING IS AN INDUSTRY."

"WHY do you ask questions about the ships?" inquired the commander. "A navy isn't ships. A navy is men."

During the winter of 1920-21 I accompanied the American battleship fleet on its winter cruise. The Atlantic fleet of seven battleships and eighteen destroyers, commanded by Admiral H. B. Wilson, with the accompanying auxiliary vessels of its train, joined the precisely similar Pacific fleet under Admiral Hugh Rodman at Panama. The combined fleets cruised together to a short distance south of Callao, Peru. There they separated, the Atlantic fleet turning back to pay a visit of ceremony to the Republic of Peru, while the Pacific fleet continued on to the Republic of Chile.

Then a juncture of the fleets was again effected and they cruised in company to Panama, where the annual inter-fleet athletic competition was held. En route literally every moment possible was occupied in the practice of maneuvers.

ON SHIP DRILL DEPENDS WAR-TIME VALUE OF THE FLEET

Upon the degree of perfection reached in these ship drills depends in great part the war-time value of the fleet. It will not have been forgotten that in the battle of Jutland Admiral Scheer saved the German fleet because he was able to execute perfectly an evolution which naval authorities had declared impracticable in the hour of battle.

Upon the second separation, the two fleets left for their winter practice grounds, on opposite sides of the continent.

MAKING A GOOD SAILOR A BETTER AMERICAN

Throughout this period of close association I was over and over again impressed with the truth of the statement quoted. It is the conviction of the leaders of the American Navy that "a navy is not ships. It is men." They bend every effort toward the production of a personnel of extraordinary intelligence. They do their best to provide them with the best ships and guns and submarines and air-craft that can be built. But they hold fast to their guiding principle, that the material elements of the navy are but the instruments through which the genius of the men can be expressed.

Because the American enlisted man does not, as a rule, care to serve more than one term afloat, the navy's efforts have been extended over a wider field than those of other countries.

The American is made into an excellent sailor, as a matter of course; but it is likewise the navy's effort to make him into a better American. With this end in view, he is offered every opportunity to gain an education; he is taken on

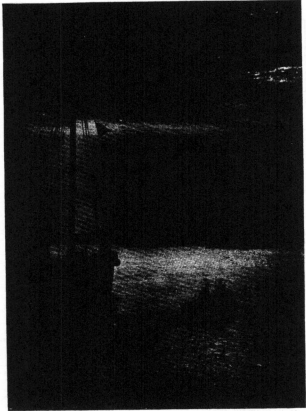

Photograph from Edwin Levick

ONE OF THE AMERICAN NAVY'S FLOATING FORTRESSES SEEN BY MOONLIGHT

Each officer of the United States Navy has a pet story to prove that the American is the
smartest sailor and best fighting man that ever crawled into a turret.

A "BOAT," ITS "GOBS" AND ITS OFFICERS

In the lingo of the sea, a destroyer is always referred to as a "boat," while the enlisted men of every naval craft are "gobs" (see text, pages 581 and 594). In the center of the front row of officers (seated) is Commander Byron McCandless, co-author of THE GEOGRAPHIC'S famous Flag Number. During the World War, Commander McCandless was in command of this destroyer, the *Caldwell*.

© Burrell Poole

SUPERDREADNOUGHTS OF THE ATLANTIC FLEET IN LINE-ABREAST FORMATION DURING MANEUVERS AT SEA

The *Arizona*, nearest the camera, and its sister ships, the *Idaho*, *Mississippi*, *New Mexico*, *Pennsylvania*, and *Tennessee*, constitute the most powerful units of the United States Navy. The first battleship equipped with 16-inch guns, the *Maryland*, goes into commission this month (June, 1921).

ONE OF THE PRIDES OF THE U. S. NAVY, THE BATTLESHIP "OKLAHOMA," IN GUANTANAMO BAY

Official Photograph, U. S. Navy

Sixty-three officers and 1,565 men constitute the complement of this fighting craft, with its main battery of ten 14-inch guns.

© Burnell Poole

AFTERNOON BAND CONCERT ON THE QUARTERDECK

Official Photograph, U. S. Navy

PLAYING BASKET-BALL ON SHIPBOARD

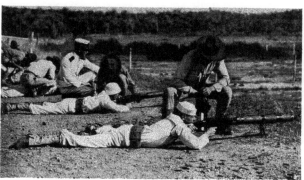

Official Photograph, U. S. Navy

LEARNING TO AIM TRUE WITH MACHINE-GUNS ON THE GUANTANAMO RIFLE RANGE

With its 264 rifle and 60 pistol targets, this is the finest rifle range used by the United States. It is a part of the 30,000 acres in the vicinity of Guantanamo rented from the Cuban Government as a naval base.

jaunts about the world; he is well fed and well clothed and his physical and moral health are guarded.

Upon his return to civilian life he has attained to a higher and more intelligent standard of citizenship.

ON BOARD THE "BLACK HAWK"

The departure of the Atlantic fleet from the Brooklyn navy yard at dawn on the morning of January 4 was unostentatious and unpicturesque. Battleships and destroyers worked singly down Ambrose Channel to effect fleet formation well out at sea.

The vessels of the train—the colliers and the supply-ships and the tenders and the cruising foundries and repair shops— smashed and wallowed in their wake.

After a time one capitalizes The Train. It is not dashing and it is a bit grubby, and no one has a kind word for it; nor has The Train a kind word for any one else. But it is what the lines of communication are to the army. Without The Train, the fighting craft had best not put to sea.

On board the *Black Hawk* one pitied the raw recruits. When the old Grace liner began to hit up her conservative twelve knots an hour, one saw the desire to see the world abate visibly in these young breasts.

The *Black Hawk* was dirty, of course. So was every other ship in the fleet. They had been lying alongside docks, submitting to repairs and taking in stores, for weeks on end. It had not been worth while to clean ship, even if it had been possible in the raw air of a northern winter. But no sooner were the ships in clean blue water than the toilet operations began.

The cargo boom of the *Black Hawk* was astraddle with boys, many of whom had never seen the sea before. The regular rise and fall of her deck sent them perishing with agony. They clung with arms and legs to the boom and put their heads down on it and cursed the sea and those who go down to it in ships. They reminded one of nothing so much as a parcel of young raccoons clinging bottom side up to tree branches.

"GOBS" COMING ASHORE AT THE RECREATION GROUNDS, GUANTANAMO

Official Photographs, U. S. Navy

AMERICAN SEAMEN IN THE MAKING ON SHIPBOARD

Soap and the safety razor are in active service with the American bluejacket. Cleanliness is one of the cardinal virtues of the men of the Atlantic and Pacific fleets.

AMERICAN BLUEJACKETS BARTERING WITH NATIVES IN ONE OF THE WEST INDIES

Official Photographs, U. S. Navy

THE "GOB'S SPECIAL," AT BRIDGETOWN, BARBADOS, BRITISH WEST INDIES

The dictionary will inform you that a 'gob" is "'a mass or lump, as of mud or meat: a large or good-sized mouthful; also a large 'sum, as of money," but not a word about sailor or seaman. How the word came to be applied to the American bluejacket is a matter of mystery, but the "gob" himself has done it, and, despite all officialdom, "gob" he is to his messmates.

ON THE BEACH AT GUANTANAMO

Official Photograph, U. S. Navy

It is here that the enlisted man begins in earnest to lose the pallor, the white knobs, and bony angles of the civilian and to take on the saddle-colored coat of tan and other attributes of husky health.

But as the worst of the sorrow passed they began to scrub the caked grime, and before the *Black Hawk* was in southern waters they were being swung alongside in cradles to paint her hull. They were already critical of other ships which had not been polished up.

THE MOTHER OF THE DESTROYERS

In the official lists the *Black Hawk* is referred to as the tender for the destroyers. On board she is known as "the mother," and this is the better title.

Almost anything that can happen inside a destroyer can be patched up by the magicians on board. The destroyers run to her with their real and imaginary troubles. They demand potatoes, brass screws, postage stamps, tooth-pulling, and sympathy.

"What's that?" I used to groan each morning in port when the steam winch over my head began to bang and clank.

"A damned destroyer," my room-mate would reply with emphasis. "A double damned destroyer. They spend their lives getting potatoes."

Eventually I learned that it was an article of the *Black Hawk's* creed that no destroyer ever goes to sea with more than a pint of oil in her bunkers; that any one of them would cheerfully start around the world on half a loaf of mouldy bread and a cigarette; that none ever carried spare parts against an engine-room breakdown or ever indented accurately for the parts when the breakdown came; that all destroyer men are idle, irreligious, improvident, and insane.

"They make us haul their condemned parts out to them in our own boats," the Black Hawk would cry, wildly: "Why? Why? Why don't they come gittim themselves? What do they think we are?"

A DESTROYER IS ALWAYS "A BOAT"

I am putting these things down for the moral betterment of the destroyers and to please the Black Hawk. Upon interrogating the boatmen, however—note: a destroyer is always a boat; never anything but a boat; to call a destroyer anything but a boat is evidence either of ignorance or malice; the Black Hawks never spoke of them except as destroyers—I learned they thought of the *Black*

Hawk precisely as of a department store and are venomous over her defects in the matter of delivery.

"What do they think their job is, anyhow?" the men of the boats inquired, acidly. "It's their business to bring our spuds."

These relations, the sort of relations that might be expected between a violent and unlovely stepmother and a horde of wild and self-willed children, are modified, however, by the contempt which all persons, even remotely connected with the boats, feel for the "inefficients" who shelter themselves from the wrath of the sea on cumbersome battleships. They say such persons might as well go to sea on dry-docks.

STRENUOUS LIFE ABOARD A BOAT

A boatman feels he is just a little better than other men. He lives harder, with less ease, fewer hot meals, more water in his boots, shorter hours of sleep, and more salt in his whiskers than any other seaman except, perhaps, the old-fashioned schoonerman who fishes for cod. He will take more chances, get away with more deviltry, and has less regard for the bones of his hands than any other man in the navy.

By the time the fleet began to buck the January seas off Cape Hatteras it had shaken down to that precision of movement which was rarely disturbed during the remainder of the cruise.

Ahead of us reached the seven battleships, plowing at one another's heels. Then came the vessels of The Train, under the guardianship of the old cruiser *Columbia*, which has a quaint habit of wallowing sidewise as she splashes along her uncertain course. On either broadside were the lines of destroyers.

One suffered for these thoroughbreds of the sea. They are at their best when permitted a speed of twenty knots or so. Thirty-five knots an hour for hours on end are nothing to them. At twelve knots they bob and roll and swing to every slant and ripple of the uneasy water.

And the fleet was pursuing its matter-of-fact twelve knots an hour. That is the economical cruising speed for bat-

Official Photograph, U. S. Navy

THE MOTHER-SHIP OF THE FLYING BOAT FLOTILLA, THE "SHAWMUT"

Once a lowly excursion boat plying Long Island Sound, the exigencies of war opened to this vessel a more serious field of usefulness. When a seaplane is in distress the *Shawmut* rushes to the rescue. Her new dignity and responsibilities, however, have not caused her to change her abominable habit of rolling unpardonably even in the most tranquil sea.

tleships. Instead of cutting knifelike through the tumble of water, the destroyers rolled and jerked and rumbled like drums to the tapping of the impertinent waves.

NO SLEEP ON A BOAT WHEN THERE'S A BIT OF A SEA

At such times, especially if there is a bit of a sea on, no one thinks of sleep on a destroyer. Old hands have an acrobatic habit of wedging their heads and necks in a corner of the berth and so securing a certain amount of rest. The cabin transom is really the best place to sleep, though. One can so pack pillows and coats that there is at least a chance of staying on.

One night on the *Black Hawk* a vicious rain squall wakened me. I had been sleeping with my state-room ports open, and as I drowsily raised myself to cut off an incipient flood I was fascinated by what I saw through the open ports.

The line of destroyers had approached our starboard more nearly than ever before, so that I could see their drunken lights reel over the sea. The waves were piling high, as they have a habit of doing when the Gulf Stream runs coun-

Official Photograph, U. S. Navy

A KITE BALLOON ATTACHED TO THE BATTLESHIP "PENNSYLVANIA," USED FOR
SPOTTING PRACTICE

ter to the wind, and the unfortunate boats, pulled down to the agonizing twelve knots an hour, were stumbling and swinging helplessly.

One could imagine how the green water spouted over their bridges and hissed down the engine-room hatches, and what a foul smell of oil and bilge and stale food and cigarettes and pipes and perhaps a few oranges (there are folk who hold that an orange is sometimes a help in seasickness) was battened inside those airless cabins; for when a destroyer gets away from port on anything but a painted ocean all apertures must be stopped against the water. It spurts through every keyhole.

LOYALTY TO HIS CRAFT A BOATMAN'S CHIEF CHARACTERISTIC

Yet no one belonging to the boats would think of leaving. The men come through a seven-day run looking as though stricken by the wrath of Heaven.

On board the larger ships one might hardly know there had been wind or sea, but the men tumbling in a destroyer have been unable to bathe or shave or sleep.

Their food has been sandwiches, which they have only been able to eat by booking one arm around a stanchion. They are covered with bruises, except where the flesh has been so indurated and woodened that it can bruise no more.

Their eyes are ringed with rime and red as those of an angry parrot. Yet they are so constituted that to tell about the boats they can go without food, sleep, or other necessaries. Their eyes shine and they reach for your arm to shake you into an appreciation of their pets.

"Come with me," said the engineer of the destroyer squadron at the end of a particularly nauseous four-day wallow on the *Black Hawk*. "I'll give you a treat."

Observation had made me somewhat skeptical of the potentialities of enjoy-

© Burnell Poole

THE ATLANTIC FLEET AT SEA

Official Photograph, U. S. Navy

A FEW UNITS OF THE ATLANTIC FLEET IN GUANTANAMO BAY

584

DIVING FOR PENNIES IN THE CLEAR WATERS OF KINGSTON HARBOR, JAMAICA

This spirited scene is duplicated in all West Indian harbors where the water is clear and free from the danger of sharks.

ment offered by the treats of boatmen. But the officer was earnest.

"We'll make a speed run on the *Crowninshield*," said he. It does not seem proper to repeat what he said of the *Crowninshield*, or of certain other boats in the flotilla. No boat could possibly be so nearly perfect. Nor do I believe that boats develop an intelligence of their own, nor that they respond to affection as so many good children might.

But one can see why the boats are so well loved. There is a freedom on them that is not to be found on the larger vessels, and snappy discipline, with few of the restraints of rank.

The least little fret in international relations may hold a chance to win a name.

The boatman feels, I fancy, as the cavalryman does with a thoroughbred horse under him, or like the fighting flyer, as he storms up in a new plane. There must always be the sense of controlled power and independence and always the hope of adventure.

ARRIVAL AT GUANTANAMO

On the morning of January 9 we woke to find ourselves in the sheltered bay of Guantanamo, at the extreme southern tip of the Island of Cuba. Two days before the men had gotten into whites and the blue uniforms had been stowed away for another year.

Now the gobs stripped off blouses and by order appeared in the sleeveless sing-

A STILT DANCE IN THE BRITISH WEST INDIES DELIGHTS THE AMERICANS

THE COLOR LINE IN JAMAICA

A COUNTRY ROAD IN JAMAICA

A BANYAN TREE IN KINGSTON, JAMAICA

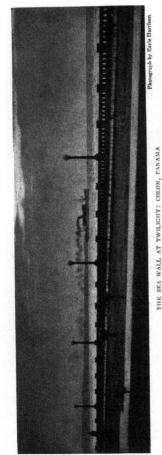

THE SEA WALL AT TWILIGHT: COLON, PANAMA

Photograph by Earle Harrison

let which is a part of the navy's working costume.

The decks began to offer sights that were at once absurd and pathetic. One thinks of a seaman as a lively person with thick shoulders and an excellent saddle-colored coat of tan, and spring in his heels; but these youngsters who timidly came out of their civilian husks offered studies in white knobs and bony angles.

Their elbows were sharp as boat-hooks and their poor little forearms were puny and pale as the stems of clay pipes, and their feet, slender and weak from years in leather boxes, stepped gingerly upon the rivet-studded deck.

Then the sunburning began. One sunburns by order in the navy, just as one gets vaccinated or shaved. The first day in the sleeveless singlets produced large blotches of bright and inflammatory red upon their virgin shoulders. By the second day some had neatly blistered.

There must be a virtue in sunburn. In another month these anemic, appealing kids had clothed their young skeletons with fine muscles and were as hard as so many sides of frozen beef. Some had begun to sprout little torpedo beards, too. But that is another story.

WHAT GUANTANAMO MEANS TO THE ATLANTIC FLEET

How many know that the United States has a plant of extraordinary value and efficiency at the Bay of Guantanamo? Or what it means to the Atlantic fleet each year? I did not. I'll confess it.

I had a vague idea that the fleet each winter visited a cactus-bordered beach on which the men walked for health's sake, and that from time to time it went outside for battle practice.

I knew there were a couple of tiny towns near by—Caimanera and Bucoron—where rum, roulette, and ruin might be had at a price. But the background to the picture was always bare white sand and cruelly hard sunlight and scrubby bushes, with a restless surf beating at an inhospitable strand.

That impression was utterly wrong.

Guantanamo Bay lies at the southern

tip of the Island of Cuba—a tip that is for the most part useless for farming purposes. It consists of a body of water almost completely landlocked, covering about 10,000 acres.

Not far away is the harbor of Santiago, where Admiral Cervera's fleet was crushed during the Spanish War. Some of the rusty hulks still lie in the water's edge.

Around the bay is a circle of low, brown hills, covered by a mesquite scrub, into which long, narrow valleys thrust like the outstretched fingers of a hand. The water's edge is bordered by the vivid green of the mangrove, except that occasional warm and sandy beaches invite swimmers.

Now and then the sharp fin of a shark is seen. Pelicans drift overhead with their air of aldermanic dignity. Fishhawks are forever circling against a sky of almost incandescent blue.

THE FINEST RIFLE RANGE USED BY THE UNITED STATES

Here the United States has rented from Cuba, "for so long a time as it desires," an area of 30,000 acres, in which the bay is included and inclosed. A rental of $2,000 annually is paid.

At the station were 1,100 men at the time of the fleet's visit. There is a sufficient number of marines to preserve order and act as first aid in the event of any near-by trouble, and the remainder are workmen.

The striking features are the rifle ranges, which have a capacity of 264 rifle targets and 60 pistol targets up to 1,000 yards range, and can be expanded indefinitely.

This is the largest and finest range used by the United States. In a month ranges could be placed up these narrow valleys, against the background of the soft brown hills, in which ten thousand men could be taught sharpshooting each day.

Once there was a golf course, but it has been made over into a landing ground for the ship-planes that were forever maneuvering overhead. They are to be carried on especially built vessels, and a maximum of skill will be required of the flyers to get away from and return to

these moving platforms without accident. Hence the continual acrobatics in the air.

LIGHTNING PLAYS HAVOC WITH BALLOONS AT GUANTANAMO

There is a balloon school, too, in which observers are taught to ascend in captive balloons—the "sausages" of the war—but Guantanamo's neighborhood seems to be dangerous to these craft. Last year three were brought down by lightning.

There are hospitals and club-houses and canteens.

On the flat land ten baseball diamonds have been laid out for the use of the fleet, and ground has been cleared for others. There are tennis grounds and handball courts and all the features that are needed in a plant which is designed not only as a training ground for 15,000 to 20,000 seamen annually, but to provide for their healthful recreation.

The wild animals of the hills have learned that this is sanctuary, so that one sees deer now and then and kicks flocks of wild guineas from beneath one's feet. The commandant has built good roads and pleasant walks and charming gardens, but the principal attraction—as soon as the gobs hear of it—is the pig-pen.

Officially, this pig-pen is accounted for by the best of utilitarian reasons. Figures prove that the animals earn hundreds of dollars for the government annually. But I shall always believe that the originator of the pig-farm was a Middle Westerner with fond recollections of the old farm.

There is something homy and comfortable under this Cuban sun about the grunt of a Duroc Red lady whose small children are gathering sustenance while she sleeps in the shade. Sailors are always standing about regarding this spectacle wistfully.

The corporal of marines who is the official custodian of the pigs has the air of proud importance which might befit a beefeater guarding the Crown jewels in the Tower of London.

CAIMANERA AND ITS ONE THOUSAND ASSORTED SMELLS

And there is, of course, always Caimanera. One thousand assorted smells as-

Photograph by H. G. Cornthwaite

A TOWING ELECTRIC "MULE" CLIMBING FROM ONE LOCK LEVEL TO THE NEXT
HIGHER: PANAMA CANAL

Ships are not allowed to use their own power, but are towed through the locks by electric
locomotives to prevent accident.

sail the nostrils when one climbs on its
rickety boat wharf. Small dogs sleep in
the sun or scratch themselves with an
irritated vigor rarely manifested by other
dwellers of the town.

Little naked gourd-shaped babies permeate the principal thoroughfares and
make excellent mudpies between showers.
At intervals the Caimanera waterworks
is dragged through the plaza on a cart
pulled by two goats. Life is leisurely,
contemplative, and eminently social.

"Sis's Place" and "The Two Sisters"
and "The American Bar" woo thirsty
callers by a display of backbars stacked
with bottles. There are no fronts to the
saloons, so that one pauses on the pave-
ment, so to speak, to wet an arid whistle.

There was a time when the bottles in
the backbar were racked with the cork
ends outward, but on one occasion an
Indian-club swinger practiced his art
upon the passers-by. Since then the bottles have been more difficult to reach.

The buildings along one side of the
principal street are half supported by
piles.

On their verandas, overhanging the
water, one sees dark-skinned women

dressed in flowing white, languidly fanning themselves as the ship's barge put-
puts in. All the way up the Guantanamo
River the atmosphere has suggested
Joseph Conrad's African backgrounds.
The dark currents, the violent green of
the contorted mangroves that curtain the
banks, the "Red Mill" at which a sugar
schooner bakes lazily in the sun and near
which a solitary saloon is thrust invitingly forward over the water—all have
a remote and exotic air.

Near the saloon an American sailor
with the black-and-green badge of the
shore patrol on his arm roasts pro-
fanely. There are various areas in Caimanera which are emphatically out of
bounds.

ONE HOUR ASHORE

At Caimanera the officers on liberty—
gobs never get ashore here—hurry to
Pablo's or the American Club. Every
moment is of value. The boats rarely
reach the dock before 5 o'clock and they
start fleetward at six sharp. Pablo and
Toney rain perspiration from their dark
brows as they shake 'em up. At either
hand stands a crockery pitcher filled with

Photograph by H. G. Cornthwaite

THE GATUN SPILLWAY WITH TWO GATES OPEN, DISCHARGING WATER AT THE RATE
OF 20,000 CUBIC FEET PER SECOND: PANAMA CANAL

The spillway regulates the height of the water in Gatun Lake and prevents it overflowing
during heavy tropical floods. At times six or more gates are opened at once to take care of
big floods. The illustration shows two of the gates open.

the seductive Daiquiri. A boy grinds at the ice-shaving machine in the rear. Pablo and Toney barely have time to make change.

So it goes down the dingy, dusty, sometimes flagrantly muddy street, with its weird multitude of vicious odors. Cubans look at the Americans with a certain reservation. It is not the Cuban temperament to hurry so over a handful of drinks. Nor does the Cuban need to hurry. Big negroes, with the strong features of the Arab, look one squarely in the eye.

Here and there one sees a conical thatch, which lends an air of distinction to the sordid streets of one-story shanties roofed with tin.

On Sundays the cockfight lasts all day long. The owners parade the streets with their sleepy birds under their arms and the laborers throng in from near-by plantations in their newest clothes.

SETTING SAIL FOR COLON

Brightwork was shining and paint was glistening when the fleet set sail for the second leg of its long cruise to Peru.

It was a different fleet entirely. The rust of the winter had been rubbed off ships and men alike. The next stop was to be at Colon, at the Atlantic end of the Panama Canal, where the Strangers' Club opens its hospitable doors.

Every writer of Central American fiction has come here to gain color for his narrative. Some have not gone beyond it. Why, indeed, should they, when any day the men drinking at the tables can offer the story of adventures that are incredible and true? One hears of the college man who has turned savage and of the lake that is paved with golden vessels sunk in it to save them from the hands of the Conquistadores——

But I temporarily abandoned the fleet at Guantanamo in favor of the seaplanes. The Atlantic flotilla of F. L. 5's had hopped down the coast from Philadelphia and was on its way to the Panama Canal.

I made my temporary headquarters on board the *Shawmut,* the mother-ship of the airboats.

The *Shawmut* has not lost the habit of rolling she acquired before the war, when

THE "NEW MEXICO" LEAVING LOCKS IN THE PANAMA CANAL

Close-up photographs of this kind constitute a part of the argument of the aviation advocates, who believe that the airplane is making obsolete the first-line battleships (see "America in the Air," by Brigadier General William Mitchell, in THE GEOGRAPHIC for March, 1921).

she used to carry excursionists down Long Island Sound. It is my conviction that she will roll her rail under if a gob misplaces a tin knife. Maybe I am wrong, but I know what I know.

A DEFENSE OF "GOB"

At any rate, this seems a good place to defend the word "gob." Every naval officer resents its use. Not one will concede there is such a thing or a word as gob. It will be admitted that it does not sound sweet and pretty, any more than "chow" or "hike" or any of the other one-syllabled words that came into use through war. But as the sailors call themselves gobs (and never by any chance refer to themselves as tars or bluejackets or seamen) and unquestionably discovered or invented this name, I submit that they are a sufficient authority.

Gold lace may fume, but between decks they will continue to be gobs.

Flying to Jamaica on board one of the seaplanes did not seem a stunt at all. There is something reassuring in the bulk and weight of these giants, just as there is in the apparent indestructibility of a battleship. They are not as large as the N. C. type, it is true, but a boat with a top wing span of 107 feet and which when loaded weighs seven tons is not as terrifying as the butterflies in which one flies over land.

LIKE FLYING ON THE VERANDA OF A COUNTRY CLUB

From my seat in the rear cockpit, with a 330-horse-power Liberty motor rasping at either ear, I looked down the twin lines of massive wing struts and felt as though I were out for a fly in the veranda of a country club.

It was only as we sat upon the *Shawmut's* emotional deck in the calm waters of Kingston harbor that I learned the day had been a bad one for flying.

BALBOA, THE SEAT OF UNITED STATES GOVERNMENT IN THE PANAMA CANAL ZONE © Publishers' Photo Service

Balboa and its adjacent community, Ancon, make an ideal residential city with wide, well-lighted streets, and spacious houses with wide, screened verandas. The United States Government operates all the commissaries and restaurants, and owns all the houses, which are occupied by the canal operatives and their families. The large building on the hill is the administration headquarters.

Official Photograph, U. S. Navy Air Service

THE AMERICAN ARMORED CRUISER "HUNTINGTON" AS SEEN FROM A KITE BALLOON

In addition to its target range, its naval and aviation base, Guantanamo has a balloon school, but the neighborhood has proved perilous for the "sausages." Three were brought down by lightning at this station last year.

Official Photograph, U. S. Navy Air Service

TWENTY-ONE AMERICAN DESTROYERS ANCHORED OFF BALBOA, PANAMA CANAL ZONE

The United States Navy's destroyer fleet consists of 281 vessels of the first line and 21 of the second line. The average destroyer of the first line is from 310 to 314 feet long, has a speed of 35 knots, is equipped with four 21-inch torpedo tubes, and has a complement of eight officers, eight petty officers, and 106 men.

on the mine-layer *Sandpiper*. Captain "Bill's" specialty is in doing things that have never been done before, which is of value in the airboat service, because things are happening there that have never happened before.

The *Sandpiper* is one of the neatest and sturdiest craft imaginable. She has a wide fantail aft and a neat round body which tapers into an impertinently sharp nose. Captain "Bill" would cheerfully undertake to drive her through the Gobi Desert if called on.

On the decks of the *Sandpiper* a miniature dry-dock is carried. That was gotten overboard and the hurt plane loaded in the well. But it did not reach port. There was no bad weather—no particularly bad weather—but the waves snapped the wings off and finally destroyed the plane. It was a mere incident of airboat flying. The flyers hope

Photograph by Earle Harrison

A CORNER OF THE CATHEDRAL PLAZA, PANAMA CITY, THE HEART OF THE REPUBLIC

It was in this plaza that the people proclaimed their independence from Spain more than a
hundred years ago, and from Colombia in 1903. The cathedral dates from 1760.

that some day they will be given a motor that will never stop, and so they will be absolved from the necessity of carrying a burdensome boat.

OLD PROVIDENCE, WHERE IT COSTS NOTHING TO LIVE

One wonders what the inhabitants of Old Providence Island thought when the seven remaining airboats of the flotilla swooped down to wait for news of the foundered plane. No airboat had ever been seen before in that part of the world.

The crews perched in their tiny cockpits, unable to go ashore because of fleet quarantine regulations, and looked longingly at the land.

Old Providence is four miles long and eleven hundred feet high and everywhere is a vivid green. Sometimes a little yawl brings mails and canned goods from Colon. This does not often happen.

"How much does it cost to live here?" the flyers asked of the port official, who was paddled in a dugout to give them pratique. Over his head he held a yellow umbrella. In one hand he fitfully agitated a small yellow flag.

"Why, nothing," said he, puzzled.

All day long the flyers sat in the shifting shade of the upper wings, as their planes bobbed on the placid water, and smoked many angry cigarettes.

But Old Providence did its best to entertain them. The brown belles and beaux paddled out to chat. Small, naked boys did perfectly unbelievable feats of swimming and diving for pennies. Large family parties were rowed out in sturdy boats. The most delicious oranges and bananas the Americans had ever tasted were offered at infinitesimal prices. But of the romantic island of Old Providence the flyers thought only as of a place in which they had been marooned.

"Gee," they reported. "We were glad to get away."

THE MARVEL OF THE CANAL TAKEN AS A MATTER OF COURSE

It is doubtful if Young America, as represented 'tween decks, fully appreciated the marvel of the Panama Canal as the warships passed through its great locks. They were immensely interested, of course. One could see the fighting tops of the great ships white with sailors. Their decks seemed to have been whitewashed, so thickly had the men gathered to gaze overside at the electric mules and the lock mechanisms (see page 592).

But most of them had been fed for the better part of their lives with stories and pictures of the canal. They took great achievements as a matter of course, as all Americans do. Nothing seems impossible or even improbable to them.

The boys commented wisely on the various mechanical gadgets used in the canal operation, but the thing that really aroused their enthusiasm was the zone's flylessness.

This was within the range of their every-day knowledge. Every boy born on a farm or reared in a village knows of the plague of flies and mosquitoes the summer brings. All knew that the tropics are the happy home of every bug that flies or creeps. Yet they saw the Canal Zone—this green ribbon which unites the two oceans in the very heart of the tropics—as clean and shaven and bugless as though it had been painted on a backdrop.

For one question I heard about the building or operation of the canal I heard a dozen about the process of getting rid of winged pests.

"We're a pretty smart people at that," was one boy's conclusion. "We can get rid of them whenever we really want to."

THE TWO FLEETS JOIN

It was at Panama that the Pacific fleet joined that from Atlantic waters, and the combined fleets steamed for Peru and Chile in company.

The greater part of each day was occupied in what were called maneuvers, but were really tactical exercises. The purpose was to accustom the deck officers of the various ships to work in concert. The spectacle of more than sixty ships of war moving in harmony was a superb one, but it was not until it was announced that the fleets were to have night battle practice that the real thrill came. That night I was leaning over the port rail in company with an officer.

The night was one of soft blackness. Not a star was reflected in the placid sur-

© Publishers' Photo Service

ABLUTIONS AL FRESCO: A STREET SCENE IN PANAMA CITY

face of the Pacific. There were no lights visible, although one knew that somewhere fourteen battleships and thirty-six destroyers were playing the war game together. There was not a sound except the rustle of the water thrust back by our blunt prow.

Suddenly my eyes became aware that a sinister shadow kept silent company with us. At first it seemed only a blob of deeper black. At last I realized that a destroyer was so near by that I might have tossed a biscuit aboard. For the first time I fully realized the deadliness of the boats.

"Why do we not fire on her?" I asked my companion.

"We were sunk ten minutes ago," said he. "We're out of action."

At Panama the heat had been blistering, as the ships lay at anchor, under the protection of the cluster of round islands

that will eventually be fortified for the protection of the canal mouth. The protection of the canal mouth. The protection is a perfect one. No ship of war can ever hope to get between the meshes of that screen of islands, although the canal remains as undefended from aircraft as is every other secret place in the world.

But the moment the start was made for waters farther south a cool breeze searched the most remote corners of the ships. Even the old-timers found it difficult to adjust themselves geographically.

We were bound "up south" and leaving the warm neighborhood of the isthmus for the delicious coolness of the Equator!

In the distance we later saw the green shores of Colombia and Ecuador, visible on the horizon as faint lines of cloud.

The farther south the fleet traveled, the more delightful became the temperature.

A BANANA-LADEN DUGOUT CANOE ON GATUN LAKE, PANAMA

The faintest of breezes ruffled the calm water, so that one might sit all day upon the quarterdeck without the pages of one's book so much as fluttering.

At night the stars were barely over the mastheads and vividly bright. A luminous lane led straight over the waters to Venus, which gleamed with such an emphasis of light that the first words of the newcomers to the deck each night were usually addressed to the Queen of Beauty.

LIGHT OVERCOATS AT THE EQUATOR

The sailors spent hours in identifying the constellations, although in northern latitudes they had not paid a moment's tribute to the heavens.

When the neighborhood of the Equator was reached, light coats were needed at the evening movie shows. The sailors who clung to singlets clotted in the neighborhood of the engine-room hatches, from which a warm and pleasant current was always flowing.

For this reversal of the expected the Humboldt Current was responsible. This may not be quite as remarkable a phenomenon as the Gulf Stream is on the other side of the continental wall, but its effects upon South American conditions are hardly less marked. It has actually shoved the South Temperate Zone north; so that while the vicinity of the Equator should be the warmest place in the world, it actually is not.

© Publishers' Photo Service

THE FRUIT MARKET ON THE BEACH AT PANAMA CITY

This deep, cold current sweeps the depths of the sea and brings up algæ that are foreign to these latitudes. The fish follow the algæ, and the birds follow the fish. Penguins which belong in the Antarctic Circle are living happily off the coast of Peru.

Much of the wealth of Peru is derived from the guano beds produced by the guano-birds, which are estimated to have an actual value to the state of $15 per pair. The giant condor, which lives at a chilly altitude of 15,000 feet, is able to swoop down to sea-level to prey on guano-birds, thanks to the temperate influence of the Humboldt Current.

But to the old-timers—the flat-feet—these are the Horse Latitudes and the Doldrums. They had many stories to tell of ships becalmed here for weeks, in spite of the whistling of the bos'ns and the throwing overboard of silver money and the affectionate pampering of such Finns as might be on board.

CROSSING THE LINE

Sometimes, though not often, an albatross was sighted. Nor were sharks as numerous as the greenhorns had supposed.

After all, the chief interest at this time was in the arrival of Neptune and his courtly party. The men who had crossed the Line before talked of little else for weeks. In every dark corner Neptune and his courtiers swore gruffly at landlubbers who tried to overhear their deliberations.

Costumes were being made and tried on behind locked doors. Some had been brought from New York and some were a part of the permanent equipment of the ship. Most of the dresses of ceremony had been bought at Panama, however, where the Chinese shopkeepers do a regular business in purveying starry crowns for Neptune himself, and the latest and most shameless lingerie for Amphitrite.

One or two of the younger and prettier seamen were seized upon for pages to be decked out in long silk stockings and permanent blushes. No one was more unhappy aboard than these pages in the days just preceding that of the ceremony. An exception may, perhaps,

THE MUNICIPAL MARKET IN PANAMA CITY

The "Ditty Box Guide Book," which was given to every member of the fleet before landing at Panama, gives this version of the origin of the nickname "spiggoty," applied by Americans to the natives of Panama: "When the Americans first arrived on the Isthmus the cab-drivers would shout, 'Me speak it, the English.' This soon changed to 'spickety,' and then to 'spiggoty.' Thenceforth the Panamanians were 'spiggoties.'"

be taken to that statement. No doubt the most unhappy man was the greenhorn who permitted himself to be frightened by the heartless old tars, who told of Neptune's severity.

"How is your heart?" they would rumble to the young man from the prairie town, who had never even heard of Neptune before he went to sea. Horny hands would feel the frightened flutterings of the victim's breast.

"You may get through," the amateur diagnosticians would say "'I dunno. It's pretty weak. Hey, Tom. C'me here and feel this guy's heart!'"

THE GREATEST LEVEE NEPTUNE EVER HELD

This was certainly the greatest levee that Neptune ever held since he began his practice of climbing over the bows of ships as they cross the Equator. Not less than 25,000 men appeared before his

courts on the sixty-odd ships. The total was probably greater, for it was estimated that not more than 5 per cent of the 31,000 men on the combined fleets had ever crossed the Line before.

On the night of January 23 the Herald came aboard. Up to this moment the promised ceremony had seemed distant and humorous. Now it took on an air of dignity. One felt that this fine old tradition of the sea would be upheld in a proper spirit. The ships had been jogging placidly along, when from our foretop came a frightened hail:

"Light ho! Light on the starboard bo-ow."

At first the men who rushed to the rail thought the light was from one of the life-rings equipped with lamps, for use when men go overboard on dark nights. But it was seen to be a flare set adrift in a tub from whatever ship was

"THE FARTHER SOUTH THE FLEET TRAVELED THE MORE DELIGHTFUL BECAME THE TEMPERATURE"

In practice cruises it is not only the constant "march and countermarch" of the ships that is watched, but the amount of smoke which issues from the funnels of each ship. Good engineering means no plumes of smokes.

Official Photograph, U. S. Navy

THE "NEVADA" FIRING HER FOURTEEN-INCH GUNS

In the extreme foreground note the heads of the men of a sister ship watching the gigantic plume of black smoke emitted from the muzzles of the great guns.

SPLASHES FROM GUNFIRE: TARGET PRACTICE AT SEA

To the layman these splashes seem uncomfortably close to the observation ship, from which
the photograph was taken. Note the muzzles of the big guns at the top of the picture.

at the head of the long line. As it fell
astern, a hoarse voice was heard from
the bridge and a masquerader in spun-
yarn wig and whiskers cried that at 8
o'clock of the following morning Neptune would come aboard.

The formula is unvarying. In all the
years and on all the thousands of ships
the wording has hardly changed.

SHIPS SURRENDERED TO NEPTUNE'S NAVIGATOR

So, at 8 o'clock of the morning of
January 24 every ship in the fleet checked
her way, while a ponderous figure, topped
with a gilded crown, climbed aboard from
the chains and marched down the deck
at the head of his crew.

The men grinned at Neptune and Amphitrite and the court jester and the rest,
of course. That is American fashion.

For all that, there was an element of
solemnity as well as of fun in the ceremonial. It seemed to link these young-

sters with Drake and Morgan and the
others who first sailed these seas.

'The captains turned over their ships to
Neptune's navigator, who took the bridge.
The crew moved to the forward decks,
where the tanks had been built. The court
criers began to read the names of those
who had never crossed the Line before.

Ordinarily this performance is somewhat drawn out. When there are but
two or three to admit to Neptune's realm
the monarch's ingenuity is taxed to give
them a fitting initiation. But Neptune
had too much work before him on this
occasion to give his humorous inclinations free rein. On some of the battleships 1,400 men were to be ducked, and
few vessels had fewer than 200.

Only the first two or three victims
heard the speeches put in Neptune's
mouth by tradition. After that quantity
production was the rule. The courtiers
and dentists and doctors rolled up their
sleeves and fell upon the neophytes.

Photograph by Herbert Corey

FATHER NEPTUNE AND HIS COURT ON BOARD THE "BLACK HAWK"

Probably never before in the history of the sea has Neptune held so great a levee as when the Atlantic and Pacific fleets of the United States Navy "crossed the Line" on the trip to Peru and Chile. Not less than 25,000 men appeared before his courts on the sixty-odd ships.

"What's your name?" Neptune would growl.

As the unfortunate opened his mouth to reply a nauseous pill of grease was slipped between his teeth. The barbers slashed at him with a huge brush covered with a lather of flour paste and shaved him with a wooden razor. He was tripped in the barber's chair and spilled backward into the tank six feet below.

The bears seized him, thrust him under water half a dozen times, swung him to the deck again, and turned, perspiring, to catch the next apprentice as he whirled head downward through the air toward the floury waters of the tank.

The extraordinary feature of the visit of Neptune was yet to be seen. In the effort to give a proper scenic investiture to their rôles, those of the flat-feet who had been chosen for the parts of courtiers had permitted their beards to grow.

Later on it seemed they could not give them up. They tortured their whiskers into new and strange shapes. They wore them as torpedoes and paint brushes and little dabs of hair high up on the cheekbones and as galways and burnsides. People turned upon the street in Panama and Lima to gaze after them, and the enraptured gobs thought of new ways in which to train their hirsute adornments. It may be they are still bearded, but I doubt if their mates have resisted temptation.

THE FLEETS SEPARATE

South of Callao, Peru, the two fleets separated for a time. The Pacific fleet went on to Chile, while the Atlantic fleet turned back to Peru. Both were to pay visits of ceremony.

The vessels comprising the Atlantic fleet coasted along this extraordinarily arid coast whose brown dryness is in such contrast to the tropical verdure farther north.

Both Chile and Peru are bordered by a belt of desert country, lying between the Andes and the sea. Here the Andes have three separate crests, which act as compressors in squeezing the last drop of

THE HARBOR OF CALLAO, PERU, WHERE THE ATLANTIC FLEET PAID ITS CALL OF COURTESY

Chief seaport of Peru, Callao is also the gateway to the capital, Lima, situated seven miles inland. It is a town of 30,000 inhabitants, mainly Peruvian, but with a goodly sprinkling of Indians, Japanese, and Chinese. The American fleet was fortunate in visiting the city sufficiently early in the year to avoid "the painter," an unpleasant odor which fills the harbor in March and April. The phenomenon is attributed by some to the remains of innumerable small dead fish brought in by the tide.

moisture out of the breezes which sweep up the Amazon Valley from the Atlantic. Once in seven years, or thereabouts, a little rain falls in the dry belt.

For years Peru accepted this condition without demur. There were advantages in the absolute dryness of the air.

Not far from Lima is a buried city, four thousand years old, in which the houses were built of unburned clay. Their fronts were molded in crude designs by the Indian architects, and the tracery has remained unharmed by the passage of years. Bodies of animals found on the plains have been mummified without decay.

One would think that some one would have reasoned that cities cannot exist without water, and that there are abundant evidences to prove that at one time this coast was extremely populous. But no one did.

PUTTING ANCIENT IRRIGATION DITCHES TO WORK AGAIN

Then a foreigner made a discovery. Wandering through the arid region, he saw what he thought might be the remains of ancient irrigation projects. A little investigation convinced him that he had under his eyes a complete irrigation system, built by the Incas or their prede-

MEN OF THE AMERICAN FLEET IN THE STREETS OF PERU'S CAPITAL

It was a cordial welcome which Lima extended to the representatives of the northern republic; even the motion-picture theaters were specializing in American films; but the great show staged for the benefit of the visitors was the bull-fight, held in the largest bull-ring in the world.

cessors, which had but to be cleared of the accumulated sand to function again as perfectly as in the Incan days.

He paused long enough to buy a large tract from the Peruvian Government, and then cleared out the old ditches and turned into them the water that had been flowing down from the hither slopes of the Andes to lose themselves in the desert sands. Today he has one of the greatest sugar estates in the world, and every main-line ditch is just as it was first surveyed and excavated thousands of years ago by a people we are apt to think of as savage.

The Peruvian Government was not slow to take the hint. It has been slow to act because of financial conditions, but today an American engineer is in the employ of the government tracing out the lines of the ancient Inca systems. They are complete and practicable, and need

only to be cleared of sand to restore to fertility the thousands upon thousands of acres which were once cultivated here.

As one ascends the Andes the rainfall increases, until on the other side of the triple summit a typical tropical jungle is to be found. But on this arid plain, for years dismissed as a useless desert, are to be founded the great farming projects of the immediate future in Peru.

A CORDIAL WELCOME FROM PERU

The Atlantic fleet entered the harbor of Callao, which is the principal Peruvian port, on January 31, and a week of good fellowship followed. One soon ceased to be cynical about the Peruvian welcome, or to look behind it for any motive of self-interest.

Every Peruvian one met said and seemed to feel that the United States is the great and good friend of his own

SPINNING YARN FOR PONCHOS IN THE HIGHLANDS OF PERU

The progressive president of the Peruvian Republic finds that his greatest difficulty in advancing the prosperity of the Indian element of his country's population is the absence of a sense of "the need for something." The Indian of Peru wants almost nothing that he cannot produce himself. His little garden supplies him with food and the wool of a sheep or two provides his wardrobe.

republic. President Augusto B. Leguia voiced what appeared to be the sentiment of the nation:

"In all our dealings with other nations we have found the United States the only one whose actions have not been dictated by selfish interest," he said.

There is a startling absence of ceremony, too, in spite of the gold lace and elaborate trappings in which the officials of state came to pay their formal call upon the officers of the fleet. This little ceremonial was quite the equal in form and color of those one is accustomed to seeing in the European states.

The green and gold carriages in which President Leguia and the members of his cabinet drove through Lima's narrow streets glittered with plate-glass. A troop of cavalry clattered in front of and behind them.

A CALL UPON THE PERUVIAN PRESIDENT

But when I desired to call upon the President of Peru I was admitted to his presence without a tenth of the delay one expects to encounter in trying to enter the office of a country banker.

"I want to go to the office of President Leguia," I told the Indian non-commissioned officer in command of the half dozen soldiers at the door of the palace, across the great plaza from the cathedral in which Pizarro's bones lie in rather doubtful state, in a glass-sided casket which every visitor to Lima sees.

The non-commissioned officer waved me on. I crossed the wide patio and found another and was again waved on.

Eventually I reached a secretary, who spoke English and left me contemplating the red Turkey carpet and gold-legged chairs which are the invariable properties of a state chamber in Latin America.

Along one side, an open gallery shaded the president's office. From it one could look into a dusty and faded little patio, in which a few discouraged bushes struggled against the prevailing lack of water.

"Come this way," said the secretary.

President Leguia seems a sort of a Hoover in his own country. He is a slender, quick-spoken, frank-eyed man who made a fortune in business before he thought of entering politics. He speaks English very well, too, thanks partly to early training and partly to the fact that after he had been "revoluted" out of office and exiled he made his home abroad for some years. He is now engaged in trying to modernize his country, and to this end is engaging school teachers and surgeons and administrators and engineers from the United States.

DESIRE THE SECRET OF PROGRESS

"I believe," said he, "that the first step in progress is to need something."

There is no more docile, biddable, kindly man in the world than the Indian of Peru. But he wants almost nothing that he cannot produce himself. He usually holds a little "tierra" which has come down to him from his forefathers and which had been assigned to them by Inca law.

The produce of his little garden feeds his family, and the wool of a sheep or two, spun into yarn and woven into cloth by his wife, provides the domestic wardrobe.

A few pots are his household furnishings, and he sleeps at night in the blanket he draws around his shoulders by day. There seems almost no reason why he should work for money.

"So I am trying to teach him to want things," explained President Leguia. "When he feels he must have a Yankee alarm-clock and his wife demands a sewing-machine, he will be willing to work regularly. Then we can make some progress."

MAKING BATHS POPULAR

He illustrated his theory by a story. In an up-country mining camp the American managers wished to encourage the employees to bathe. They built a bathhouse and equipped it with hot and cold water sprays, but the Indians regarded it as a further evidence of the incurable folly of all foreigners.

Then the Americans announced that those who bathed would be given tickets, and these tickets could be cashed in for five cents each. Soon every Indian was taking his daily bath.

"Now the Americans have reversed the process," said Mr. Leguia with a dry chuckle. "The Indians have formed the habit of taking a daily bath and cannot

© Publishers' Photo Service

VESPER HOUR IN LIMA

From the bell tower of the Church of San Domingo the priest looks over the lovely city to the twin-towered cathedral in the distance. Pizarro gave to Lima the sobriquet ' City of the Kings."

© Publishers' Photo Service

THE CATHEDRAL, LIMA, WHERE REST THE BONES OF PIZARRO

The reputed mummified body of the Spanish conqueror of Peru is contained in a glass casket in this handsome building of semi-Moorish architecture. It was Pizarro who founded the cathedral in 1535. Little remains of the original building, however.

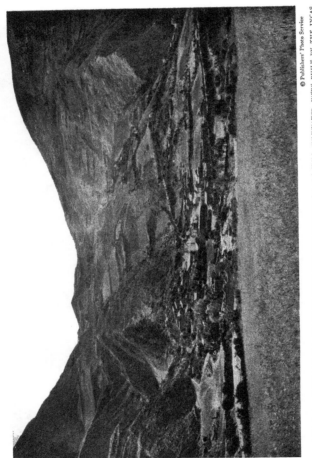

WHEAT-FIELDS AMONG THE MOUNTAINS OF PERU: THE TERRACES, WHICH ARE STILL CULTIVATED, WERE BUILT BY THE INCAS
CENTURIES BEFORE THE ARRIVAL OF COLUMBUS IN THE WESTERN HEMISPHERE (SEE TEXT, PAGE 612)

It was in this country and among these terraces that the potato was born into the vegetable kingdom (see "The Staircase Farms of the Ancients,"
by O. F. Cook, in The Geographic for May, 1916).

stop. So they work in order to get the five cents."

It is eight miles from Callao, the seaport, to Lima, the capital of Peru, but the sailors thought nothing of that. They had been provided by the Navy Department with concise little manuals, edited with the coöperation of the National Geographic Society, in which the things worth seeing had been set forth, a plan of the city given, and the monetary system explained. Armed with these, the men went where they pleased.

There were 13,000 men on the vessels of the fleet and most of them had several days' liberty, yet not one unpleasant incident marred the visit. It is true that now and then one met a sailor who was irrepressibly musical, and it is likely that the masters-at-arms had occasion to put a man or two in the "lucky bag" on each ship.

THE CONDUCT OF THE MEN OF THE FLEET INCREDIBLY GOOD

But the conduct of the men of the fleet was incredibly good. Pisco is a white whisky which plays havoc with the unaccustomed, but after one trial the gob let pisco alone.

Every man gazed at Pizarro's bones and most of them had a working version of the Conquistador's life and achievements, and all seemed to know the story of his death.

"He'd a-killed all seven," one heard them say, "only his sword got stuck in the wishbone of one of 'em."

They walked endless miles through Lima's dusty streets and regarded openeyed the Indian women who jounced in astraddle of little horses with milk-cans slung over the saddle bows, and winked harmlessly at the giggling girls who sheltered themselves behind iron grilles at the second-story windows.

At least every fourth man had a camera and they bought picture postcards tirelessly. All Peruvians smiled at them in a comradely way and the Americans were inspired to try their phrase-book Spanish on their hosts. Often it worked, too.

In every café one saw little parties about the round tables, laboriously work-. ing out an *entente cordiale*. Between times they grinned at each other and made motions to the brown boys who served drinks.

THE GREAT EVENT OF THE VISIT—THE BULL FIGHT

So far as the gobs were concerned, however, the great event of the week was the bull-fight. This was given by the Peruvian Government in honor of the American visitors, and Belmonte, who is one of the most celebrated Spanish matadors, played his spectacular part.

It was a superb bull-fight, as bull-fights go. The Lima ring is the largest in the world and the bull has more of a chance than he has elsewhere. The bulls were savage and active, too, so that in the very first fight an important section of Belmonte's skin-tight breeches was torn away and a red line appeared across his white skin.

The sailors gasped. It was more than they had counted on, apparently. Then a horse was gored, and from that moment the sympathy of the sailors was definitely against the men who took part in this abhorrent torture.

"Go it, bull," they cried whenever a bull charged. "Go get him, bull."

A little brown bull was the real hero of the day, in spite of the gold-laced fighters and the splendor of the spectacle and the age-old ceremony with which the game of butchering bulls and torturing horses was carried out.

This animal gave some evidence of a reasoning faculty. He refused to charge the blindfolded horses, although they were wheeled broadside on to him by the red-capped attendants, and the heavily armored picadors yelled at him insultingly. He paid no attention to the red mantles that were flaunted in his face. But when he charged—and he charged freely—he went straight for the man who waved the flag.

In three minutes he had the ring cleared. The expensive matadors were sitting down behind the shelters smoking cigarettes. The men with the banderillas had climbed to the safety of the balconies. The picadors were silent and the poor old horses were asleep with their chins resting on their knees. So the door was opened and the little brown bull went triumphantly out.

© Publishers' Photo Service

A BIRD'S-EYE VIEW OF LIMA, SHOWING THE GREAT BULL-RING, THE LARGEST IN THE WORLD

In this great arena a special bull-fight was staged for the entertainment of the men of the American Navy (see text, page 619).

"He was a cross-eyed bull," was the explanation that came up to us.

THE TAXI CHAUFFEURS WENT ON A STRIKE

Even the strike of the taxi chauffeurs did not disturb the happiness of the sailors. Ordinarily the chief delight of the seaman ashore is to pack as many exemplars of his profession as possible in some sort of a cab and then go careering. But the pilots of the four-seatos, who had contentedly served the Limans at a rate of one dollar and a half an hour, American money, for any sort of an automobile that would go, took umbrage at the act of the government in advertising in all the papers that the rate was one dollar and a half and no more, and advising the Americans to resist four-seato aggression.

Therefore they went on strike. They were still willing to haul a Peruvian at the old rate, but if they could not gouge the Yankees they would have nothing to do with them. It did not annoy the boys. "We've seen taxi chauffeurs before," they said.

Hard work began with the departure of the fleet from Callao and the resumption of maneuvers, when the Pacific fleet joined those of the Atlantic squadron in those tragic waters which saw the crushing of Craddock's gallant British squadron by Admiral Von Spee's ships.

But it was not only the constant march and countermarch of the ships that was watched by the admirals of the fleet. The radio was continually crackling with admonitions to the engineer officer of this ship or that.

ENGINEERING WITHOUT SMOKE

First-rate engineering to a landsman means engineering without a breakdown, but on the sea it likewise means engineering without smoke. A plume of black can be seen by an alert enemy a score of miles away.

In the thirty-odd days I spent on board the flagship *Pennsylvania* I did not see enough smut come from her funnels to daub a lady's handkerchief. There were other ships which smudged the skies like so many factories.

Everything else is subordinate to shooting, on board an American battle-

ship, however. No one can know whether our men are actually better marksmen than the gunners of other navies, but they assuredly believe they are.

The gun crews watch and gossip at target practice, and between times practice speed in loading and firing without orders from the officers.

One of the ships of the American battle fleet was once classed as a bad ship because her captain was a notorious sundowner—which may mean that he was tyrannical and unfair and may only mean that he was disliked.

The captain who replaced him established a "meritorious mast," so that men might be called in front of their fellows for praise as well as for blame. Then he began to preach better shooting.

"There isn't a better ship in the fleet," said the man who ought to know. "His men work with the guns all day long."

WHERE THE PROBLEM OF AIMING THE BIG GUNS IS WORKED OUT

Therefore I found occasional entertainment in that perspiring center known as the plotting-room. Here young mathematicians juggle with curves and logarithms to discover precisely how to hit the enemy's ship with the first salvo.

If they work out this problem accurately by the aid of adding-machines and established formulas, in the midst of jangling telephones and whistling speaking-tubes, the enemy's ship goes out of action in a beautiful burst of black and red. Otherwise the spotters in the fighting tops and in the ship-planes must take up the work of finding the range.

In getting the range, practically the only factors really considered are the presumed distance between the two ships, the speed at which each is going, the nature of the wind and the density of the atmosphere, and, last of all, the temperature of the powder-magazine. It seems almost too easy.

Upon returning to Panama the fleet paid its visit of ceremony to the ruins of Old Panama, precisely as it had to Pizarro's bones in Lima. The navy's looksee manuals informed the men that Sir Henry Morgan sacked the town in 1671, after struggling through the jungle for days.

Photograph from Herbert Corey

A CRITICAL MOMENT, WHEN THE "CROSS-EYED" BULL CHARGED (SEE TEXT, P. 619)

The members of the Atlantic fleet, for whose edification the spectacle was staged, occupy
practically all the seats in that portion of the structure shown in the photograph.

Photograph from Ernest T. Fauroat

THE BULL-FIGHT AT ITS HEIGHT: LIMA, PERU

From the moment the first horse was gored the sympathies of the American guests were
with the bull rather than with his human torturers. "Go it, bull!" was the cry which echoed
from every tier of seats, as the animal would charge the gold-laced matador and the
resplendent picadors.

"MICKY" RECEIVING HIS MORNING BATH

On their way home after their visits to Peru and Chile, the men of the Atlantic and Pacific fleets acquired an amazing assortment of pets and mascots at Panama City. Pigs, ducks, parrakeets, dogs, rabbits, and even a tiger's cub were included in the floating menageries.

He seems to have been thoroughly modern in his financial theories. After capturing many mule-loads of gold, he had it all loaded on board his own ship. Then he set sail for the nearest port in which a responsible official could be found, bought a pardon and the knighthood which gave him his title, and then blithely set about hunting down his old comrades for the bounty which was then paid for buccaneering scalps.

One visit to Old Panama was enough, however, and once was enough over the rough country road that led to it through fields that gave evidence of the fertility of the Isthmus when the jungle is once cleared away, and between rows of tinroofed shacks in which brown men and women mixed relentless drinks for all comers.

Some of the sailors carried their antiquarian researches so far as to visit the golden altar which had been taken from Old Panama in time to save it from Sir Henry Morgan's hands.

PETS OF ALL SORTS COLLECTED

Most of the men devoted themselves to the investigation of the town. Many bought the most extraordinary assortment of pets that even a longsuffering American battle fleet ever carried. Pigs and dogs and parrakeets and ducks and rabbits were taken aboard. One man even brought on board the *Pennsylvania* a tiger's cub, the tiger being of the Isthmian variety, of course. Then they were ready for the inter-fleet competition.

One result of the boxing matches was the development of perhaps the first Filipino champion in the person of Javier, a mess attendant, who won the flyweight boxing contest without the slightest difficulty and seemed competent to win a score more in the same ring. The Atlantic fleet won most of the boxing and wrestling and aquatic events, but lost the three baseball matches.

The fact is worthy of note, because it afforded opportunity for one of the finest exhibitions of sportsmanship one could ask to see.

TRUE SPORTSMANSHIP IN THE NAVY

The Atlantic fleet had a band of 195 pieces—the sort of band that made the listeners want to throw their hats and cheer when it played patriotic pieces—and a cheering section under the leadership of a young ensign, who once made New York talk about him instead of the Army and Navy football game, when he led his cheerers there. It also had an immense banner, one hundred odd feet long, with the letters A-T-L-A-N-T-I-C printed on it six feet high.

So the Atlantics were beaten. Through the series the gobs had rooted for their players as best they knew how. They had cheered defiantly and insultingly and they had cheered when cheers were needed to restore morale. Curly had thrown handsprings and turned flipflops and danced with the music. They had clapped their hands—a thousand of them, like this—tatatata, tatatata, tatatata, tata, tata, and they had driven the Pacifics utterly wild by wailing "Margie" at them. No one knows why the Pacifics so hate and utterly detest this harmless song of "Margie"; but they do.

Then the Atlantics were beaten. And that huge band crashed into a chorus that made the cold chills run up and down one's spine and brought the spectators yelling to their feet. The big banner with "Atlantic" on it, supported on boathooks by forty sailors, was paraded about the field, preceded by a cheering section of five hundred men and followed by a solid square of loyal Atlantics. Homage was paid to the victors, and then the Atlantics marched to the box where the admirals sat in state, and cheered while admirals and staffs stood in salute.

It was a fine and sportsmanlike thing to do.

Then the fleets parted for the winter's work with the guns.

Notice of change of address of your GEOGRAPHIC MAGAZINE *should be received in the office of the National Geographic Society by the first of the month to affect the following month's issue. For instance, if you desire the address changed for your August number, the Society should be notified of your new address not later than July first.*

FAMILIAR GRASSES AND THEIR FLOWERS

By E. J. Geske and W. J. Showalter

(With Illustrations in Color from Paintings by E. J. Geske)

THE dynasty of the grass family dates back to the days of the forefathers of the horse, the camel, and many others of the important herbivorous animals of the present day, and there is little doubt that the evolution of many animals into orders and forms of today was greatly facilitated by the advent of the grass family in the vegetable kingdom.

Today, of all the plants that cover our earth, grasses rank second to none in importance. In the matter of utility to man and beast, no plant or group of plants has ever played so great a part in the history of the world, and we may well say with Solon Robinson that "Grass is King."

The 10,000 species of the order, of which 1,300 are indigenous to the United States, are distributed throughout all the zones of the earth, and range in size from a few inches in height to veritable forest trees towering sixty feet and more.

Wherever rainfall sufficient to sustain plant life occurs, and at intervals of time not too distant, and with temperatures above freezing at least part of the year, some members of the family will be found. They readily adapt themselves to soil and conditions and flourish and propagate their kind.

Regions that afford ideal conditions are the great prairies of the United States and Canada, southern Russia, Siberia, the grassy plains of South America, and Africa.

THE BAMBOO IS A GIANT GRASS

Wherever the rainfall is insufficient for forests and the climate is not too arid, grasses prevail over all but the hardiest vegetation. In these areas often more than 90 per cent of the indigenous plant life belongs to the grass order, and, except where cultivation of some species has excluded its rivals, it is not uncommon to find from twenty to sixty distinct species inhabiting almost any locality.

Rice, wheat, corn, oats, barley, and rye are grasses. They enter so largely into the relations of mankind that the country which is best able to supply the world with these necessary articles of food commands the destinies of nations.

Several groups of grasses, like sugarcane, furnish sugar and its by-products. Brooms, paper, rugs, hats, and innumerable articles of commerce are made of grasses, and even houses are built and furnished with their products, not only in darkest Africa, but in many civilized countries.

The giants of the order are the bamboos, the great trunks of which furnish material for an endless number of articles of commercial importance. The pigmies are the various forage grasses, which furnish pasturage for domestic animals and beautify our parks and lawns.

Nor is the story of the merit of the grasses more than half told when it is related that they are "Man's bread and meat; many things good, and most things sweet."

GRASSES GUARD THE SOIL

Grasses are the overseers of the soil. What is more irresponsible than the sands of the seashore and of the desert? Driven hither and thither by every shifting wave and wind, they now drift here and lodge there. Now they bury forests, now they expose the bones of those who lie asleep in God's Acre, while in the waste spaces of earth the sand-storm overwhelms the traveler and his caravan.

If it were not for the grasses, the soil of hillside and plain would be as shifting as the sands of seashore and desert. Every raindrop would be a vehicle on which a grain of the soil would steal a careless ride down to the sea.

But the grasses pin the soil down to its duty. The barren hillside may become a mass of gullies and gulches, but where the grass is master, the soil bécomes the faithful servant of man. Even the trees and the shrubs would not possess a sure footing, did not the grasses help hold down the soil around them.

Though of all plants the most common, the grasses are of all common plants the least known.

The story of the grasses begins long before the age of man. In the geologists' Book of Nature, there are records of grasses that gladdened the face of the earth in the days of the tiny eohippus, from which the horse is sprung.

Some of the grasses have served man so long and in turn have been served by him that they have become as powerless to live without him as he is unable to get along in comfort without them. Imagine corn and wheat fighting their own way through the years. How soon they would fail without a plow and a harrow and a cultivator to prepare their beds and fight their battles with the weeds.

But other grasses have fought for themselves so many generations that they ask naught of any one. These travel along the roadsides of the world, sending their seeds hither and yon, until they have effected a foothold by "peaceful penetration" in a thousand communities.

HOW GRASSES SEND THEIR SEED ABROAD

The bur grass sends its thorny seed burs far and wide, attached to some passing animal or human being. The terrell grass produces seeds encased in cork-like hulls, which float to new fields on the waters of the brook beside which it grows. Couch grass grows from the root as well as from seed, and sends its spear-pointed rootstalks up through many a new foot of soil.

The beach grass has long since learned the tricks of the sand in attempting to bury all that would bind it, and has worked out a plan for circumventing the resourceful wanderer. It rises out of the sand as fast as the dune can build itself up, and at the same time sends its roots downward until it clinches the dune to solid earth.

Many grasses spread like the couch grass, by runners or rootstocks, having a succession of joints, from each of which arises a shoot that shortly takes root on its own account and in its turn sends out other runners.

Others develop only fibrous roots, and usually, like orchard and panic grasses, are found in bunches or tufts.

The seeds of some species are undestroyed and undigested by the animals that feed on them and get their chance to build a new colony through these carriers.

GRASS SEEDS WITH BARDS

Some other kinds of seeds are provided with novel weapons for forcing themselves into the soil. They have a prickly callus which bears stiff hairs growing away from the point like the barbs from the arrow's head. Once the prickly point has penetrated the soil, to draw it out is difficult, since the stiff hairs, rubbed the wrong way, interpose a strenuous objection.

A strong, bent contrivance, known as an awn, and twisted like a rope, is used by some grasses to bury their seeds in the ground. The rope-like twist is influenced by dampness and dryness—it uncoils when damp and coils again when dry. This acts as a motor to drive the seeds into the ground.

In high latitudes and corresponding altitudes, where the ripening of the seeds is uncertain, entire spikelets are transformed into leafy shoots, provided at the base with rootlets ready to grip the ground and grow wherever they fall.

What wonder that, in view of all these devices, one feels like saying of them as Darwin said of the schemes to which the plants resort in order to insure cross-fertilization: "They transcend in incomparable degree the contrivances and adaptations which the most fertile imagination of the most imaginative man could suggest, with unlimited time at his disposal."

WHY GRASSES HAVE JOINTS

Every one has noted what are popularly known as the "joints" of grass, but which are technically described as "nodes." Their mission is not, as most people believe, to give strength to the stem, but rather to help it always to stand upright.

The cells of these "nodes" are known as "geotropically sensitive"—attracted to the earth or driven from it. A wind comes along and bends the grass, so that it becomes unable to resume its upright position. Thereupon the cells in the "node" on the side of the stem that inclines toward the earth begin to lengthen, thus, by imperceptible degrees, lifting the stem to its upright position again.

The Illustration Magnifies 30 Diameters

BARNYARD GRASS (*Echinochloa crusgalli*)

The familiar barnyard grass is a strong-growing annual, one to four feet in height, having wide leaves. It grows in moist or well-manured soil. The flowers occur in a compound raceme, or cluster, of spikelets and have long and very rough awns (beards or bristly appendages). The feathery stigmas are crimson and showy. Flowers appear from August to October.

The Illustration Magnifies 50 Diameters

TIMOTHY (*Phleum pratense*)

Timothy is said to have derived its name from Timothy Hansen, of Maryland, who introduced it into the American Colonies from England in 1720. It is a native of Europe, but is now widely distributed throughout the world. It is the standard hay grass of temperate regions. The flowers are arranged in a dense spike about three or four inches long and containing several hundred flowers.

II

The Illustration Magnifies 50 Diameters

KENTUCKY BLUEGRASS (*Poa pratensis*)

Kentucky bluegrass is a meadow grass of high order and unequaled for pasturage. It is also a favorite lawn grass where the soil is fertile, there is plenty of moisture, and the sun is not too hot. It blooms in May and June and the flowers occur in a loose panicle, or tuft, of spikelets, each spikelet having three to six flowers.

III

The Illustration Magnifies 25 Diameters

PURPLE-TOP (*Triodia flava*)

Purple-top is a perennial growing to the height of 3 to 5 feet, with smooth, flat leaves. It is found in
dry fields from southern New York, the Ohio Valley and Missouri southward.   It is a very showy grass with
large panicles of flowers which occur in 5- to 7-flowered spikelets.   The glumes, or husks, are deep purple
and shining, and the feathery stigmas almost brown, standing out in strong contrast against the bright green
of the stem and foliage.

IV

The Illustration Magnifies 50 Diameters

YELLOW FOXTAIL (*Chaetochloa lutescens*)

Yellow foxtail is a smallish plant commonly found in cultivated fields as a weed. The flowers grow in a dense cylindrical spike of a yellowish color. A cluster of bristles accompanies each floret, the coloring of which is delicate and beautiful. The perfect flower is transversely wrinkled, and surmounted by stigmas of a ciliate character (marginally fringed with hairs), beautifully colored. "Pigeon grass" is a name used in England for this plant.

RYE-GRASS (*Lolium perenne*)

Rye-grass is a perennial with flowers occurring 8 to 15 in spikelets situated alternately on the rachis, or axis. It grows in fields and lots and is commonly considered a weed. Its blooming period is in June. Found mostly in the eastern part of the United States, it was introduced from Europe, where it occurs in numerous varieties.

REDTOP (*Agrostis palustris*)

Redtop is a favorite perennial forage grass of the northern States, where it is much cultivated. Its flowers occur in a cone-shaped open panicle. The glumes are green and whitish or oftener with a reddish blush, which has given rise to the name "redtop."

The Illustration Magnifies 25 Diameters

ORCHARD GRASS (*Dactylis glomerata*)

Orchard grass is a stout perennial growing in fields and yards, and attaining a height of three feet. The flowers occur in 3- and 4-flowered spikelets forming a dense, branching panicle. In early spring it affords good pasturage. Flowers appear in June, the flowering stem growing out of a dense tuft of broad leaves.

VIII

Nowhere else will one get a more striking picture of why botany seems a hard study to the layman than in the names of the different parts of a stalk of corn. The tassels are the stamens, the ear is a "spike borne in the axils of the leaves," and the grains are "the fertile flowers." The corncob is a "thickened rachis." The chaff covering the cob is "the flowering scale and palet," and the silk forms "the elongated pistils of the flower."

One of the most interesting of all the wars that Nature stages is the struggle between the grasses and the trees. Go where the forest and the prairie meet and watch the efforts of the timber to drive wedges into the grasses' lines; and observe the counter-offensives of the grasses in gaining footholds on the treeward side of the No Man's Land of the battle zone.

Strategy and tactics alike enter into the struggle. There is nothing of the barbarity of the frost and forest struggle, for neither attempts to dislodge the other; each seeks only to outlast the other and to prevent the other from bringing in reinforcements.

UNDER THE EYE OF THE MICROSCOPE

Under the eye of the microscope one may see something of the true glory of the unpretentious grasses, and in the accompanying color series the power of our eyes has been multiplied so that we may discern something of the beauty that lies hidden so deep that the cursory glance is not privileged to behold.

We are told a great deal about the beautiful green of lawn and meadow, but little is known by the layman of the gorgeous and often grotesque flowers, for they can only be seen to advantage under the microscope.

The flowers of most of the smaller grasses are perfect, and their component parts are readily comparable to the larger and better-known flowers.

In grass flowers the petals and sepals are replaced by glumes; there are usually three anthers and an ovary or pistil surmounted by one or two stigmas, these latter often branched and feathery in form.

When seen under the microscope at moderate enlargement and with reflected lighting, the color effect and structure suggest a delicate piece of beaded work profusely jeweled and built into fantastic forms and designs.

From their extreme delicacy it follows that the superficial parts of the flowers are semi-transparent, the colors being iridescent or those of one part glowing faintly through the translucent cell walls of the overlying parts. Words are inadequate to do justice to their beauty.

In the smallest enlargements of the accompanying color series of illustrations, one square inch becomes slightly more than four and one-half square feet; in the largest, we are able to look at the grasses under a magnification which is equal to stretching a square inch into twenty and one-quarter square feet, or to expanding one square yard into something more than half an acre.

If one could see familiar objects on a like scale of magnification, a normal man would be half as high as the Washington Monument and a mouse would become almost as big as a horse.

BARNYARD GRASS (Echinochloa crusgalli) [Plate I]

The familiar barnyard grass sometimes known as cockspur grass is a strong annual, growing from one to four feet tall, possessing wide leaves, and having an affinity for moist, rich soil. The flowers appear from August to October. The plant flourishes throughout North America except in the extreme north.

TIMOTHY (Phleum pratense) [Plate II]

There is no grass of the field more familiar than the timothy, with its tall, bright green stalk, its succulent blades, and its cattail head. It is the most prized among the haying crops, bringing a higher price per ton than any of the other grasses. In good land, timothy often grows to a height of five and six feet and its heads attain a length of from six to ten inches. The average timothy head is, perhaps, four inches and contains several hundred tiny flowers. The pollen, dashed with lavender, disperses with the slightest touch and is borne on the wings of the wind on its mission of fertilization.

Timothy is a provident grass. During the days of abundant moisture in the ground it stores up nutriment in bulbous thickenings at the base of the stems, which enables it to survive periods of drought better than the majority of its companions of the field.

Timothy seed is one of the lightest of the grass family, and many a farmer allows acres to pass the haying time in order that the seeds may develop. Usually the timothy harvest comes after wheat and rye are in the barn.

It is believed that timothy gets its name from a Maryland planter by the name of Hansen—Timothy Hansen—who is supposed to have imported the grass from England in 1720.

KENTUCKY BLUEGRASS (Poa pratensis) [Plate III]

Though attaining its most luxuriant growth in the far-famed bluegrass region of Kentucky, whose limestone soils also produce Burley tobacco, fat cattle, and fleet-footed thoroughbreds, Kentucky bluegrass is by no means limited in its habitat to the State that was once the "Dark and Bloody Ground." Indeed, it is one of the most common of American grasses and claims for its domain almost every limestone area from the Atlantic to the Pacific.

The habit of the bluegrass in spreading by sending up a running rootstock renders it an ideal lawn grass, since it so readily forms a fine turf. It blossoms in June, ahead of the summer grasses, the flowers occurring in a loose panicle of spikelets, each spikelet possessing three or four flowers. In dry or sandy soil the grass is unprosperous-looking and harsh, but where the limestone pasture-land has sufficient moisture it grows from two to four feet tall and makes that happy time which is known as "knee-deep in June."

PURPLE-TOP (Triodia flava) [Plate IV]

Purple-top is a perennial, growing from three to five feet in height, with smooth flat leaves. It is found in dry fields from southern New York and Missouri southward. It blooms in August and September, along with the purple eragrostis, and towers above its associates, the busy panic grasses and the slender paspalums. It comes at a time when it can share the sunshine with the pennyroyal and the other mints which are so often found in its neighborhood. The flowering head of the purple-top is somewhat sticky to the touch.

YELLOW FOXTAIL (Chaetochloa lutescens) [Plate V]

Belonging to the foxtail group, which includes the millets, this grass is widely distributed. It is very attractive when studied carefully, for the dense, yellowish cylindrical spike is full of florets, each accompanied by a cluster of bristles the coloring of which is delicate and beautiful. The perfect flower is transversely wrinkled and surmounted by beautifully colored stigmas.

The millets, cousins of the yellow foxtail, were among the most ancient of cultivated grains. Even the lake dwellings of the Stone Age reveal such quantities of these grains as to lead to the conclusion that they must have yielded the principal bread supply of prehistoric men.

RYE-GRASS (Lolium perenne) [Plate VI]

Rye-grass is a perennial, growing in fields and lots, and is commonly considered a weed. Its blooming period is in June. Rye-grass is found mostly in the eastern part of the United States and is probably an emigrant from Europe, where it occurs in numerous varieties.

Rye-grass has the reputation of being probably the first of the grasses cultivated as a forage plant, and since the days of Charles II has been held in high esteem in England. In America other grasses have answered so well the needs of the farmers that the rye-grass, does not figure in his cropping system.

A cousin of Lolium perenne—Lolium temulentum—is supposed by some to have been the tares among the wheat mentioned in the Gospel according to St. Matthew.

In Scotland the seeds of the Lolium temulentum, commonly called darnel, bear the name of "sleepies," on account of what was supposed to be the narcotic effect of its seeds. Scientific investigation has revealed the fact, however, that this effect is produced only by those grains which have become diseased through the attack of a fungus.

REDTOP (Agrostis palustris) [Plate VII]

There are few more interesting grasses than the redtop. It belongs to the bent grasses, which are a group made up of hundreds of species scattered throughout the temperate zones. They monopolize the field and wayside in midsummer as thoroughly as the goldenrod rules the landscape of autumn.

The redtop clothes the land in iridescent tones of reddish purple. One variety used to be known as "bonnet grass" and is found extensively along the reaches of the Connecticut River. It derived its name from the thrifty habits of the New Englanders of yesteryear, who braided the stems into hats.

The flowers of the redtop occur in cone-shaped panicles, while the plumes are green and whitish with a reddish blush reaching its deepest note in the redtop. The illustration shows an albino form of this species.

ORCHARD GRASS (Dactylis glomerata) [Plate VIII]

One of the earliest of the grasses that gladden the springtime is the orchard grass. The English call it cocksfoot grass because of a fancied resemblance of the branching panicle to the rooster's foot.

This plant is a living example that even among the grasses the prophet usually receives his first recognition abroad; for, although it was brought from England to America, it was never appreciated in the mother country until it acquired its abode here.

The orchard grass spreads its flowering panicles to the winds in the days when the odor of new-blown clover sweeps through the land, and with its anthers ranging from purple and yellow to terra-cotta and pink, depending upon the quality of the soil and the quantity of light, it is no mean rival of the clover for recognition.

This species ranks high as a farm grass, since it offers the husbandman pasture for his herds in the springtime and is one of the last to retire before the cavalry of Jack Frost, which precedes the infantry and artillery of Winter.

The flowers of orchard grass are in their glory in June, the stem growing out of a dense tuft of broad leaves.

A NEW NATIONAL GEOGRAPHIC SOCIETY EXPEDITION

Ruins of Chaco Canyon, New Mexico, Nature-Made Treasure-Chest of Aboriginal American History, to be Excavated and Studied; Work Begins This Month

(With illustrations from photographs by Charles Martin, of the National Geographic Society Reconnaissance Party of 1920)

THE National Geographic Society announces the sending out this summer of an expedition to undertake extensive excavations and studies in the Chaco Canyon of northwestern New Mexico. This expedition hopes to discover the historic secrets of a region which was one of the most densely populated areas in North America before Columbus came, a region where prehistoric peoples lived in vast communal dwellings whose ruins are ranked second to none of ancient times in point of architecture, and whose customs, ceremonies, and name have been engulfed in an oblivion more complete than any other people who left traces comparable to theirs.

Through the findings of this expedition, which begins its work this month (June), the National Geographic Society expects to reveal to its members a shrine of hidden history of their own country.

Both in scientific value of its findings and by adding a new chapter to the progress story of the human race, this project promises to rival such expeditions of The Society as that which dug out the marvelous city of the Andean ancients, Machu Picchu, or that which added to North America's known spots of majesty and seeming magic, the now famous Valley of Ten Thousand Smokes, in the vicinity of Titan Katmai.

A POPULOUS ISLAND IN SEA OF SAND

Chaco Canyon is that segment of the Chaco River which is cut out near the borderland of San Juan and McKinley counties, New Mexico. Its sheer, but sometimes crumbling, walls of sandstone rise from its floor anywhere from 100 feet to a height nearly equal to the United States Capitol dome. From their upper ledge stretch semi-desert wastes, making for an isolation which adds another mystery to the bygone metropolis of the canyon's maw: Whence came the lumber to build and whence the water to cultivate the corn, beans, and squash of these aboriginal farms?

To answer questions like these, the expedition not only will include archeologists, who will study periods of habitation and the origin of the tribes, but also will have agriculturists and geologists, who hope to patch from a crazy-quilt of half-submerged ruins a complete picture of the lives, customs, and culture of these early Americans.

A GIANT CANAL CARVED BY NATURE

From an airplane this gash in the desert surface might resemble a magnified sector of the Panama Canal. Closer inspection would disclose, however, not an expanse of water, but an unwatered canyon, in or bordering which are a dozen huge ruins that look to the casual observer like remains of giant apartment-houses, containing hundreds of rooms, with associated temples or sanctuaries, known as kivas, and lesser dwellings, the true significance of which is not yet known.

More astounding still, some of these larger structures, such as the Pueblo del Arroyo (arroyo—wash), one of the two ruins upon which The Society's expedition is to concentrate its investigations, are built after the familiar E-shaped ground plan of the modern office building,

PUEBLO BONITO FROM THE NORTHEAST: CHACO CANYON

NORTH WALL OF THE PUEBLO BONITO AS SEEN FROM THE NORTH:
CHACO CANYON

Large beams have been torn from the round holes at the top. Today no timber of this size is to be found within forty miles of the canyon. The openings at the ground level have been cut by vandals.

HOPI GIRL OF WALPI (PLACE OF THE NOTCH)

A HOPI "UNIVERSITY" AT WALPI

Garfield's ideal of Mark Hopkins on one end of a log with himself on the other was anticipated by the American Indian, whose tribal leaders are teachers in such primitive but effective agricultural "universities" as this scene discloses.

TWO BEAUTIES OF THE ZUÑI

The Zuñi women are attractive Indian types when young; the Zuñi pottery still is made according to aboriginal methods and patterns. The latter is as distinctive for its coloring as is the exquisite ware of the Chaco Canyon for its black and white design.

summer made a report which, when examined by The Society's research committee, bristled with such interesting scientific problems that authorization and appropriation were made for the expedition which begins its work this summer, under the leadership of Neil M. Judd, curator, American Archeology, U. S. National Museum.

Within an area less than half that of the District of Columbia there are eighteen enormous community houses having from 100 to 800 or more rooms. There also are other structure types, such as the three- to twelve-room dwellings, groups of "talus pueblos" under the wall of the canyon, in the immediate vicinity of the large buildings, and tiny cliff houses and storage cists under the canyon wall itself.

Then there are circular structures, adjacent to both large and small dwellings, and a semi-subterranean home built of mud instead of stone—the last mentioned found by The Geographic's reconnaissance party—which points to possibility of other ruins of greater antiquity that will be invaluable in tracing the development of this aboriginal civilization. The existence of these last mentioned in the Chaco Canyon region had not previously been suspected.

If the major groups were inhabited simultaneously, it is estimated the canyon population could not have been less than 10,000. This Indian city lay in a region so unfriendly that even the nomadic Navajo has not attempted to

with the addition of a curved wall binding the ends of the E projections and forming inner courts. The other ruin to be studied, Pueblo Bonito (bonito—beautiful), is a D-shaped building, with its curved wall 800 feet long.

Archeologically this ancient Island of Manhattan, surrounded by a sea of sand, may accurately be described as "a hundred miles from anywhere"; for it is 100 miles north to the cliff dwellings of the Mesa Verde, 100 miles south to the ancient Zuñi towns, and 100 miles west to the ancestral site of the Hopis.

A reconnaissance party dispatched last

cultivate it. Hence the question, What has happened there? Did the climate change? Were the surrounding arid wastes once fields of cotton, corn. squash, and beans? Or did these aborigines of northwestern New Mexico have an irrigation system akin to that of the Ifugaos of the Philippines or the rice terraces of China? Was the American Indian independent of any Nile, toward whose delta such an ingenious people as the Egyptians tended; and did he build apartments no less colossal and of more immediate service than the Egyptian "race of undertakers" constructed for their dead?

One fact is fairly certain, that this people of a period variously placed between the time of Julius Cæsar and William the Conqueror had a democratic form of government and elected a governor every year.

To the explorer, the Sherlock Holmes of ancient annals, equipped with pick and shovel, even a cursory inspection reveals clues that point to the recovery of buried treasure of history. For example, attention was attracted by masonry reinforced by timbers beneath the precipitous rocks that frown over the Pueblo Bonito. This represents a naive effort to support a huge mass of solid rock weighing thousands of tons which threatened to topple on the great building beneath. This child-like engineering experiment was surprising, in view of the architectural skill disclosed in the construction of buildings which are superior

NEIL M. JUDD, LEADER OF THE NATIONAL GEOGRAPHIC SOCIETY EXPEDITION, AND SANTIAGO NARANJO, GOVERNOR OF A SANTA CLARA PUEBLO

Having been a member of numerous expeditions sent to study the natural wonders of the Southwest, Mr. Judd is admirably equipped by experience to undertake the work entrusted to him by the National Geographic Society. He was one of the members of the Utah Expedition, which, with a surveying party of the General Land Office, in 1909 discovered the Nonnezoshi, greatest of nature's stone bridges (see "The Great Natural Bridges of Utah," in THE GEOGRAPHIC for February, 1910).

in masonry to any other aboriginal structures in the United States.

The Society's reconnaissance party examined and reported upon the availability of 16 of the canyon's 18 major ruins. The Pueblo Bonito and the Pueblo del Arroyo were selected as promising the richest rewards. These two ruins lie in the very heart of the Chaco Canyon National Monument.

Pueblo Bonito has been called the fore-

PUEBLO BONITO AS SEEN FROM THE CLIFF OF CHACO CANYON, LOOKING TOWARD THE SOUTHEAST

Excavation will outline the structure more clearly and uncover, in addition, the rare treasures of exquisite jet and turquoise ornaments, utensils, and tools of the 1,000 or more aboriginal tenants of this mammoth D-shaped apartment-house, which had an acreage and height comparable to the United States Capitol, disregarding its dome.

most prehistoric ruin in the United States. It is the largest of the ruins, the most complex in design, the most impressive. It seems to tell most clearly the unwritten story of the forgotten people who once dwelt within its silent walls. It covers an area approximately that of the United States Capitol. Its 800 rooms probably sheltered from 1,000 to 1,200 souls.

These mysterious tenants tilled the soil of the broad, level canyon floor; they hunted deer and antelope on the mesas overlooking the valley; they probably waged war on the Navajo, the Philistines who pressed upon them from the north.

RUINS ARE DEPOSIT VAULTS OF CERAMIC TREASURES

Already ceramic remains of rare artistry have been taken from Pueblo Bonito, exquisite ornaments of jet and turquoise mosaic, tools and utensils of bone, stone, and wood. Tons of earth and stone have been removed in search of material. Yet the great ruin still guards priceless secrets. The architecture remains to be studied and further evidence of the pursuits of its people needs to be found and interpreted.

Less than a city block west of Pueblo Bonito is Pueblo del Arroyo, occupying a perilous position, as indicated by its name; for the wash, or arroyo, which passes the structure threatens to cut away the bank upon which the ruin is situated. The pueblo virtually is virgin soil for the investigator. It probably stood four stories high. The upper story is gone, the first is buried, leaving only the second and third exposed.

It possesses characteristics that make all the ruins noteworthy and one, in addition, of paramount importance. Beneath the pueblo, exposed only by the caving of the arroyo bank, is a dwelling of the "small-house" type noted above. It is considered that two periods of occupancy at one site, each with its distinctive remains, offers an unparalleled opportunity for study of culture sequence. So far as is known, this is the only instance in Chaco Canyon where such superposition occurs. The fortunate proximity of Pueblo del Arroyo and Pueblo Bonito affords one advantage to the expedition in a region where many handicaps must be overcome.

Geographically the Chaco Canyon ruins have a special interest. They denote admirably the exceptional characteristics that result from an exceptional environment. Being a people hemmed in by natural barriers, their area of activity was restricted.

They were able to meet their material needs by expending only a fraction of their energy. Hence the surplus found expression in religious ritual, attested by the great ceremonial chambers; in architectural monuments, as did that of the European cathedral builders of the Middle Ages; and in ceramics, which flourished there as never before or since, for the black and white ware of the Chaco Canyon has been cited as marking the high point of this art in the Southwest.

The Chaco Canyon is a desert today, unwatered except by floods in the rainy season. The geologist must be relied upon to describe conditions of water supply and crops when the great houses were occupied. Specialists in desert flora must coöperate with the geologists in an effort to picture the economic life of these ancients. Only by the combined findings of these various experts can it be determined whether the inhabitants left because natural changes threatened their food supply, or whether falling cliff masses impressed their superstitious minds as being omens of evil.

Being within the Chaco Canyon National Monument, the Pueblo Bonito and Pueblo del Arroyo ruins are reserved and protected for the American people. The National Geographic Society's investigations, made possible under a permit granted by the Secretary of the Interior, therefore constitute a gift to the public.

The excavations, at the expense of The Society, should solve many of the problems now apparent. Repairs will prevent rapid disintegration of the walls and insure longer life to the ruins.

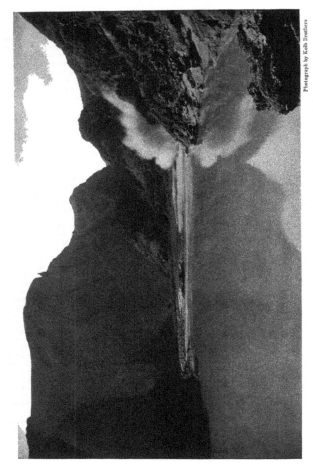

BLASTING ROCK TO MAKE PLACE FOR THE FOUNDATION OF THE GRAND CANYON SUSPENSION BRIDGE

The completed canyon bridge will be 420 feet along the roadway and is suspended 60 feet above the river in normal flow, but only 13 feet above the rushing torrent when it is at its greatest flood. This is the only bridge across the Colorado above Needles, California, which is 360 miles by river course to the south.

THE GRAND CANYON BRIDGE

By Harriet Chalmers Adams

THE suspension bridge over the Colorado River in the Grand Canyon is practically completed. Late this summer it will be possible to ride from El Tovar, on the south rim of the stupendous chasm, to the Kaibab plateau, on the north rim.

The bridging of the Granite Gorge of the Colorado opens up a new wonderland in the Grand Canyon National Park. From the Kaibab plateau, which averages 1,000 feet above the better-known south rim of the canyon, new and amazing panoramas are presented.

Last month I rode down to the river over a trail not yet opened to tourists, messed with the bridge crew, and spent the night in the gorge. The bridge is 11 miles by trail from El Tovar and 4,700 feet below Yaki Point, on the Coconino plateau. The saddle trail, following the Bright Angel and Tonto trails to the river, and up Bright Angel Canyon to the Kaibab forest, is about 31 miles in length. Rim-to-rim travelers will spend the night in a camp near Ribbon Falls, about eight miles beyond the river.

It was a chilly morning when we started for the bridge camp. The wind surged through the pines and pinyons, and twisted the gnarled cypress trees overlooking the chasm. It is the Rim of the Eternal, to be approached with awe; but people differ.

I heard a stout woman, standing by the lookout, say to her daughter, "Oh! Clara, I'm terribly disappointed. We've come at a time of year when there's no water in the canyon!"

A tall man, with a red face, was explaining to a thin man in a plaid suit that, in contour, the canyon was exactly like the doughnuts his mother used to make.

SPRINGTIME ON THE TRAIL

Once down the trail it was springtime. Shimmering blue-jays chattered among the Douglas firs and emigrant butterflies zigzagged by. High in the cliff a canyon-wren piped up a love ditty.

The "expedition" consisted of the Chief Ranger of the Grand Canyon Na-tional Park, the wandering lady he escorted, and our two mules. The ranger, whose first love was the Yellowstone, has been many years in the park service and regards our national playgrounds with reverence. He is of the opinion that all those caught carving their names on rocks or trees should be lined up and shot at sunrise.

Down we dropped to the Tonto plateau, the green shelf on the canyon wall lying between the ruby-stained limestone and the gray Archean granite. Here winds a trail of romance.

ONCE THE HIGHWAY OF THE CLIFF-DWELLERS

In the shadowy past this was the highway of the Cliff-dwellers. Here, in later years, Spaniards whose names are not written on the historic page adventured. There came occasional fur trappers from lands far to the north; the first of those great explorers who dared the descent of the river; hardy miners, whose half-hearted workings still border the Tonto trail.

We counted seven wild burros descended from pack animals abandoned by the miners. Deer were recently seen in this part of the canyon. Mountain-sheep hide on ledges high up the wall. Many other wild creatures still find refuge in this vast wilderness.

The only animals that we saw, besides the burros, were woodrats nearly as large as squirrels. These "trade rats" accumulate great mounds of rubbish. From a camp they walk off with the soap and the spoons, leaving pebbles and sticks in exchange.

The pack-train, carrying the bridge material from railroad to river, made its half-way camp at Pipe Creek. Here only a lonely black kitten greeted us. The pack-train was "on the job." It has been a tremendous undertaking to move the lumber, cement, and cables down the 11 miles of steep, winding trail to the bridge site. Many are the exciting tales told by the packers. On one trip a horse

645

Photograph by Kolb Brothers

PACK-TRAIN CARRYING LUMBER DOWN TO THE BRIDGE SITE

"It has been a tremendous undertaking to move the lumber, cement, and cables down the 11 miles of steep, winding trail to the bridge site."

THE PACK-TRAIN MAKING ITS TORTUOUS WAY TOWARD THE RIVER BED

In transporting the material from the rim of the Grand Canyon to the suspension bridge camp site many difficulties were encountered. On one occasion one of the pack-horses went over the cliff, carrying two other animals with it. Only the resourcefulness and daring of one of the men saved the remainder of the train by cutting the ropes.

went over the cliff, carrying two others with him; but a resourceful lad cut the rope and saved the remainder of the train.

Since January these pack-trains have been steadily trudging up and down between the hidden river and the railroad on the rim.

A REHEARSAL FOR CARRYING THE CABLES

The transportation of the 1,200-pound cables alone marks an epoch in bridge-building. The superintendent of the Grand Canyon National Park, who supervises the bridge-work, is an engineer whose varied experience ranges from setting the official height of Mount McKinley, in Alaska, to locating a Patagonian railroad. He conceived the idea of "rehearsing" the carrying down of the cables by estimating, with ropes, just the proper length of line necessary between each mule, as the train swung around the curves. The 1,200-pound cable was then loaded on to eight mules roped together,

with the weight evenly divided, a man walking at the head of each mule.

The sun was high in the heavens as we made the final drop down the newly cut trail in the granite wall to the bridge camp by the river. There were three sleeping tents in the camp, a dining-room tent, and a kitchen. The cook played star rôle. It is he who makes or breaks a camp. This particular cook put Broadway chefs to shame, in spite of the fact that everything but the water had to be packed down from above.

A 420-FOOT BRIDGE

I was fortunate in having the contractor himself explain the bridge to me.

The completed bridge will be 420 feet along the roadway, with a span of 500 feet from center to center of the bearings. The two main steel cables are placed about 10 feet apart and are anchored to the canyon walls 80 feet above the floor level, by means of sections of 80-pound railroad iron set into the rock with concrete.

Photograph from Harriet Chalmers Adams

SWINGING ACROSS THE COLORADO RIVER GORGE ON A WIRE CABLE

This method of crossing the canyon was employed during the work of constructing the suspension bridge.

Americans, brawny and bronzed, not a pound overweight. One used to be a lumber cruiser in Alaska; another has mined in southern Chile; a third was a cowboy "before they fenced in the whole bloomin' Southwest." One is an amateur astronomer, who spends his evenings with his telescope under the stars. He says you can see the stars better from the depths of a canyon. Several go in for photography. One has a gift for whistling and can imitate the bird calls. There is a good bit of poetry and adventure nailed into the Grand Canyon bridge.

Night in the Granite Gorge of the Colorado! They gave me the tool and meat tent for an abode. I recalled a game we played in childhood, "Heavy, heavy, hangs over your head!" It turned out to be the bacon. The framework of my tent was formerly the iron cage in which the infrequent traveler crossed the river by cable. Colonel Roosevelt crossed in this way on his ride up to the Kaibab forest.

A DEEP, MASTERFUL, SULLEN RIVER

When the camp slept and moonlight flooded the gorge, I slipped out of my sleeping-bag and walked to the river. The Colorado is a deep, masterful stream, sullen, unfriendly. No habitations border its canyon shores. It has a flow of 20,000 cubic feet per second, reaching a maximum of 200,000 cubic feet. By day its walls take on a strange, reddish-purple glow, but by moonlight they were softly pink. A weird rock,

Hanging galvanized steel cables, clamped to the main lines above, carry the wood floor of the bridge. A seven-foot wire meshing is strung along the sides as a protection for animals and pedestrians.

The bridge is 60 feet above the river in normal flow and 13 feet above the highest known water-mark in June floods. This is the only bridging of the Colorado above Needles, California, 360 miles to the south by river curve, as you "step it off" on the map.

Now for the bridge crew. Never have I seen a finer-looking lot of men—typical

which they call the Temple of Zoroaster, dominated the scene. Jupiter rode high in the heavens.

Across the river lay the ruins of an ancient Indian village, its broken stone walls strewn with prehistoric pottery — coils and Greek-key patterns— such as are found among the Mesa Verde cliff-dwellings (see illustration, page 652). Perhaps it was never a permanent settlement, only a temporary winter refuge of some peaceful plateau tribe driven down from the heights by the warring Utes. The early chroniclers of the canyon did not mention these Indians.

Who will write the long-ago romances and tragedies enacted within this mighty gorge?

A chill wind swept down the canyon and I crept back to my tent.

Next morning, when the 10 o'clock sun looked over the cliff, we crossed the river in a canvas boat, rowing well upstream and coming back with the current to the landing beach. The boat leaked. It is difficult to swim the river because of the heavy sand and silt; but in case of an upset one would probably be tossed up on the rocks before reaching the rapids.

Photograph from Harriet Chalmers Adams

THE FLYING-MACHINE, THE GRAND CANYON BRIDGE CREW CALLED IT

They went down by gravity and pulled up by hand on the other side.

LITTLE BRIGHT ANGEL, THE BRIDGE MASCOT

We climbed the bed of Bright Angel Creek, which here enters the Colorado, to the clump of cottonwoods still called "the Roosevelt camp." Here we discovered the bridge mascot, Little Bright Angel, a gray burro who lives in Elysian Fields, with clear water, plenty of grass, and a care-free life. We fed him pancakes sent by the cook, his favorite dish.

There are 113 crossings of the creek on the trail up Bright Angel Canyon to the north rim, and the little burro knows every one of them. Not long ago he guided the foreman of the bridge-crew up to the plateau, showing him just where to cross the stream.

I had heard that a distinguished American from Philadelphia, an enthusiast over the Grand Canyon, was to be the first to cross the Grand Canyon bridge; but the foreman told me, somewhat confidentially, that Little Bright Angel would be the first fellow across.

Photograph from Harriet Chalmers Adams

THE PACK-TRAIN ARRIVING AT THE BRIDGE CAMP, ON THE BANKS OF THE COLORADO

The camp consisted of five tents. Three were used as sleeping quarters for the bridge crew, one for a dining-room, and the fifth for a kitchen.

"You see," he explained. "Bright Angel has stood so long on the north shore of the river hoping to get across. He can't swim over, and he doesn't like the canvas boat."

Up in the Kaibab forest—"the island forest," a great naturalist has called it—live wild animals which have developed on original lines. The Kaibab squirrel and its cousin, the Albert, with their broad feathery tails, are the only American squirrels with conspicuous ear-tufts.

The herd of deer, variously estimated at from 12,000 to 15,000, are the mule deer, with large, broad ears and rounded, whitish tails, tipped with black. Where there are deer, there are pumas, or mountain-lions. They call them cougars in this part of the country. Uncle Jim Owens, an old-timer on the north rim, has hung out a sign: "Cougars killed to order." He has a record of 1,100 skins. His cabin walls are covered with them.

Other beasts of prey are the big gray timber-wolf, the coyote, and the fox. A man who lives here and explores unfrequented cliffs tells me there are antelope on the green shelf under the north rim. "Uncle Jim" has a promising buffalo herd, 64 in all. Isolated on a promontory and protected, the herd is sure to increase.

TINY "WARRIORS" OF THE "PEACEFUL PEOPLE": ARIZONA

These Hopi youngsters are "children of the sun" in a double sense. They are dedicated to that luminary deity by their parents, and are completely exposed to its penetrating desert rays until they are well in their 'teens. As their name signifies, the Hopi Indians are pacific; but they are far from being pacifists. Their "standing army" consists of a clan in which every mother raises her son to be a warrior, just as certain other clans are entrusted with the perpetuation of the Snake Dance and similar customs. The extinct Chaco Canyon pueblo people (see "A New National Geographic Society Expedition," pages 637 to 643) may have had an even more complex social organization, for they dwelt in apartment-houses compared to which many pueblo dwellings are but groups of cottages.

CLIFF PALACE: MESA VERDE

This most celebrated of Mesa Verde ruins is an example of a pueblo type in contrast to that of Bonito. The forgotten people of the Chaco found in canyon depths a refuge which the Mesa Verde dwellers utilized a cliff to attain. Recently a subterranean entrance was discovered to this "palace" of 200 rooms.

Photograph by Charles Martin

WHERE NATURE UPSET HER PAINT-POT: CANYON DE CHELLY

Bright red sandstone cliffs, piercing the sky to heights ranging between that of the Washington Monument and of the Eiffel Tower, sheltered a prehistoric people, probably of the same general period as those of Chaco Canyon. This most brilliantly colored of all the canyons of the Southwest lies in the heart of the Navajo Desert, northeastern Arizona.

653

Photograph by Frederick I. Monsen

HOPI BOYS OF WALPI, ARIZONA

Walpi has a cliff-top location comparable to the monasteries of Meteora, Greece. Climbing along a steep trail, where ladders formerly had to be used at some stages, has developed a lithe, agile people. Hopi children are among the handsomest of the Pueblo Indians, though their symmetry sometimes is marred when their heads have been flattened by the cradle-board.

HOPI POTTERS OF ARIZONA ENGAGED IN AN ART THAT SURVIVES THE CENTURIES

Photograph by Charles Martin

Compared with these Indian workers of 1921, Cape Cod fishermen are followers of an infant American industry. Ceramics have more than an esthetic significance. Pottery making, for example, is indicative of a pueblo people, for the ware is too fragile for nomad use.

Photograph by Charles Martin

EL RITO DE LES FRIJOLES (LITTLE CANYON OF THE BEANS) : NEW MEXICO

The honeycomb circlet in the foreground is the pueblo ruin of Tyuonyi. This photograph was taken from the top of a cliff along whose base for three miles stretches a series of "talus pueblos," a type of dwelling also found in Chaco Canyon (see text, page 640).

Photograph by Charles Martin

CANYON DE CHELLY MONUMENT: ARIZONA

In the shadow at the base to the right a cliff dwelling was found. On a ledge just above is a man, whose form is a tiny speck against this lone sentinel among the fantastic "back-drops" of multihued canyon walls.

"BRACED-UP" CLIFF AT PUEBLO BONITO, CHACO CANYON

The scattered stones at the bottom of this leaning tower of Chaco are an enigma. They represent a naïve effort to prop up a massive cube of solid rock on the part of these aboriginal engineers, who exhibited contrasting skill and acumen in the construction of Pueblo Bonito, to the left (see also text, page 641).

CANYON DEL MUERTO, A BRANCH OF CANYON DE CHELLY : ARIZONA

Photograph by Charles Martin

Cliff dwellings abound on nearly every ledge. It is not strange that peoples living in such an environment should conceive man to have emerged from a vent in the earth's surface.

EVENING EFFECT: WALPI, ARIZONA

Photograph by Frederick I. Monsen

Like the people of San Marino, who climbed a mountain to live in liberty and serenity, the Hopi, self-styled "People of Peace," took refuge in the cliffs of northeastern Arizona to avoid constant warring with cruder tribes. Walpi is on the summit of a sheer cliff.

THE PAINTED DESERT: ARIZONA

Photograph by Frederick I. Monsen

Photograph by Charles Martin

GIRL OF ORAIBI, THE METROPOLIS OF THE HOPI

Among the Hopi, famous for their snake dances, skill in weaving, dyeing and embroidery, and complex mythology, may be found lore which will provide clues to the Chaco Canyon people.

Photograph by Frederick I. Monsen

THREE LITTLE MAIDS AWAY FROM SCHOOL: HOPI INDIANS, ARIZONA

The white man is proud of his juvenile courts; the Hopi red.man is proud that he has no need for such institutions. A Hopi father considers it an essential duty to teach his children to abhor lying and stealing, to respect and obey their elders, and to be self-supporting.

Photograph by Frederick I. Monsen

AN OLD WAR CAPTAIN OF LAGUNA PUEBLO, NEW MEXICO

The bow and arrow today are relics of bitter tribal wars of long ago. A more potent
mace is a cane, prized by many council chiefs, who hold this symbol of prestige because of a
visit to the "Great White Father" in Washington. Some of these canes have been handed
down from patriarchs who made the cross-continent journey during Lincoln's administration.

As summer breezes, stealing
Fill woods and fields with song.
This soup will keep you feeling
Light-hearted all day long.

The natural food

A whiff of the savor from off the fire! A plate set out before you, steaming its invitation! The first delicious, invigorating spoonfuls! And then the glow of pleasure and satisfaction that comes over you! Right from nature comes

Campbell's Tomato Soup

Your outdoors appetite tells you how good it is for all the year. The pure juices of luscious tomatoes, after the sun has ripened them to a glowing red are blended with creamery butter, pure granulated sugar, tempting herbs, spices and other ingredients. Just so much sheer enjoyment and tonic healthfulness!

Campbell's Bean Soup

The old-fashioned flavor makes this bean soup as delicious as it is nourishing. You are sure to like it. Include a can or two in today's grocery order.

21 kinds   **15c a can**

Campbell's Soups

LOOK FOR THE RED AND WHITE LABEL

CRANE SERVICE

Enables You to Select Complete Heating, Plumbing and Sanitation Equipment through One Central Source of Supply

WHEN you are ready to consider such installations for a building of any size or character, visit the nearest Crane branch with your architect, where you will find it a simple and pleasant matter to fill all of your requirements.

Crane Service provides a wide variety of designs from which you can choose precisely the types you want—and, above all, it safeguards you with uniform quality throughout the entire installations. It is complete, convenient, reliable.

Call on any Crane Branch for the fullest co-operation

Partial View of Crane Exhibit Rooms in New York

We are manufacturers of about 20,000 articles, including valves, pipe fittings and steam specialties, made of brass, iron, ferro-steel, cast steel and forged steel, in all sizes, for all pressures and all purposes, and are distributors of pipe, heating and plumbing materials.

Actual Size

Here is Efficiency

A Focusing Model
of the

Vest Pocket
Kodak *Special*

with Kodak
Anastigmat Lens *f*.6.9

In this new camera the focusing is accomplished by slightly turning the lens flange, the focusing *scale* appearing on the shutter itself.

So effective is this manner of focusing that the lens may be brought to perfect focus for subjects as close as three feet, thus doing away with all necessity for the use of a Portrait Attachment in making "close ups".

The shutter is the Kodak Ball Bearing with speeds of 1/25 and 1/50 of a second, the usual time and "bulb" exposures, and of course the full range of stops from *f*.6.9 down to *f*.32.

A remarkably compact camera—likewise an unusually efficient camera—autographic, and richly finished.

The Price, $21.00
includes the Excise War Tax.

At your dealer's,

EASTMAN KODAK COMPANY, Rochester, N. Y., *The Kodak City*

The Chronometer:

MODERN Navigation dates from 1762, when John Harrison's Chronometer reached the West Indies, after a voyage of sixty-one days, with an error of only five seconds.

The rich prize which Parliament had offered for half a century —twenty thousand pounds sterling—went to Harrison. His victory, after thirty years of struggle, hinged on his previous invention of the Compensating Pendulum.

Unlike the modern ship's-watch, his timepiece was not suspended in gimbals but carried on a pillow.

The world war set new standards in naval timekeeping. The torpedo boat, with its terrific vibration, baffled America's experts till Elgin railroad watches were adapted to the service. And the first acceptable ship's-watches supplied our navy in quantities sufficient to equip the U. S. Emergency Fleets were—as might have been expected—

The $150 Corsican in yellow gold ∞ ∞ Three-fifths actual size ∞ ∞ An unretouched photograph

Elgin Watches

PRESIDENT

PLANT ENGINEER

CONSULTING ENGINEER PRODUCTION MANAGER

55 Acres of Borrowed Trouble

"The first tanker will dock here in just four months," declared the President, looking across the fifty-five-acre site of the gigantic new oil works. "By then we simply must be in full running order."

"And that includes piping," mused the Production Manager.

"It can't be done," exclaimed the Plant Engineer. Why, there's more than 68 miles of piping required here."

"I don't care if there's a hundred and sixty-eight," flashed the President, "this job's got to go through on schedule."

"But sixty-eight miles of piping—" the Plant Engineer came back, "high and low pressure steam lines, acid, air and water lines, besides connections to stills and coking plants—why there's over fifty thousand joints to make trouble."

The Consulting Engineer turned—started to reply. But again the President broke in—"Not more than four months, remember."

"All right," persisted the Plant Engineer, "but if you rush construction like that, you can figure on acres of leaky joints after the construction army is gone—."

"You're borrowing trouble, old man," smiled the Consulting Engineer. "I'll bet you a suit of clothes there won't be a hundred leaks in the whole job when it's tested."

"Only a hundred leaks in 68 miles of rush piping? I'll go you."

It was hardly a fair bet. For the Consulting Engineer knew the service he would get from Grinnell Company. And his confidence wasn't misplaced. The job was done on time, and after the test only six leaks were reported. He won his suit of clothes with ninety-four leaks to spare!

GRINNELL
INDUSTRIAL PIPING
Automatic Sprinkler Systems, Heating, Power and Process Piping
Your kind, their kind, every kind of Piping

How Pretty Teeth

are ruined during sleep

When you retire with a film on your teeth, it may all night long do damage.

Film is that viscous coat you feel. It clings to teeth, gets between the teeth and stays. The tooth brush does not remove it all.

That film causes most tooth troubles. So millions find that well-brushed teeth discolor and decay.

How film destroys.

Film absorbs stains and makes the teeth look dingy. It is the basis of tartar. It holds food substance which ferments and forms acid. It holds the acid in contact with the teeth to cause decay.

Millions of germs breed in it. They, with tartar, are the chief cause of pyorrhea.

Few escape its damage. So dental science has for years been seeking a film combatant.

New Methods found

Now ways have been found to fight film and film effects. Able authorities have proved them. The ways are combined in a dentifrice called Pepsodent. Leading dentists everywhere advise it. And millions of people every day enjoy its benefits.

Watch it for ten days

This offers you a 10-Day Tube. Get it and watch its effects.

Each use of Pepsodent brings five desired effects. The film is attacked in two efficient ways.

It multiplies the salivary flow. It multiplies the starch digestant in the saliva, to digest starch deposits that cling. It multiplies the alkalinity of the saliva, to neutralize the acids which cause tooth decay.

It also keeps teeth so highly polished that film cannot easily adhere.

These five effects, attained twice daily, have brought to millions a new era in teeth cleaning.

Send the coupon for the 10-Day Tube. Note how clean the teeth feel after using. Mark the absence of the viscous film. See how teeth whiten as the film-coat disappears.

Judge by what you see and feel. Our book will tell the reasons. This is too important to neglect. Cut out the coupon now.

PAT. OFF.

REG. U.S.

The New-Day Dentifrice

A scientific film combatant combined with two other modern requisites. Now advised by leading dentists everywhere and supplied by all druggists in large tubes.

"We Never Had Such Comfort—

and besides, we have burned one-third less coal with complete relief from work and worry."

The IDEAL Type "A" Heat Machine

thrives on the blustery storms of winter in any climate. In mild weather it is equally economical because of its automatic control. Great fuel economy, cleanliness, and minimum of caretaking are advantages enjoyed by thousands of owners. Provide your family with complete protection from winter ills and make this dividend-paying *investment* in an IDEAL TYPE "A" HEAT MACHINE for your home.

Write for catalog with test-chart records of efficiency and economy

AMERICAN RADIATOR COMPANY

Dept. 55 · NEW YORK and CHICAGO
Sales Branches and Showrooms in all large cities
Makers of the world-famous IDEAL Boilers and AMERICAN Radiators

A Food for Growth

Boys and girls, on the way to manhood and womanhood must have food of sturdy building qualities.

Grape=Nuts

is exceptionally rich in the elements needed to build young bodies strong and well; and it has a natural sweetness and charm of flavor for young and old.

"There's a Reason" for Grape=Nuts

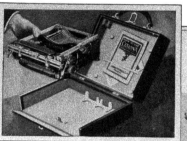

Now you will understand
the reason why Corona folds

YOU wouldn't carry a jack-knife open; nor a book, nor a traveling bag, nor a fountain pen. When you want to use them they're open, when you carry them they're closed.

Merely as a contributor to your convenience Corona's patented folding feature would be tremendously worth while. But there is another and more fundamental reason why it is of so much importance to you.

In no other way could you have *all* of the advantages of a normal, practical typewriter without the weight and bulk of a "standard" machine.

Take the type-bar as a specific instance. It is the same length as the type-bar of "standard" machines and it travels through the same 90 degree arc. Hence the same speed as a bulky machine, and the same lightness of touch.

The Corona dealer can show you a half dozen advantages of the same character.

Ask him to show them. Open the case, unfold Corona and write —how responsive its action, how easy its touch. Fold it up, slip it back in its case and close the cover —now you'll understand the reason why Corona folds.

The price of a brand new Corona, including the carrying case, is only $50.00. You can rent a Corona for a small monthly sum, or you can buy one on easy payments.

"Fold it up—take it with you—typewrite anywhere"

CORONA
The Personal Writing Machine
TRADE MARK

Built by CORONA TYPEWRITER COMPANY, Inc., Groton, N.Y.

There are more than 1000 Corona Dealers and Service Stations in the U. S.

The Personal Writing Machine

Corona
Typewriter
Company,
Inc., Groton,
New York

Send me your interesting booklet No. 66 about Corona.

Name..........................

Address.......................

Rarely does an indisputably safe investment yield so high a dividend, over as long a term of years, as an investment in a Hoover for your home. For this efficient cleaner actually returns its full cost every twelve months through the savings it effects. It pays this 100% annual dividend by reducing your present carpet-cleaning expense, by lowering the cost of household help, by cleaning so dustlessly that less laundering of curtains is required, by bettering the family health through fostering greater sanitation, and by very materially prolonging the life of your valuable rugs. Only The Hoover gently beats out all nap-wearing, embedded grit, as it electrically sweeps and suction cleans. *Invest* in a Hoover.

The HOOVER
It Beats — as it Sweeps — as it Cleans

Write for booklet, "How to Judge an Electric Cleaner," and names of Authorized Dealers licensed to sell and service Hoovers bearing our guarantee

THE HOOVER SUCTION SWEEPER COMPANY, FACTORIES AT NORTH CANTON, OHIO, AND HAMILTON, ONTARIO

The Hoover lifts the rug from the floor, like this — gently beats out its embedded grit, and so prolongs its life

Service Beyond Price

The ears of the people are within your call; their voices are within your hearing. From near neighbor to distant cities and villages, millions of slender highways made alive by speech converge within the small compass of your telephone.

Telephone service cannot be estimated by usual values. Imagine how complete a revision of our methods of living and working would have to be made if the telephone ceased to operate.

Disasters, both personal and to whole communities, are daily prevented by the telephone. And this guardianship is a part of its ceaseless service.

Glad tidings are forever streaming over the telephone. The meeting of national crises, the accomplishment of vast business undertakings, the harmonizing of a nation's activities; these compose a portion of the telephone service which is beyond price.

But the miracle of the telephone is realized in the emergency when it is so vital to health, happiness and success as to put its value beyond price.

AMERICAN TELEPHONE AND TELEGRAPH COMPANY
AND ASSOCIATED COMPANIES

One Policy   *One System*   *Universal Service*
And all directed toward Better Service

AN UNUSUAL ELK PICTURE   Photograph by W. J. Stroud

Sure Ways to Enrich Summer Days

Complete your vacation kit and summer equipment with these two closer-to-nature books. They are *fascinating—useful—beautiful.* They delight the outdoor minded—child or adult, sportsman, scientist, teacher, camp leader. The little folks never tire of their pictures and "truly" stories. Ideal for the porch reading table—indispensable at summer hotel or camp.

Wild Animals of North America	**The Book of Birds**
200 pages 300 illustrations	240 pages 308 illustrations
127 full-color plates by Louis Agassiz Fuertes	250 full-color bird portraits by Louis Agassiz Fuertes
The unique plates enable grown-up or child to identify instantly and accurately large and small animals seen in yard, park, zoo, or woods. Edward W. Nelson, Chief of the U. S. Biological Survey, with the art of the born narrator, admits us into the inner lives of the animals and reveals their habits and the important parts they play in our existence.	Those attendants at the summer bird concerts who would like to know the names of the feathered songsters or would identify winged creature seen in meadow, thicket or water, will find this book invaluable. Henry W. Henshaw, the great protector of bird life and famous ornithologist, charms every reader as he discloses the hidden beauties and romances of our neighbors in the trees.
Royal Buckram, $3.00, postpaid in U. S. A.	Royal Buckram, $3.00, postpaid in U. S. A.

------------------------------ CUT ON THIS LINE ------------------------------

Dept. H, National Geographic Society,
    Washington, D. C.:
                                                                    1921

Enclosed.................Dollars

Please send...........copies THE BOOK OF BIRDS ....copies WILD ANIMALS OF NORTH AMERICA.

NAME...

ADDRESS...

(6-21)   **Obtainable only from the Society's Headquarters**

Chin Golf! *Out in 38 — and coming easy!*

DO you play Chin Golf? It is the latest popular game. Play it Winter or Summer; at home or at your club.

Chin Golf is not a 19th hole proposition —nothing like stove baseball or conversational tennis, but a regular indoor sport.

Any man who shaves himself can play it. Count your razor strokes when you shave, and see how low a score you can make. It puts fun and friendly rivalry into shaving.

If you are a golfer, you will get the idea at once; but, even if you never have schlaffed with a driver, nicked with a niblick, or been bunkered, you may be a winner at Chin Golf.

You are sure to like the course and have a good score if you use Colgate's "Handy Grip" Shaving Stick.

Fill out the attached coupon, mail it to us, with 10c. in stamps, and we will send you a "Handy Grip," containing a trial size Colgate Shaving Stick. Also we will send you, free, a score card, the rules for playing Chin Golf, and a screamingly funny picture made especially for Colgate & Co. by Briggs, the famous cartoonist.

The picture is on heavy paper, suitable for framing or tacking up in locker rooms. It will help you to start every day with a round of fun.

"And then he took up Chin Golf"

COLGATE & CO.
Dept. 66
199 Fulton Street, New York

COLGATE & CO.
Dept. 66
199 Fulton St., New York

Enclosed find 10c., for which please send me Colgate's "Handy Grip" with trial size Shaving Stick; the Briggs's Cartoon, score card, and rules for Chin Golf.

Name..

Address..

....

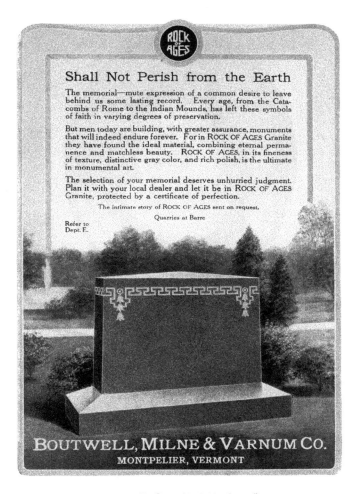

Shall Not Perish from the Earth

The memorial—mute expression of a common desire to leave behind us some lasting record. Every age, from the Catacombs of Rome to the Indian Mounds, has left these symbols of faith in varying degrees of preservation.

But men today are building, with greater assurance, monuments that will indeed endure forever. For in ROCK OF AGES Granite they have found the ideal material, combining eternal permanence and matchless beauty. ROCK OF AGES, in its fineness of texture, distinctive gray color, and rich polish, is the ultimate in monumental art.

The selection of your memorial deserves unhurried judgment. Plan it with your local dealer and let it be in ROCK OF AGES Granite, protected by a certificate of perfection.

The intimate story of ROCK OF AGES sent on request.

Quarries at Barre

Refer to
Dept. E.

BOUTWELL, MILNE & VARNUM CO.
MONTPELIER, VERMONT

LaFayette

This cameo, which is mounted on the radiator
of every LaFayette, is the seal we place upon
our work in witness that we have held fast to
our purpose: to produce with honest metal and
unhurried skill the very finest car we could.
And the car's deportment in the service of its
owners has so impressed motorists who have
been schooled in long association with fine
cars, that they see in the cameo a symbol of
unmatched worth and constant satisfaction.

LaFAYETTE MOTORS COMPANY at *Mars Hill* INDIANAPOLIS